Explosives
Engineering

Explosives Engineering

Paul W. Cooper

WILEY-VCH

New York • Chichester • Weinheim • Brisbane • Singapore • Toronto

The procedures in this text are intended for use only by persons with prior training in the field of explosives. In the checking and editing of these procedures, every effort has been made to identify potentially hazardous steps and safety precautions have been inserted where appropriate. However, these procedures must be conducted at one's own risk. The authors and the publisher, its subsidiaries and distributors, assume no liability and make no guarantees or warranties, express or implied, for the accuracy of the contents of this book or the use of information, methods or products described in this book. In no event shall the authors, the publisher, its subsidiaries or distributors, be liable for any damages or expenses, including consequential damages and expenses, resulting from the use of the information, methods or products described in this book.

Paul W. Cooper
424 Girard Blvd., SE
Albuquerque, NM 87106

This book is printed on acid-free paper. ∞

Library of Congress Cataloging-in-Publication Data
Cooper, Paul W., 1937–
      Explosives engineering / Paul W. Cooper.
         p.   cm.
      Includes bibliographical references (p.   –   ) and index.
      ISBN 0-471-18636-8 (alk. paper)
      1. Explosives.   I. Title.
   TP270.C7438 1997
   662'.2—dc20

Printed in the United States of America

ISBN 0-471-18636-8 Wiley-VCH, Inc.

25 24 23 22 21

Dedicated to my Dad,
the late Nathan Cooper,
a helluva engineer!

Preface

The field of explosives engineering incorporates a broad variety of sciences and engineering technologies that are brought together to bear on each particular design problem. These technologies include chemistry, thermodynamics, fluid dynamics, aerodynamics, mechanics, electricity, and electronics, and even meteorology, biology, and physiology. Although excellent textbooks and research papers are found in each of these areas, there has been little, if any, literature available that ties all these diverse technologies together into a unified engineering discipline for this complex field of explosives engineering.

The purpose of this text is to attempt to fill that gap. It is based, in large part, upon engineering philosophies and approaches I have developed during my career to solve numerous design problems. The text is broken into six general areas, each of which is bound together by a common technical thread.

Section I deals with the chemistry of explosives. It starts with definitions and nomenclature of organic chemicals, based on molecular structure, which is included to bring nonchemists up to speed on being able recognize and describe pure explosive compounds and mixtures and not to be intimidated by chemists' jargon. It then describes the many forms in which these explosive chemicals are used. Using molecular structure as the common thread, the text then goes into the estimation of the stoichiometry of oxidation reactions, the prediction of explosive detonation velocity and pressure properties, and the quantitative analysis of thermal stability.

Section II deals with the energetics of explosive reactions: Where does the energy come from, and how much do we get out of a particular explosive reaction? This section also uses molecular structure as the common thread tying together the thermophysical and thermochemical behavior of these reactions. In this section the thermochemical properties of the materials are used to predict the explosive properties.

Section III deals with nonreactive shock waves. The thread here is composed of three simple equations that describe the conservation of mass, momentum, and energy across the shock front. In this section we learn how to deal quantitatively with shock waves interacting with material interfaces and other shock waves.

Section IV combines the thermochemistry from Section II with the shock behavior of Section III to describe detonation (reactive shock waves). This section begins with simple ideal detonation theory and then goes on to quantitative calculations of detonation interactions at interfaces with other materials, and then deals with nonideal effects, those that cannot be predicted by ideal theory, such as the effects of size and geometry.

Section V describes the initiation of explosive reactions and the application of initiation theory to the design and analysis of initiating devices such as nonelectric, hot-wire, and exploding-bridgewire igniters and detonators. The thread that sews together all initiation phenomena is an energy-power balance, which describes the rate at which energy is deposited in an explosive and the rate of energy lost from the explosive through heat transfer.

Section VI takes all the previous information and, hanging that on a common thread of dimensional analysis, goes into the development of design scaling and scaling databases. Scaling theory and data are used here to predict the formation and flight of fragments generated by explosive devices; the production and behavior of air- and water-blast waves; the formation of craters from aboveground, ground-level, and buried explosive charges; the formation of material jetting and how that is applied to the design and behavior of lined cavity-shaped charges, as well as to the process of explosive welding.

Missing from this text is any mention of the computer codes and programs that may be used for the solution of many explosive design problems. That is an intentional omission. This text is intended to give the reader the basic understanding and working tool kit to deal with various explosive phenomena. When computer codes are used, this basic understanding of the phenomena provides a reality check of the output of computer-derived solutions.

Acknowledgments

I wish to acknowledge and thank the following people who helped with bringing this book to completion: Glenda Ponder for the editing and typing and formatting of the original manuscript; Dr. Olden L. Burchett (Sandia National Laboratories, retired), Dr. Brigita M. Dobratz (Lawrence Livermore National Laboratory, retired), and Stanley R. Kurowski (Sandia National Laboratories, retired), who devoted so much time and work in the editing and checking of the final manuscript.

My sincere thanks and appreciation also to the following people who reviewed the manuscripts and provided many excellent comments and improvements: John L. Montoya (Sandia National Laboratories), Dr. Gerald Laib (Naval Surface Warfare Center White Oak), Dr. James E. Kennedy (Los Alamos National Laboratory), Dr. Carl-Otto Lieber (Bundesinstitut fur Chemisch-Technische, BICT, Germany), Dr. Hugh R. James (Atomic Weapons Establishment, England), Dr. Pascal A. Bauer (Professor, Ecole Nationale Superieure de Mecanique et d'Aerotechnique, Paris, France), Dr. Eric J. Rinehart (Field Command, U.S. Defense Nuclear Agency). Mr. J. Christopher Ronay (Institute of Makers of Explosives), and Dr. Ronald Varosh (Reynolds Industries Systems, Inc.).

Paul W. Cooper
Albuquerque, NM

Contents

CHEMISTRY OF EXPLOSIVES

1

Organic Chemical Nomenclature

1.1 Basic Organic Structures

The carbon atom is the basic building block of organic molecules. A brief look at the carbon atom reveals that its atomic number is six, which means that it has six protons in its nucleus and six electrons around its nucleus. Its atomic weight is 12, which means that it must have six neutrons as well as six protons in its nucleus. The first electron shell is complete with two electrons, which leaves four more electrons for the second or outer shell. The second electron shell needs eight electrons to be complete, and thus the carbon atom can either gain or lose four electrons to have a complete outer shell. In other words, the carbon atom has a valence of four. In organic chemicals, the carbon atom fills the outer shell by sharing electrons with other atoms forming shared pairs of electrons or covalent bonds.

The four bonds with which carbon attaches to other atoms are equally distributed in a singly bonded carbon atom. Picture, then, that the bond sites of carbon are like the corners of a tetrahedron. Organic molecules, therefore, are three dimensional. Because it is difficult to draw complex, three-dimensional figures, we represent organic molecules by convention with a two-dimensional system of notation.

Carbon, with nothing bonded to it, is represented in Figure 1.1(a). Each dot represents one of the four electrons in the outer shell. Carbon can share its electrons with the electrons of other carbon atoms to form complex chains. If there is one shared pair of electrons between two carbon atoms, it is a single bond [Figure 1.1(b)]. Each shared pair of electrons can also be represented by

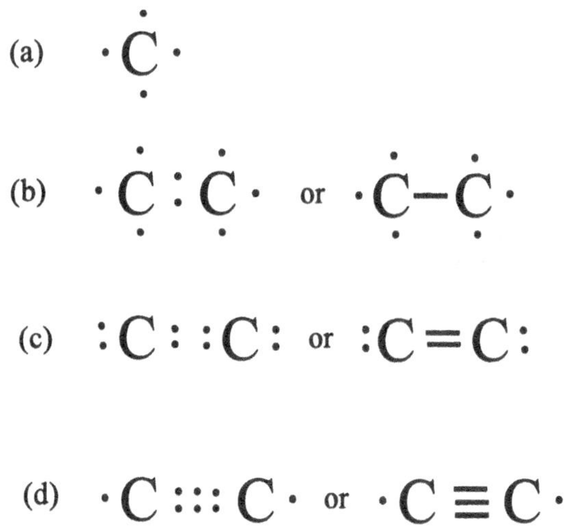

(a) $\cdot \overset{\displaystyle \cdot}{\underset{\displaystyle \cdot}{C}} \cdot$

(b) $\cdot \overset{\displaystyle \cdot}{\underset{\displaystyle \cdot}{C}} : \overset{\displaystyle \cdot}{\underset{\displaystyle \cdot}{C}} \cdot$ or $\cdot \overset{\displaystyle \cdot}{\underset{\displaystyle \cdot}{C}} - \overset{\displaystyle \cdot}{\underset{\displaystyle \cdot}{C}} \cdot$

(c) $: \overset{\displaystyle \cdot}{C} : : \overset{\displaystyle \cdot}{C} :$ or $: C = C :$

(d) $\cdot C ::: C \cdot$ or $\cdot C \equiv C \cdot$

Figure 1.1. (a) Carbon; (b) single-bonded carbons; (c) double-bonded carbons; and (d) triple-bonded carbons.

a line. If there are two shared pairs of electrons between two carbon atoms, it is called a double bond [Figure 1.1(c)]. A triple bond, shown in Figure 1.1(d), consists of three shared pairs of electrons between two carbon atoms.

If all the remaining electrons each form a covalent bond by sharing with the electron of a hydrogen atom (hydrogen has one available electron to form a covalent bond), then a molecule of hydrogen and carbon, or a hydrocarbon, is formed. Some examples are shown in Figure 1.2. Remember that in stable organic molecules, carbon has four covalent bonds and hydrogen has one.

1.2 Alkanes

Hydrocarbon molecules in which the carbon atoms are attached to each other only by means of single bonds are called *saturated*. Open-chain, saturated hydrocarbons form the group called *alkanes*, shown in Figure 1.3. Their names all end with the suffix *ane*.

The names of the four hydrocarbons of the alkane chains shown in Figure 1.3 are derived from the Latin named numbers as shown in Table 1.1. If one bond is not attached to hydrogen, thus leaving it open to attach to some other atom, the name can end with *yl*, instead of *ane*. Two different structures of butylbromide are shown in Figure 1.4(a) and (b). Each carbon in the chain is numbered starting from the end nearest the heteroatom.

Note that a shorthand version of the structure, $-CH_x$, can be used where there is no ambiguity caused; thus the 1-butylbromide in Figure 1.4(a) could be written

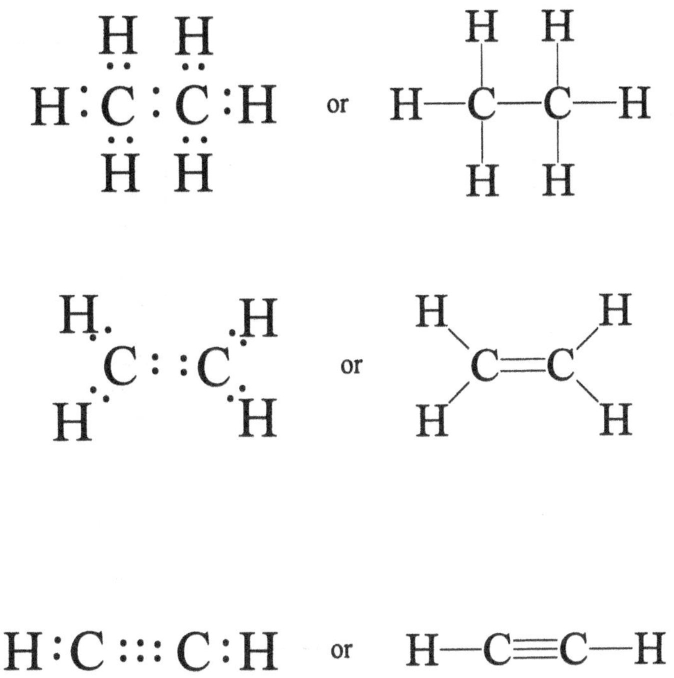

H H
H:C:C:H   or   H—C—C—H
H H

H.   .H
  C::C   or
H   .H

H:C:::C:H   or   H—C≡C—H

Figure 1.2. Three simple hydrocarbon molecules.

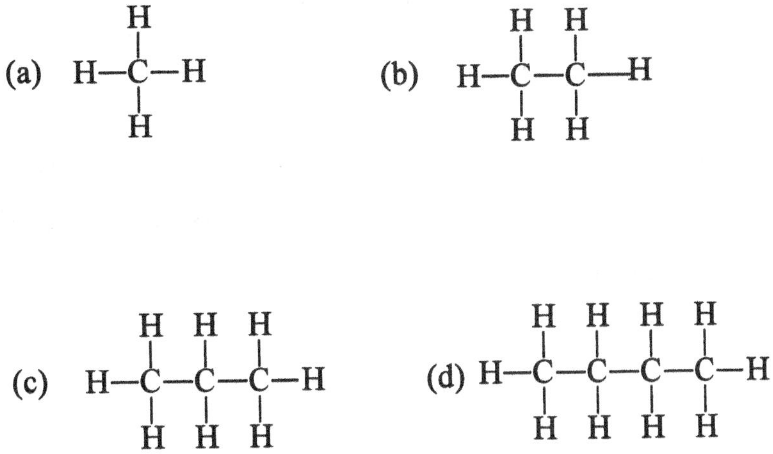

(a) H—C—H   (b) H—C—C—H

(c) H—C—C—C—H   (d) H—C—C—C—C—H

Figure 1.3. Alkanes (saturated hydrocarbons): (a) methane, (b) ethane, (c) propane, and (d) butane.

Table 1.1 Alkanes

Carbons in Chain	Name
1	Methane
2	Ethane
3	Propane
4	Butane
5	Pentane
6	Hexane
7	Heptane
8	Octane
9	Nonane
10	Decane
11	Undecane
12	Dodecane
13	Tridecane
14	Tetradecane
15	Pentadecane
16	Hexadecane
17	Heptadecane
18	Octadecane
19	Nonadecane

as shown in (c). The ending *ane* can also be retained, as shown in the same two structures of bromobutane in Figure 1.5.

If another shorter alkane is attached to one of the nonterminal carbons, forming a branched alkane, the longest carbon chain forms the basis of the name, and the attached alkane is the prefix as shown in Figure 1.6. Figure 1.7 shows the structural formula of 2-methyl-2,3-dibromopentane in four steps.

(a)
$$\begin{array}{cccc} H & H & H & H \\ | & | & | & | \\ H-C-C-C-C-Br \\ | & | & | & | \\ H & H & H & H \end{array}$$

(b)
$$\begin{array}{cccc} H & H & Br & H \\ | & | & | & | \\ H-C-C-C-C-H \\ | & | & | & | \\ H & H & H & H \end{array}$$

(c) $H_3C-CH_2-CH_2-CH_2-Br$

Figure 1.4.   (a)–(c) Butylbromide.

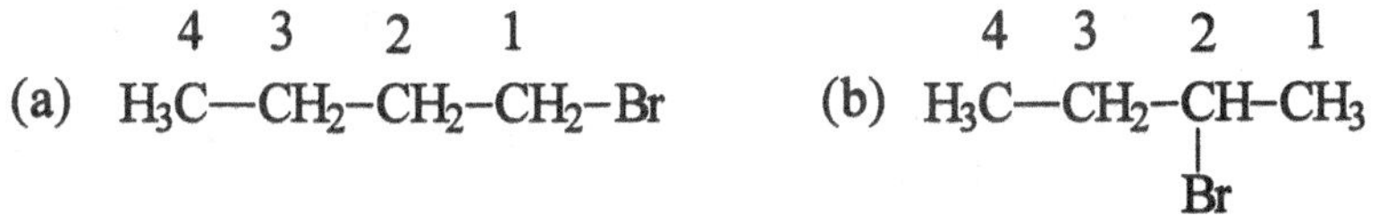

$$\underset{\text{(a)}}{\overset{4\quad3\quad2\quad1}{H_3C-CH_2-CH_2-CH_2-Br}}$$

$$\text{(b)}\quad\overset{4\quad3\quad2\quad1}{H_3C-CH_2-\underset{\underset{Br}{|}}{CH}-CH_3}$$

Figure 1.5. (a) 1-Bromobutane; (b) 2-bromobutane.

1.3 Alkenes

If there are one or more double bonds in a hydrocarbon, it is unsaturated. Unsaturated, straight-chain hydrocarbons with one double bond are called *alkenes*. Their names are identical to the alkanes, except they end with *ene* instead of *ane*. An example is shown in Figure 1.8. If there are two double bonds, the chain is called an alkadiene, and the names end in *adiene*, instead of *ene*. An example is given in Figure 1.9.

If three double bonds exist, the group is called alkatrienes, with the names ending in *atriene*. Exceptions are the compounds ethylene ($CH_2{=}CH_2$) and allene ($CH_2{=}C{=}CH_2$), which retain their common names.

1.4 Alkynes

When there is a triple bond in the chain, it is referred to as an *alkyne*. The names end with *yne* instead of *ane*, but otherwise are named similarly to the alkanes and alkenes. Chains with multiple triple bonds are likewise called alkadiynes, with names ending in *adiyne*; alkatriynes, with names ending in *atriyne*; and so forth. The exception is that the compound acetylene ($CH{\equiv}CH$) retains its common name. Unsaturated hydrocarbon chains are numbered starting at the end of the chain that gives the double or triple bonds the lowest numbers. See Figure 1.10.

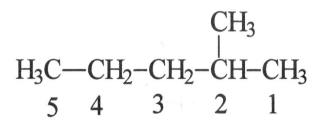

Figure 1.6. 2-Methylpentane (this material is also called isohexane).

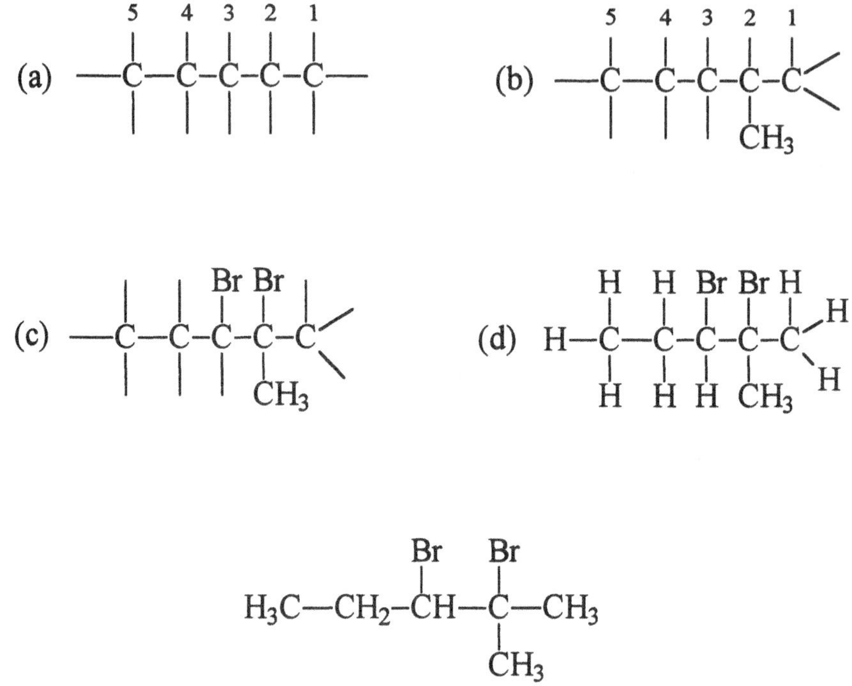

Figure 1.7. Structural formula of 2-methyl-2,3-dibromopentane: (a) pentane is the major chain; therefore, there is a straight saturated five-carbon chain as the major backbone; (b) 2-methyl-; there is a methyl group on the number two carbon; (c) -2,3-dibromo; dibromo means two bromine atoms, and they are on the number 2 and 3 carbons; (d) the rest of the bonds are not specified; therefore, they are all bonded to hydrogen; thus we have 2-methyl-2,3-dibromopentane.

$$5 \quad\; 4 \quad\quad 3 \quad\quad 2 \quad\quad 1$$
$$H_3C-CH_2-CH=CH-CH_3$$

Figure 1.8. 2-Pentene.

$$6 \quad\; 5 \quad\quad 4 \quad\quad 3 \quad\quad 2 \quad\quad 1$$
$$H_3C-CH=CH-CH_2-CH=CH_2$$

Figure 1.9. 1,4-Hexadiene.

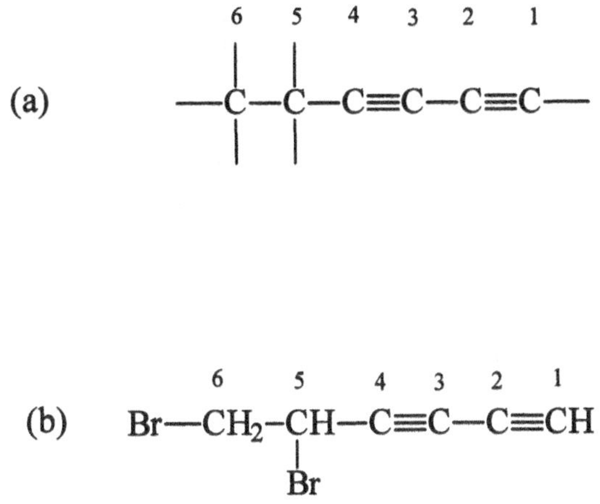

(a)

$$\overset{6}{|}\ \overset{5}{|}\ \overset{4}{}\ \overset{3}{}\ \overset{2}{}\ \overset{1}{}$$

$$-\overset{|}{\underset{|}{C}}-\overset{|}{\underset{|}{C}}-C\equiv C-C\equiv C-$$

(b) $\overset{6}{Br}-\overset{5}{CH_2}-\overset{4}{\underset{\underset{Br}{|}}{CH}}-\overset{3}{C}\equiv\overset{2}{C}-\overset{1}{C}\equiv CH$

Figure 1.10.   Structural formula of 5,6-dibromo-1,3-hexadiyne: (a) the hexadiyne ending means that the major chain has six carbons and that there are two triple bonds in the chain. Since it is 1,3-hexadiyne, the triple bonds must be between the number 1 and 2 carbons and between the number 3 and 4 carbons. (b) The 5,6-dibromo, of course, indicates two bromine atoms, one each bonded to the number 5 and 6 carbons.

1.5 Cyclic Forms

Most of the chains mentioned with three or more carbons can be bent around and formed into a ring. Such ring compounds are named similarly to the straight chains, except that their name starts with the prefix *cyclo*. Cyclopropane and cyclohexane are shown in Figure 1.11.

We thus have the families cycloalkanes, cycloalkenes, and cycloalkynes, as well as the multi-double and triple-bond variants such as cycloalkadienes and -atrienes, and cycloalkadiynes, -atriynes, etc. Naming the cyclo compounds corresponds to the naming of the straight-chain forms except that carbon atoms are numbered such that substituents are on the lowest numbered carbon atoms. This

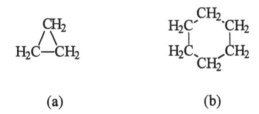

(a)   (b)

Figure 1.11.   (a) Cyclopropane; (b) cyclohexane.

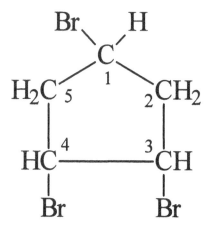

Figure 1.12. 1,3,4-Tribromo-cyclopentane.

is shown in the 1,3,4-tribromo-cyclopentane (Figure 1.12) and in 1,3-cyclo-hexadiene (Figure 1.13). In the latter case (Figure 1.13), the carbon atoms are numbered so that the double bonds receive the lowest possible numbers. Figure 1.14 shows the structural formula for the compound named 3,5-dibromo-1-cyclopentene.

The compounds we have looked at so far (alkanes, alkenes, and alkynes—open chain or cyclic) are called aliphatic compounds.

1.6 Aromatics

A special ring compound, the six-carbon ring with three double bonds, is known by its common name *benzene* (Figure 1.15). This particular arrangement has a

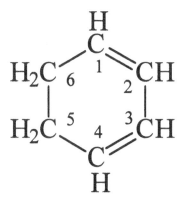

Figure 1.13. 1,3-cyclohexadiene.

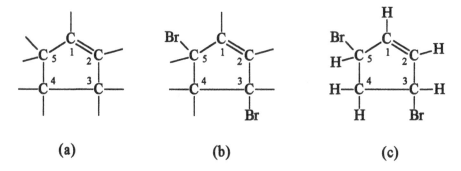

(a)   **(b)**   **(c)**

Figure 1.14. 3,5-dibromo-1-cyclopentene: (a) the 1-cyclopentene indicates that this is a five-carbon ring with one double bond in it, and that bond is between the number 1 and 2 carbons. (b) 3,5-Dibromo means that there are two bromine atoms, one each bonded to the number 3 and 5 carbons. (c) The rest of the bonds are to hydrogen; thus we have the complete formula.

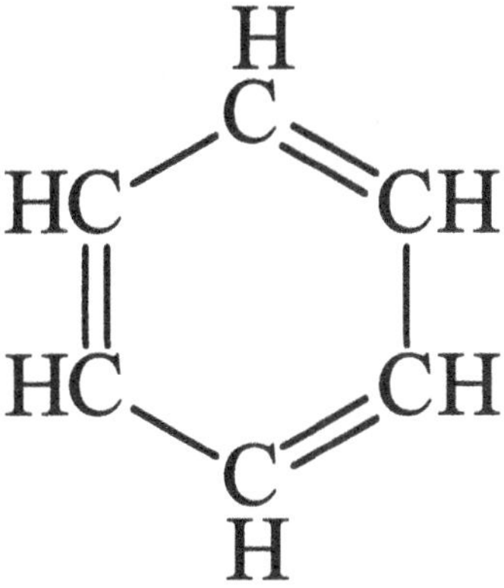

Figure 1.15. The benzene molecule.

Figure 1.16. Symbol for the benzene molecule.

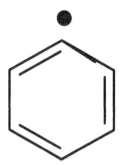

Figure 1.17.   Symbol for phenyl, the benzene molecule with one hydrogen removed.

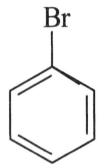

Figure 1.18.   Phenylbromide molecule.

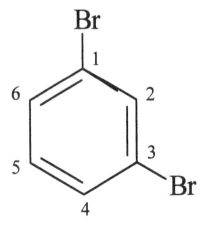

Figure 1.19.   Phenylene-1,3-dibromide or 1,3-dibromophenylene.

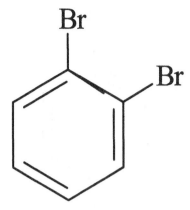

Figure 1.20.   o-Dibromobenzene (1,2-dibromobenzene).

special stability that makes this ring the basis of a different class of compounds than cycloalkatrienes. All organic compounds that contain this benzene ring are included in a class called aromatic compounds. For simplicity the benzene ring can be represented by the symbol shown in Figure 1.16.

Each corner represents a carbon atom, and if not otherwise indicated, each carbon is bonded to a hydrogen atom. If one hydrogen is removed, the resulting radical is named phenyl and is represented as in Figure 1.17. Therefore, the compound represented in Figure 1.18 is called phenylbromide. If two hydrogen atoms are removed, the resulting diradical is called phenylene. Thus the compound shown in Figure 1.19 is a phenylene-1,3-dibromide, or 1,3-dibromophenylene.

Alternatively, the name benzene may be retained. In that case, this same compound may also be called 1,3-dibromobenzene. The carbons in the benzene

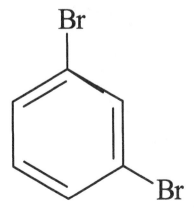

Figure 1.21.   m-Dibromobenzene (1,3-dibromobenzene).

Figure 1.22.   p-Dibromobenzene (1,4-dibromobenzene).

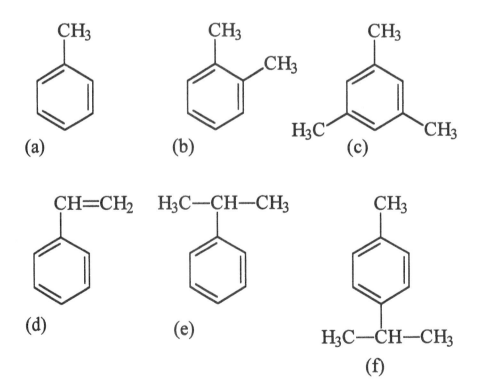

Figure 1.23.   (a) Toluene; (b) xylene (o shown); (c) mesitylene; (d) styrene; (e) cumene; and (f) cymene (p shown).

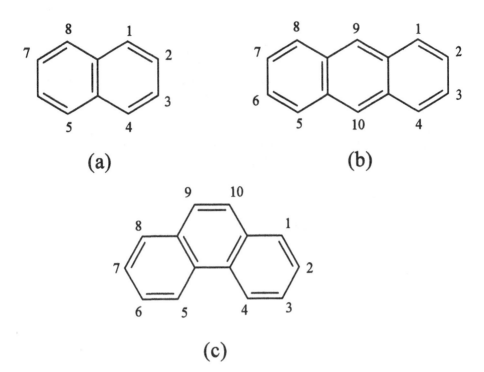

Figure 1.24. Fused polyclics: (a) naphthalene; (b) anthracene; (c) phenanthrene.

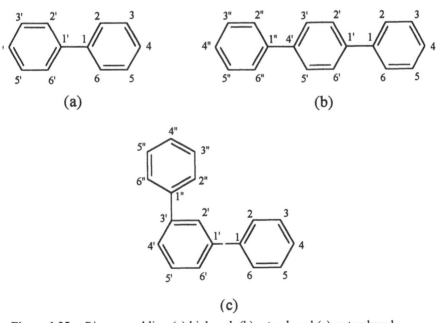

Figure 1.25. Ring assemblies: (a) biphenyl; (b) p-terphenyl (c) m-terphenyl.

ring, like all of the cyclo compounds, are numbered such that the substituents are on the lowest-numbered carbon atoms. In lieu of numbering the carbons, there is also a system of naming relative positions of substitution on the ring when there are two identical substituents. Sometimes this method is clearer to use; however, both the numbering and naming systems are used. If two like substituents are on adjacent carbons of the benzene ring, they are in the *ortho* form, as in Figure 1.20. If the two substituents are on alternate carbons, they are in the *meta* position, as shown in Figure 1.21. If the two substituents are on opposite carbons, they are in the *para* position (Figure 1.22).

The compound of which common moth balls are made is paradichlorbenzene. Certain substituted benzene compounds retain their common names. Some of these are shown in Figure 1.23.

1.7 Polycyclic Aromatic Structures

When more than one benzene ring are in the same compound, they may be joined together in different ways. If both rings share common carbon atoms, they are called *fused polycyclics*. Examples of this are the compounds shown in

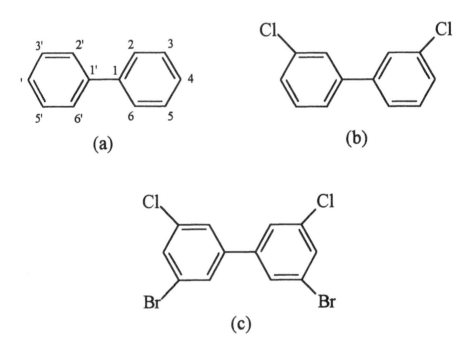

Figure 1.26.   3,3′-Dichloro-5,5′-dibromobiphenyl. (a) Biphenyl (a two-ring assembly); (b) 3,3′-dichloro indicates two chlorine atoms, one each on the number 3 and 3′ carbons; (c) 5,5′-dibromo indicates two bromines, one each on the number 5 and 5′ carbons.

Figure 1.24. Since the two common carbons have all four bonds already committed, they are not numbered.

When rings are joined such that they are not sharing common carbon atoms, they are called *ring assemblies*. Three examples are shown in Figure 1.25.

The structural formula of 3, 3'-dichloro-5, 5'-dibromo-byphenyl is shown in Figure 1.26.

For more extensive rules in organic chemical nomenclature, consult Ref. 1.

CHAPTER

2

Oxidation

2.1 Oxidation Reactions

When explosives react they produce energy by a process called *oxidation*. In this chapter we will examine this process and see how it is affected by the composition of the explosive. We will learn how to predict the composition of the products of oxidation and how to quantify the degree of oxidation.

An oxidation reaction is the chemical reaction that occurs when a fuel is burning or an explosive is detonating; it is the same in both cases. Oxidation reactions produce heat because the internal energy of the product (final) molecules is lower than the internal energy of the reactant (starting) molecules. This difference between the internal energies of the reactants and products is called the *heat of reaction*. When a fuel burns with oxygen completely to its most oxidized state, the heat of reaction is called the *heat of combustion*. When an explosive material detonates to form its products, the heat of reaction is called the *heat of explosion*. The heat of reaction (or of combustion or explosion) per unit weight of reactants (fuel plus oxidizer) is greatest when there is just enough oxidizer to burn all the fuel to its most highly oxidized products. The highest oxidation state is also the lowest internal energy state. Most explosives are made of carbon, hydrogen, nitrogen, and oxygen (CHNO). The highest oxidation state of carbon when burned with oxygen is carbon dioxide (CO_2); the highest oxidation state of hydrogen when burned with oxygen is water (H_2O). Nitrogen molecules (N_2) are at a lower state of internal energy than the oxides of nitrogen (NO, NO_2, N_2O_3, etc.); therefore, any nitrogen in the reactants forms N_2 in the products. (There is always some trace amount of NO_x, but that trace is always very small.)

Examining a simple burning or oxidation reaction, we find that 1 mole of methane, a gas, burns with 2 moles of oxygen to form 1 mole of carbon dioxide and 2 moles of water.

$$CH_4 + 2O_2 \rightarrow CO_2 + 2H_2O$$

The heat of reaction is 212.8 kcal (per mole of methane). Since 1 mole of methane weighs 16.042 grams and 2 moles of oxygen weigh 64.0 grams, the heat of reaction per unit weight is as follows:

$$\frac{212.8 \text{ kcal}}{16.042 \text{ g} + 64.0 \text{ g}} = 2.659 \text{ kcal/g}$$

2.2 Effects of Stoichiometry

If more than 1 mole of methane for every 2 moles of oxygen were present, there would still be fuel remaining at the end of the reaction. Since that extra fuel would not have burned, it would not have contributed to the production of heat, but would have added to the total weight of the combination. Therefore, even though the heat evolved by the reaction would remain the same, the heat evolved per unit weight of reactants would be lower. The same is true when there is excess oxidizer. The fuel-to-oxidizer ratio that is precisely balanced (no excess of either fuel or oxidizer) is called the *balanced stoichiometric* ratio. For the reaction just discussed, the burning of methane and oxygen to form water and carbon dioxide, the balanced stoichiometric ratio (fuel to oxidizer) is 2.0. Figure 2.1 shows the heat evolved per unit weight of reactants for this reaction as a function of the stoichiometric ratio.

The oxidizer need not be from a source separate from the fuel as shown in Figure 2.1. It can exist as part of the same molecule as the fuel; this is the case in most explosives. For nitroglycol, shown in Figure 2.2, the oxidizer is in the two substituent groups ($—ONO_2$), and the fuel is the hydrocarbon portion ($—CH_2—CH_2—$).

2.3 Reaction Product Hierarchy

As mentioned previously, most explosives consist of carbon, hydrogen, nitrogen, and oxygen and are called CHNO explosives. The general formula for all CHNO explosives is expressed as $C_xH_yN_wO_z$, where x, y, w, and z are the number of carbon, hydrogen, nitrogen, and oxygen atoms, respectively, in the explosive molecule. The simplest picture of how this reaction takes place is to visualize that, in the zone where a propellant is burning or an explosive is detonating, the reactant molecule is completely broken down into its individual component

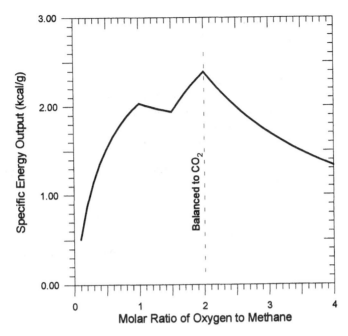

Figure 2.1. Specific energy output from burning methane with oxygen as a function of the molar ratio of oxidizer to fuel.

atoms; that is, $C_xH_yN_wO_z \rightarrow xC + yH + wN + zO$. These atoms then recombine to form the final products of the reaction. The typical products formed are as follows:

$$2N \rightarrow N_2$$
$$2H + O \rightarrow H_2O$$
$$C + O \rightarrow CO$$
$$CO + O \rightarrow CO_2$$

In the case of nitroglycol (Figure 2.2), there was exactly enough oxygen to burn all the carbon completely to CO_2. This is not the case with all explosives. Some explosives have more than enough oxygen to burn all the carbon to CO_2. These explosives are overoxidized or fuel lean. The explosive compounds that do not have enough oxygen to burn all the carbon to CO_2 are underoxidized or fuel

$$O_2NO-CH_2-CH_2-ONO_2$$

Figure 2.2. Nitroglycol.

rich. In all cases, the reaction hierarchy of the products formed can be estimated by using the following "rules of thumb."

1. All the nitrogen forms N_2.
2. All the hydrogen is burned to H_2O.
3. Any oxygen left after H_2O formation burns carbon to CO.
4. Any oxygen left after CO formation burns CO to CO_2.
5. Any oxygen remaining forms O_2.
6. Traces of NO_x (mixed oxides of nitrogen) are always formed.

This set of rules is called the simple product hierarchy for CHNO explosives (and propellants). If the explosive had contained any metal additives, these would probably not oxidize until all the above oxidation steps were completed. By traces of NO_x, we mean less than 1%. An example of this is TNT, detonated in the open air, where measurements have shown from 0.2 to 0.5% total NO_x in the original undiluted products. Of this NO_x, approximately half was NO. Some examples of oxidizing reactions are shown in Figures 2.3, 2.4, and 2.5.

For nitroglycerine (Fig. 2.3) the oxidizing reaction is as follows:

$C_3H_5N_3O_9 \rightarrow 3C + 5H + 3N + 9O$
a. $3N \rightarrow 1.5N_2$;
b. $5H + 2.5O \rightarrow 2.5H_2O$ (6.5 O remaining);
c. $3C + 6O \rightarrow 3CO_2$ (0.5 O remaining);

(8.5 of the 9 O atoms available have burned all the H to H_2O and all the C to CO_2. There are still 0.5 O atoms remaining.)

d. $0.5O \rightarrow 0.25O_2$.

The overall reaction is $C_3H_5N_3O_9 \rightarrow 1.5N_2 + 2.5H_2O + 3CO_2 + 0.25 O_2$

For RDX (Fig. 2.4) the oxidizing reaction is as follows:

$C_3H_6N_6 O6 \rightarrow 3C + 6H + 6N + 6O$
a. $6N \rightarrow 3N_2$;
b. $6H + 3O \rightarrow 3H_2O$ (3 O remaining);

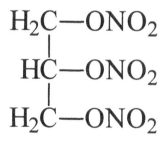

Figure 2.3. Nitroglycerine (overoxidized).

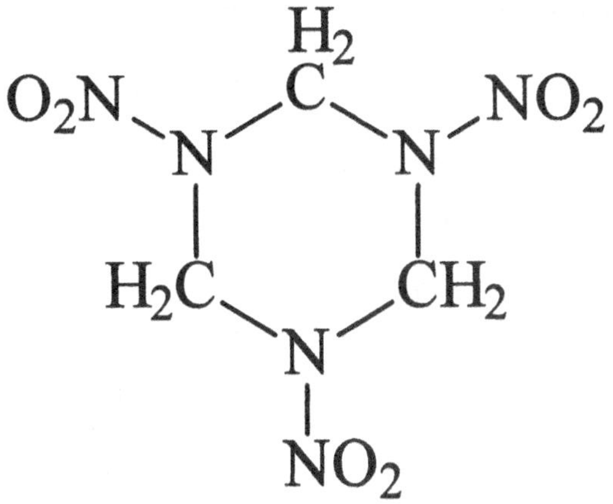

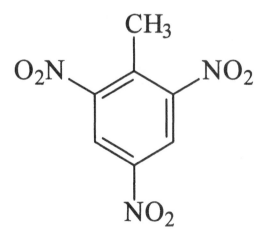

Figure 2.4. RDX (underoxidized).

c. 3C + 3O → 3CO (all the O is used up at this point; therefore, no CO_2 is formed).

The overall reaction is $C_3H_6N_6O_6 \rightarrow 3N_2 + 3H_2O + 3CO$.

In Figures 2.4 and 2.5, some of the products themselves are fuels, specifically the free carbon, C, and the carbon monoxide, CO. After the burning or detonation reaction is complete, the products may be free to expand into air. As they do this, they may mix with the oxygen in the air, burst into flame, and burn to CO_2 when the proper mixture with the air is reached. This second reaction is called a secondary *fireball*. Such fireballs can also be fueled from other burnable mate-

Figure 2.5. TNT (very underoxidized).

rials, such as casings, glues, binders, and colorants, that have been mixed with the explosive. Since very underoxidized explosives produce free carbon (which can form black smoke), the presence of black smoke is a crude indication of severe underoxidation.

For TNT (Fig. 2.5) the oxidizing reaction is as follows:

$C_7H_5N_3O_6 \rightarrow 7C + 5H + 3N + 6O$
a. $3N \rightarrow 1.5N_2$;
b. $5H + 2.5O \rightarrow 2.5H_2O$ (3.5 O remaining);
c. $7C + 3.5O \rightarrow 3.5CO$ (all the O is used up) + 3.5C.

The overall reaction is $C_7H_5N_2O_6 \rightarrow 1.5N_2 + 2.5H_2O + 3.5CO + 3.5C$.

We now know that the relative amount of oxygen in an explosive is quite important. When an explosive is exactly oxygen balanced, neither rich nor lean, it produces the maximum energy output per unit weight of that explosive. The relative amount of oxygen with respect to the oxygen required to oxidize the fuel completely in an explosive (or propellant) is expressed quantitatively as "oxygen balance" (OB).

2.4 Oxygen Balance

Referring back to the general formula for a CHNO explosive or propellant, $C_xH_yN_wO_z$, we see that, if all the carbon could be burned to carbon dioxide, we would need twice the number of oxygen atoms as we have carbon atoms (or $2x$ oxygen). Similarly, to burn all of the hydrogen to water, 1 oxygen atom for every 2 hydrogen atoms (or $y/2$ oxygen) is required. To be exactly balanced, this compound would need $(2x + y/2)$ atoms of oxygen. It has z atoms of oxygen. Therefore, the quantity $(z - 2x - y/2)$ is a measure of the OB balance for this molecule. When that number is negative, $z < (2x + y/2)$, it means z is less oxygen than needed for complete combustion; therefore, the material is underoxidized. When z is greater than $(2x + y/2)$, the quantity is positive, which means there is more than enough oxygen, and the molecule is overoxidized.

It is customary to express the OB in terms of the weight percent of excess oxygen compared to the weight of explosive. This is done by multiplying the expression $(z - 2x - y/2)$, which is in atom numbers, by the atomic weight (AW) of oxygen, and dividing by the molecular weight (MW) of the explosive material. Then it is put in percent terms by multiplying by a hundred. Thus

$$OB\% = 100 \frac{AW(O)}{MW\,(\text{explosive})}(z - 2x - y/2)$$

The atomic weight of oxygen is 16.000; therefore,

$$OB\% = \frac{1600}{MW\,(\text{explosive})}(z - 2x - y/2)$$

Table 2.1 Atomic Weights for Elements in CHNO Explosives

Chemical Element	Atomic Weight
Carbon	12.010
Hydrogen	1.008
Nitrogen	14.008
Oxygen	16.000

The MW of the explosive molecule is the sum of the weights of all the atoms. Since we know the formula is $C_xH_yN_wO_z$, deriving the molecular weight is simple. Table 2.1 gives the atomic weights for the four elements used in CHNO explosives.

Therefore, the molecular weight of the explosive is

$$12.01x + 1.008y + 14.008w + 16z$$

As examples, let us calculate the oxygen balance for the four explosives examined earlier.

a. Nitroglycol, $C_2H_4N_2O_6$:

$x = 2, y = 4, w = 2$, and $z = 6$,

$MW_{exp} = 12.01(2) + 1.008(4) + 14.008(2) + 16.000(6) = 152.068$,

$OB = \dfrac{1600}{152.068}[6 - 2(2) - 4 / 2] = 0\%$.

b. Nitroglycerine, $C_3H_5N_3O_9$:

$x = 3, y = 5, w = 3$, and $z = 9$,

$MW_{exp} = 12.01(3) + 1.008(5) + 14.008(3) + 16.000(9) = 227.094$,

$OB = \dfrac{1600}{227.094}[9 - 2(3) - 5 / 2] = 3.52\%$.

We determined earlier qualitatively that nitroglycerin was slightly overoxidized, and now we have the quantitative term that describes that fact.

c. RDX, $C_3H_6N_6O_6$:

$x = 3, y = 6, w = 6$, and $z = 6$,

$MW_{exp} = 12.01(3) + 1.008(6) + 14.008(6) + 16.000(6) = 222.126$,

$OB = \dfrac{1600}{222.126}[6 - 2(3) - 6 / 2] = -21.61\%$.

d. TNT, $C_7H_5N_3O_6$:

$x = 7, y = 5, w = 3$, and $z = 6$,

$MW_{exp} = 12.01(7) + 1.008(5) + 14.008(3) + 16.000(6) = 227.134$,

$OB = \dfrac{1600}{227.134}[6 - 2(7) - 5 / 2] = -73.97\%$.

CHAPTER

3

Pure Explosives

3.1 Grouping Explosives by Structure

We have seen how organic compounds are categorized by their structure. In this chapter we will see that explosives occur in all the major organic structural groups. We will describe these explosives and discover how they are synthesized and how some are used. We shall also examine some explosives that are not organic.

In the previous chapter, examples were given of the different organic structural families of explosives; that is, nitroglycol and nitroglycerine are aliphatics, RDX is a cycloaliphatic, and TNT is very obviously an aromatic. The two different oxidizer subgroups or substituents, —NO_2 and —ONO_2, were also presented. These two oxidizers are the major sources of oxygen in organic explosive compounds. Other substituent groups may be found on or in various explosive molecules. Some of these contribute oxidizer; some contribute fuel. Some, like the azides, contribute neither fuel nor ozidizer, but do contribute energy to the detonation process when their high-energy bonds are broken. Table 3.1 shows a number of substituent groups that may be found in pure explosive compounds.

Given the many forms or structures in which hydrocarbons can occur, along with the fairly broad number of substituent groups that can be attached onto or into them, it is apparent that an almost limitless variety of explosive compounds are possible. Of this almost limitless variety, surprisingly few are commonly used. The number is severely limited by several factors. The cost of raw materials and processing is one of the major factors. Thermal stability, chemical compatibility, toxicity, physical form, handling sensitivity, and explosive output prop-

27

Table 3.1 Molecular Substituent Groups Common to Many Explosives

Name	Formula
Oxidizer contributors	
Nitrate	$-ONO_2$
Nitro	$-NO_2$
Nitroso	$-NO$
Alcohol (Hydroxyl)	$-OH$
Acid (Carboxyl)	$-COOH$
Aldehyde (Carbonyl)	$-CHO$
Ketone (Carbonyl)	CO
Chloro	$-Cl$
Fluoro	$-Fl$
Difluoramine	$-NF_2$
Fuel contributors	
Methyl	$-CH_3$
Ethyl	$-CH_2-CH_3$
Butyl	$-CH_2-CH_2-CH_3$
Other Hydrocarbons	$-C_xH_y$
Imino	$-NH$
Amino	$-NH_2$
Ammonium	$-NH_4$
Combined fuel and oxidizer contributors	
Fulminic	$-ONC$
Nitramine	$-NHNO_2$
Other bond energy contributors	
Azides	$-N_3$
Diazo	$-N_2-$

erties are other factors. Explosives that are in fairly common use are classified into all the structural categories that we discussed earlier.

Therefore, for both simplicity and continuity, explosive compounds will be divided here into groups or families according to their organic structure. The few inorganic explosives will be put into a separate group. Figure 3.1 shows this organizational arrangement.

3.2 Aromatic Pure Explosive Compounds

Molecules that contain a benzene ring are aromatic. The simplest of the aromatics, and structurally the basis for all of the others in this family, is TNB (Figure 3.2).

Notice that the three oxidizing substituents are nitro groups ($-NO2$). Virtually all the aromatic explosives are oxidized principally by this substituent group, with the occasional addition of one or more of the others.

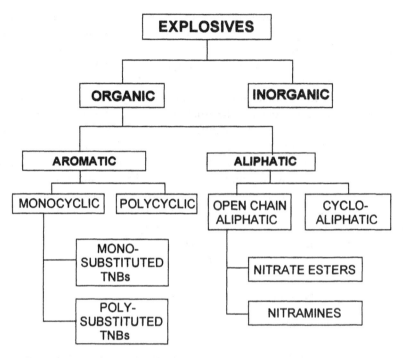

Figure 3.1. Structural organization of pure explosive compounds.

TNB is an excellent explosive. It has almost all of the virtues wanted in an explosive: stability, low toxicity, nonsensitivity, relatively high output velocity and pressure. We use very little of this explosive, however, because it is extremely difficult to synthesize and hence is very expensive.

The aromatic nitro compounds, in general, are made by direct nitration with

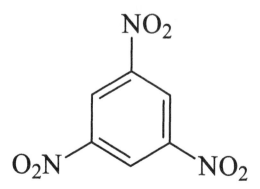

Figure 3.2. Trinitrobenzene, TNB.

nitric acid, usually mixed with sulfuric acid. The nitration is accomplished by initial attack of the ring by the nitronium ion $(NO_2)^+$. The sulfuric acid promotes and maintains the presence of the nitronium ion in the mixed acid. The first nitration step in the reaction with benzene goes quite easily to mononitrobenzene, as shown in Figure 3.3.

This step is done at about 60°C and with ordinary concentrated nitric and sulfuric acids. The nitro groups on the aromatic molecule, however, exhibit great electron-withdrawing power, and therefore, the mononitrated benzene is very resistant to further nitration because of local suppression of the nitronium ion.

To accomplish the next step, to dinitrobenzene, the temperatures must be raised to around 95°C, and fuming nitric acid must be used along with the concentrated sulfuric acid. Direct nitration to the trinitro form is virtually impossible. TNB must be made, therefore, by an indirect (and hence expensive) route. To produce laboratory quantities, it is made by starting with trinitrotoluene (TNT), then oxidizing the methyl group to the acid (carboxyl) form, and decarboxylating in hot aqueous solution, as illustrated in Figure 3.4.

3.2.1 Monosubstituted TNBs

In toluene, as well as with many other monosubstituted benzenes, the substitution group (methyl, in the case of toluene) acts as an electron donor. This counters the electron-withdrawing effect of the previously substituted nitro groups and allows higher local nitronium ion activity, thus allowing a much easier trinitration step. This is the key, then, to inexpensive synthesis of trinitro-aromatic explosives. Figure 3.5 is the TNT molecule, the first, and most important (as far as quantity of production goes) of the monosubstituted TNBs.

TNT is soluble in benzene, toluene, and acetone. It is slightly soluble in alcohol and virtually insoluble in water. Because of its moderate melting point (approximately 80°C) and the fact that it does not decompose upon melting, it is used most often in melt-cast form.

It is produced, as was earlier discussed, by direct nitration with nitric and sulfuric acids. The nitration takes place in several steps. The last step, which is trinitration, uses free SO_3 gas bubbled through the highly concentrated acids. Both batch and continuous synthesis processes are used for TNT production. It is preferred to produce the pure 2,4,6-form; the other isomers are separated out by various techniques. Purity is tested by measuring the solidification point. The

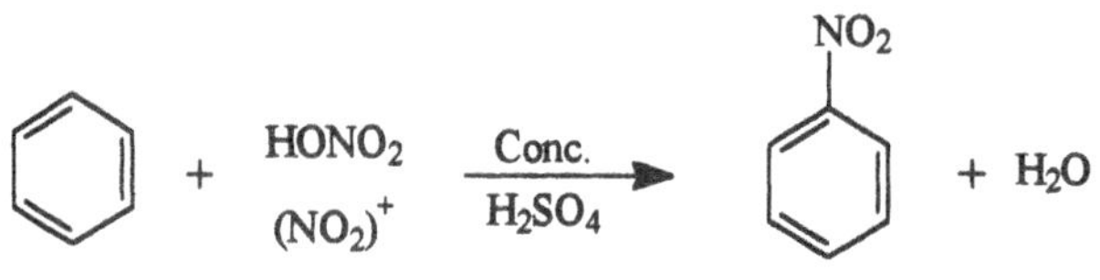

Figure 3.3.   First nitration step to mononitrobenzene.

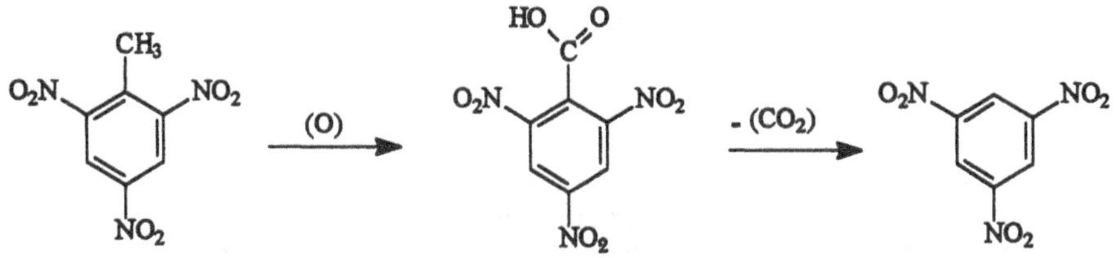

Figure 3.4. Nitration steps to make TNB.

higher the solidification temperature, the purer the 2,4,6-TNT. The pure form of 2,4,6-TNT solidifies at 80.8°C; the other isomers solidify a few degrees lower. U.S. Military Specifications call for a minimum of 80.2°C. The U.S. Department of Energy specifications for TNT used for Composition B call for 80.6°C. Physical properties of TNT, along with those of several other aromatic explosives, are listed in Table 3.2.

In its role as an intermediate in synthesizing TNB, TNBA (Figure 3.6) is made by oxidizing TNT with a solution of $KClO_3$ in nitric acid, or with a chromic acid mixture. Although it is only slightly soluble in cold water, it decarboxylates (loses the —COOH group by giving up CO_2 gas) in hot aqueous solutions, and yields TNB when exposed to water vapor.

TNBA and TNT are the two major monosubstituted TNBs that have a carbon linkage to TNB. Four major monosubstituted TNBs with a nitrogen linkage to TNB are shown in Figures 3.7 and 3.8.

Tetranitroaniline is also called picramid. It is prepared by straightforward, step-by-step nitration of aniline, or in small batches by treating trinitrochloro-

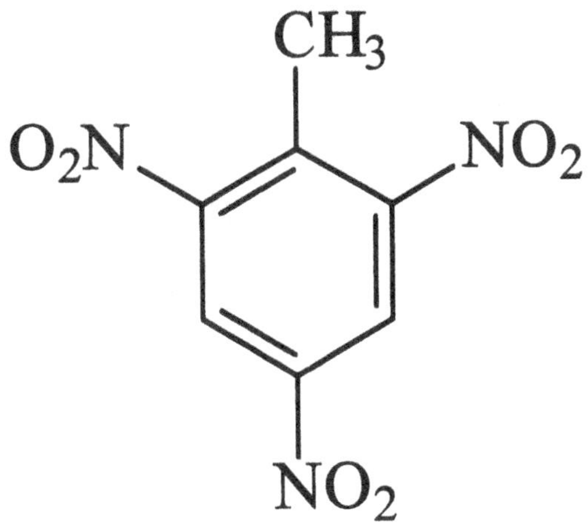

Figure 3.5. Trinitrotoluene, TNT.

Table 3.2 Physical Properties of Some Aromatic Explosives

Name	Color	Crystal Density	Melting Point	Detonation Velocity at Density
Trinitrobenzene (TNB)	Light green-yellow	1.76	123	7.30 @ 1.71
Trinitrotoluene (TNT)	Light yellow	1.654	80.8	6.90 @ 1.60
Trinitrobenzoic acid (TNBA)	Yellow			
Trinitroaniline (TNA)	Orange-red	1.762	188	7.30 @ 1.72
Tetryl	Yellow	1.73	129.5	7.57 @ 1.71
Ethyl tetryl	Green-yellow	1.63	05.8	
Picric acid	Yellow	1.767	122.5	7.35 @ 1.7
Ammonium picrate	Yellow	1.72	280	7.15 @ 1.6
Methyl picrate	Pale yellow	1.61	68	6.80 @ 1.57
Ethyl picrate	Light yellow		78	6.50 @ 1.55
Picryl chloride	Light yellow	1.797	83	7.20 @ 1.74
Trinitroxylene (TNX)	Light yellow		182	
Trinitrocresol	Yellow	1.68	107	6.85 @ 1.62
Styphnic acid	Yellow to red brown	1.83	176	
Lead styphnate	Orange yellow to brown	3.0	d[a]	5.20 @ 2.9
Triaminotrinitrobenzene (TATB)	Bright yellow	1.93	350	7.35 @ 1.80
Hexanitroazobenzene (HNAB)	Orange red		221	
Hexanitrostilbene (HNS)	Yellow	1.74	318	
Tetranitrodibenzotetrazapentalene	Orange red	1.85	378 w/d	7.25 @ 1.64
Tetranitrocarbazole (TNC)	Yellow		296	

References 3, 4, 5.
[a] Decomposition

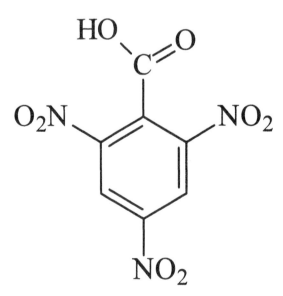

Figure 3.6. Trinitrobenzoic acid, TNBA.

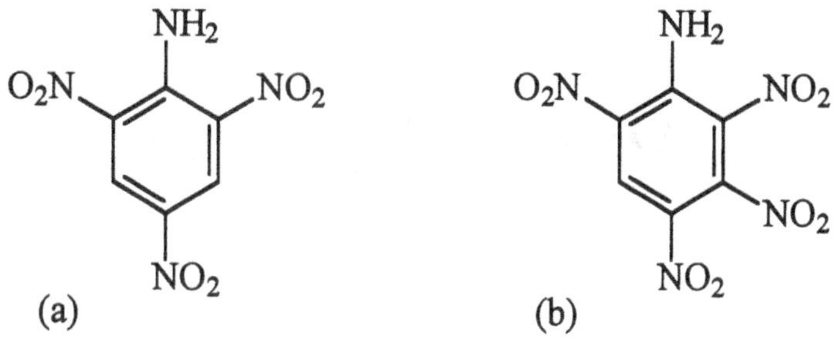

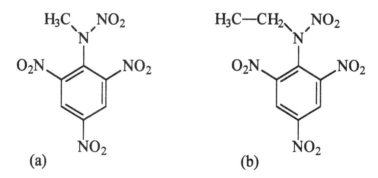

Figure 3.7.   (a) Trinitroaniline, TNA; (b) tetranitroaniline.

benzene with ammonia. Nitration can be carried a step further with moderate yields of tetranitroaniline.

Tetryl is prepared differently than the TNBs we just scanned. It is normally made by dissolving methylaniline in sulfuric acid, and then slowly pouring (while cooling) into nitric acid. The bright yellow crystals then immediately precipitate out. Tetryl is a very common booster explosive, used as an intermediate charge in many military explosive trains, and also as an output charge in some blasting caps and exploding bridgewire (EBW) detonators. Its major drawback is that it is somewhat toxic.

Ethyl tetryl, shown in Figure 3.8, is prepared the same way as tetryl, except that ethylaniline is used as the starting material. In the next group of monosubstituted TNBs, Figures 3.9 through 3.12, the substituent group is linked to the TNB through oxygen.

Picric acid (Figure 3.9) is made by dissolving phenol in sulfuric acid, then nitrating with nitric acid. Another route is from nitration of dinitrophenol, which is made from dinitrochlorobenzene. Picric acid was used extensively in World War I as a bomb and grenade filler both by itself and in mixtures with other explosives. Its major drawback (besides its toxicity) is that it reacts with metals

Figure 3.8.   (a) Tetryl; (b) ethyl tetryl.

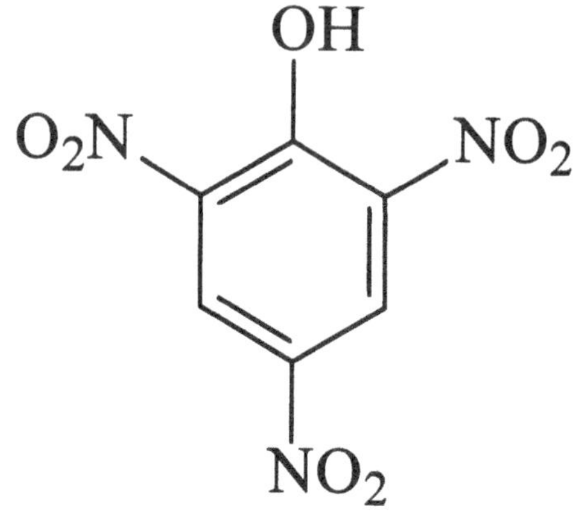

Figure 3.9. Picric acid.

in the presence of moisture, forming salts. The crystalline metal picrate salts are extremely sensitive to impact and friction. Some can detonate by themselves just sitting on the shelf. This has happened occasionally in industries that normally have nothing to do with explosives. For example, picric acid is often mixed with an aqueous solution of ferric chloride to make a cleaning solution for certain types of metals. They are safe as long as the ingredients remain in solution, but when the water evaporates, iron picrate crystallizes out. Such ''dried-out'' cleaning solutions have been known to detonate while being stored (Ref. 2).

Ammonium picrate (Figure 3.10), sometimes called Explosive D, is made by

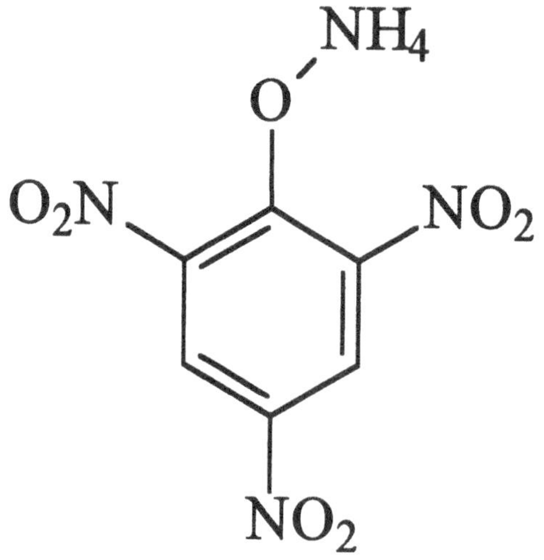

Figure 3.10. Ammonium picrate.

bubbling ammonia gas through an aqueous solution of picric acid. This explosive was also used as a bomb filler during World War I. It suffered compatibility problems similar to picric acid.

The explosive shown in Figure 3.11, also called methylpicrate, is toxic. It is made by direct nitration of dinitroanisol, which is made by reaction of dinitrochlorobenzene with methanol (methyl alcohol) in the presence of an alkali. This explosive is very insensitive to shock and friction, even less sensitive than TNT. It found some limited use during World War I as a bomb filler.

Ethylpicrate (Figure 3.12) also is quite toxic, causing skin irritation upon contact. It is prepared the same way as methylpicrate except for the use of ethanol (ethyl alcohol) in lieu of methanol.

The explosive shown below in Figure 3.13 is trinitrochlorobenzene, also called picryl chloride. It is made by direct nitration in the classical sense, by the mixture of nitric and sulfuric acids. The last nitration step is very difficult. It requires maximum acid concentrations and has a relatively low yield. This explosive, therefore, is rather expensive. It is as insensitive as TNT and has a somewhat higher output in terms of both detonation velocity and pressure. The dinitro form is more important because it is used as the starting material in the synthesis of several other explosives, as was shown previously.

3.2.2 Polysubstituted TNBs

The next group of explosives, depicted in Figures 3.14 through 3.18, is the polysubstituted TNBs, in which more than one substituent group has been added to the basic TNB molecule. TNX (Figure 3.14) is similar to TNT, except of course for the second methyl group substitution. It has a considerably lower

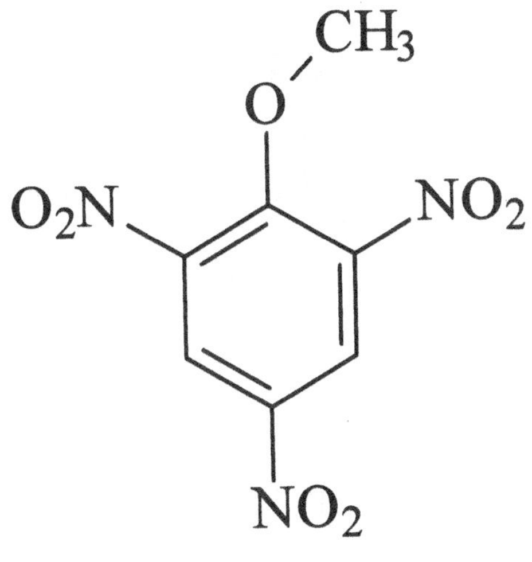

Figure 3.11.   Trinitroanisol.

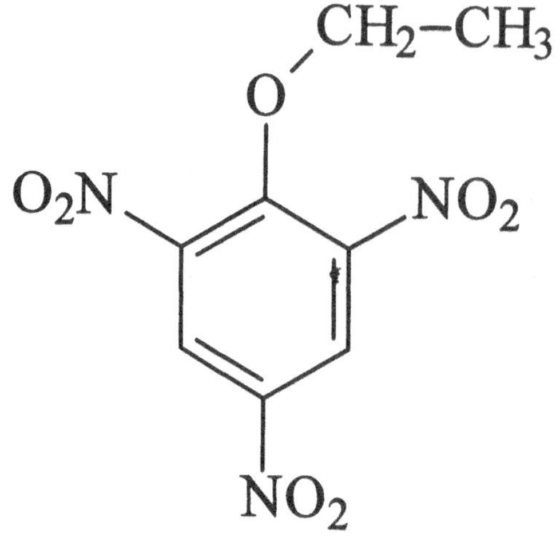

Figure 3.12.   Ethylpicrate.

oxygen balance, and the nitration to the trinitro form is difficult. It is rather expensive and has limited practical use.

Trinitrocresol (Figure 3.15) is made by nitration of m-cresoldisulfonic acid. The Germans used it extensively as a bomb and grenade filler in World War I, where it was usually mixed with picric acid in a 60:40 (by weight) mixture. The low melting point (80°C) of the mix made it an easy material to cast.

Also called styphnic acid, trinitroresorcinol (Figure 3.16) is similar to picric acid except it has two hydroxyl groups. It is made in a similar manner to several

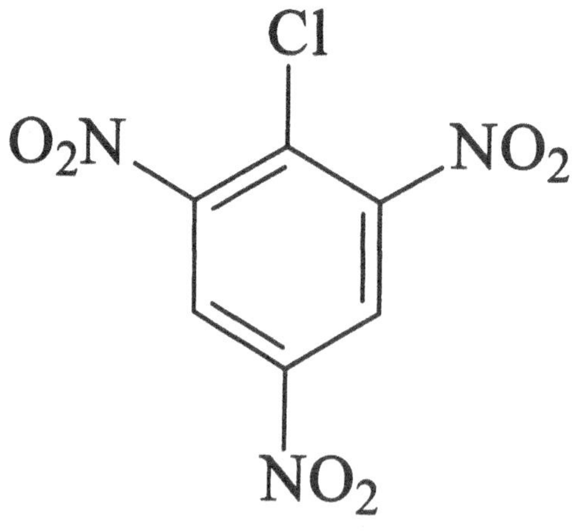

Figure 3.13.   Trinitrochlorobenzene.

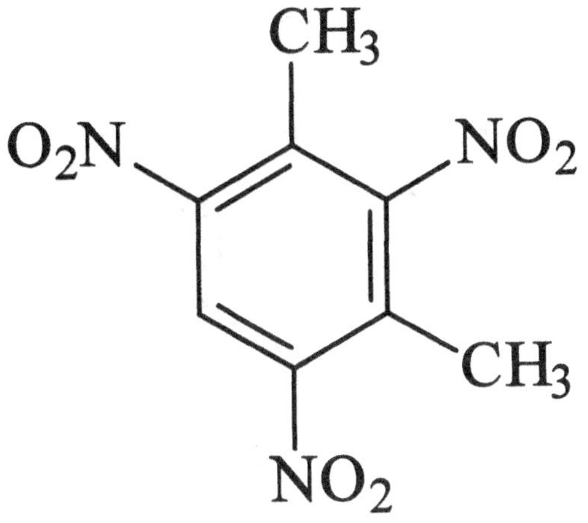

Figure 3.14. Trinitroxylene, TNX.

of the monosubstituted TNBs. Resorcinol (m-dihydroxyl benzene) is dissolved in sulfuric acid and then nitrated by adding concentrated nitric acid. It is not a particularly good explosive as far as output goes, and finds its main use as the material from which we make metal styphnate salts such as magnesium, barium, and lead styphnate.

Lead styphnate (Figure 3.17) is a primary explosive (very sensitive) widely used as the ignition element in many hot-wire detonators. In that application it is often mixed with lead azide, another primary explosive. It is also used as one of the major ingredients in modern noncorrosive percussion primers. It is pre-

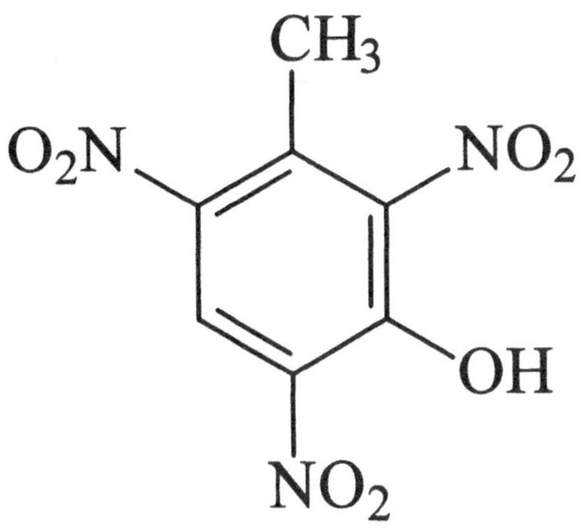

Figure 3.15. Trinitrocresol.

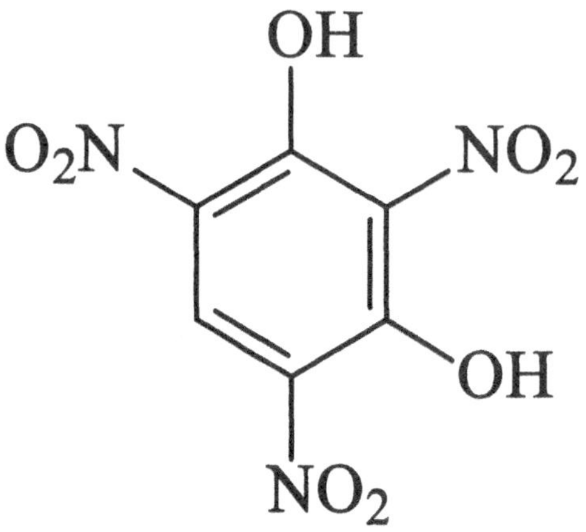

Figure 3.16. Trinitroresorcinol.

pared by reacting lead nitrate in solution with magnesium trinitroresorcinate, the magnesium salt of styphnic acid.

Going from one of the most sensitive of the aromatic explosives to one of the very least sensitive, we have TATB (Figure 3.18). This is a new, very insensitive, high explosive that is finding broad use in nuclear weapons development. The extreme degree of insensitiveness boosts the safety in handling and in accident situations, which is so crucial in that particular application. It is made by direct nitration of 1,3,5-trichlorobenzene to 1,3,5-trichloro-2,4,6-trinitrobenzene. This, in turn, is then converted to the 1,3,5-triamino- by amine substitution of the three chlorine atoms.

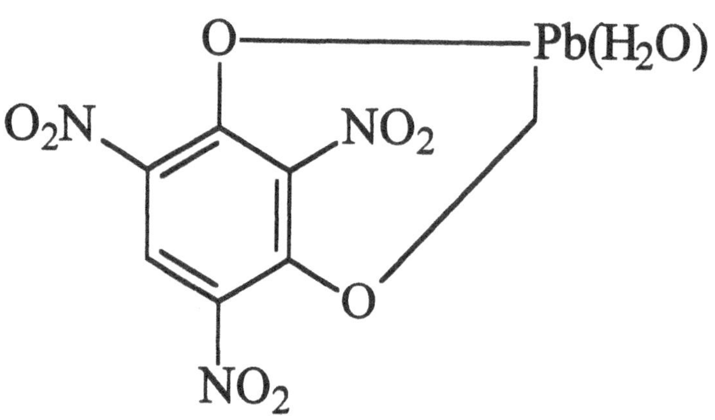

Figure 3.17. Lead styphnate.

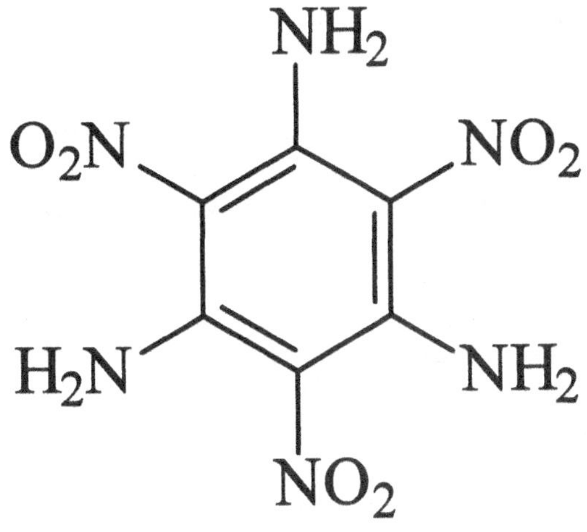

Figure 3.18.   Triaminotrinitrobenzene, TATB.

3.2.3 Polycyclic Aromatic Explosives

Referring back to Figure 3.1, we have now seen some of the monocyclic aromatic explosives. Starting with TNB, the basic structural building block, we looked at monosubstituted TNBs, those explosives made of TNB with one additional substituent group. Then we looked at polysubstituted TNBs, those explosives made of TNB with two or more additional substituent groups. To complete our view of the aromatic explosives, let us next look at the polycyclic aromatics (Figures 3.19 through 3.22), those that contain more than one benzene ring.

The molecule in Figure 3.19 resembles two trinitroanilines joined by a double bond. The —N═N— linkage is the "azo" group in this molecule's name. It is made by reacting hydrazine (H_2N—NH_2) with dinitrochlorobenzene to form

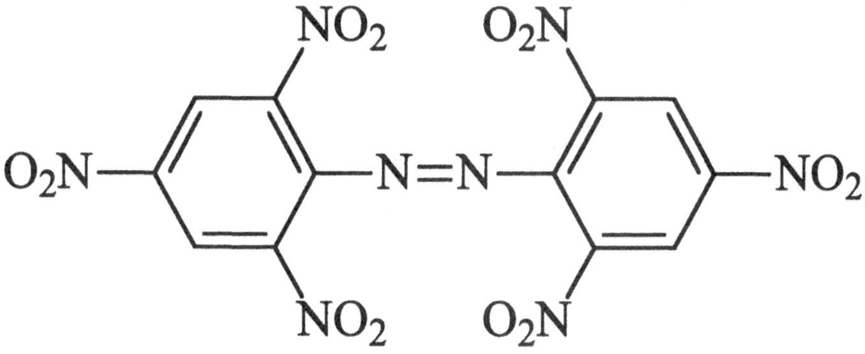

Figure 3.19.   Hexanitroazobenzene, HNAB.

tetranitrohydrazobenzene. This intermediate compound is then treated with mixed acid (nitric plus sulfuric), which oxidizes the —HN—NH— bond to the azo form —N=N—, and at the same time pushes the nitration up by two more nitro groups to the hexanitro form. This explosive is fairly stable at high temperatures (up to around 200°C) and is used in special components that must survive that kind of environment. All the polycyclic aromatics shown here are good high-temperature explosives.

HNS (Figure 3.20) is prepared from TNT and various oxidizing agents that attack the methyl group, leaving the bonds free to join as shown. The details of the various preparative routes are proprietary. HNS is used in many high-temperature components, including mild detonating fuses (MDF) and several detonators. The HNS MDF is used in emergency canopy deployment systems on several different military fighter aircraft.

TACOT (Figure 3.21) is a commercial high-temperature explosive manufactured by Dupont. It can survive temperature soaking at 275°C for 3 or 4 weeks and still be serviceable. It is made by direct nitration in mixed acid with dibenzotetrazapentalene as the starting material.

TNC (Figure 3.22) is made by direct nitration of carbazole with mixed acid. Prior to the nitration, the carbazole is treated with sulfuric acid to make it soluble in water.

3.3 Aliphatic Explosive Compounds

As you now know, aliphatic organic compounds belong to the alkane, alkene, and alkyne classes of compounds. Aliphatic explosives fall into both the open-chain and cycloaliphatic groups. The major sources of oxidizer in most aliphatic explosives are from the nitrate ester group (—ONO$_2$) and the nitramine group (—NH—NO$_2$). The nitrate esters are usually made by direct nitration of an

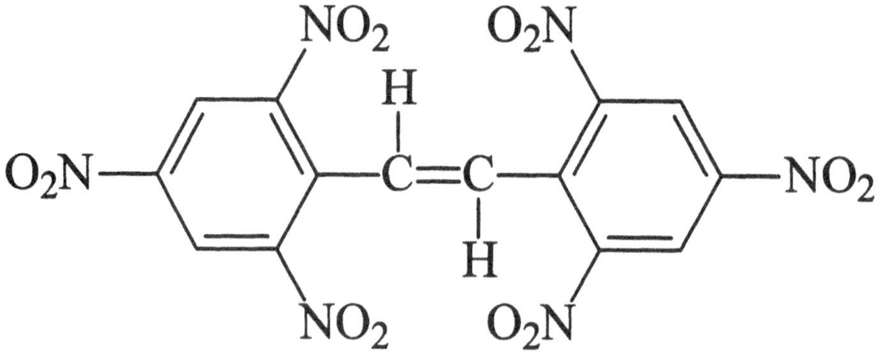

Figure 3.20.   Hexanitrostilbene, HNS.

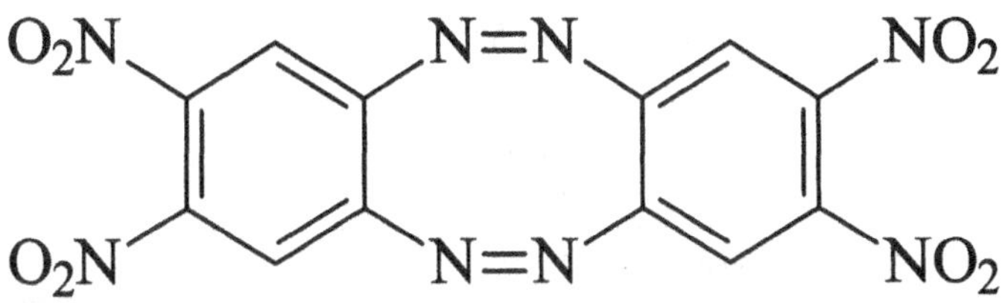

Figure 3.21. Tetranitrodibenzotetrazapentalene, TACOT.

organic alcohol by mixed acid. R— is any organic radical, such as methyl, H_3C—, etc.

$$R\text{—OH} + HONO_2 \quad R\text{-}ONO_2 + H_2O$$

The nitramines are arrived at by various routes, but generally, the reaction is similar to the direct nitration of an organic amine with nitric acid.

$$R\text{-}NH_2 + HONO_2 \quad R\text{—NH—NO}_2 + H_2O$$

3.3.1 Nitrate Esters

The first and simplest group of open-chain aliphatic explosives is that which is derived from the paraffinic, or alkane, polyalcohols. The physical properties of some aliphatic explosives are given in Table 3.3. Figures 3.23 through 3.29 show the molecules for some of these explosives.

Methyl nitrate (Figure 3.23) is very volatile and is therefore rarely used. The vapors of methyl nitrate are both flammable and explosive. Methyl nitrate is poisonous, like all the nitrated alkane polyalcohols, which can be absorbed through the skin, causing increased heartbeat, severe headache, and even nausea.

Nitroglycol (Figure 3.24) is also called ethyleneglycoldinitrate (EGDN). It is made by treating, as in the classic nitration, ethylene glycol with mixed acid. It

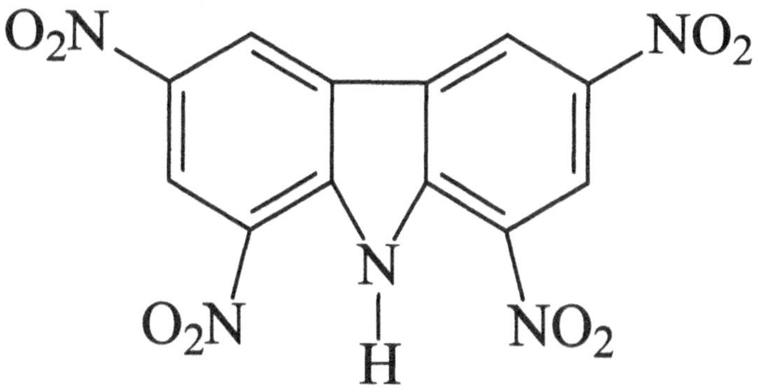

Figure 3.22. Tetranitrocarbazole, TNC.

Table 3.3 Physical Properties of Some Aliphatic Explosives

Name	Form	TMD (g/cm³)	Melting Point (°C)	Detonation Velocity at Density (mm/μs) at (g/cm³)
Methyl nitrate	Liquid	1.217		6.30 @ 1.217
Nitroglycol	Liquid	1.48	−20	7.30 @ 1.48
Nitroglycerine	Liquid	1.591	2.2	7.60 @ 1.59
Erythritoltetranitrate	Solid	1.6	61.5	
Mannitol hexanitrate	Solid	1.604	112	8.26 @ 1.73
PETN	Solid	1.76	141.3	8.40 @ 1.7
PETRIN	Solid	1.54		
EDNA	Solid	1.71	176.2	7.57 @ 1.65
Nitroguanidine	Solid	1.71	232	8.20 @ 1.7
Nitrourea	Solid		159	
RDX	Solid	1.82	204	8.75 @ 1.76
HMX	Solid	1.96 (B)	275	9.10 @ 1.9
Sorguyl	Solid	2.01		9.15 @ 1.95

References 3, 4, 5.

is used mixed with nitroglycerine in low-temperature dynamites because it suppresses the freezing point of nitroglycerine. It is an oily liquid, has a fairly high vapor pressure, and is very sensitive to impact.

Nitroglycerine (NG) (Figure 3.25) is also called glyceroltrinitrate. This oily liquid was the first modern, large-scale, commercial explosive. It replaced black powder as the main blasting agent in mining, quarrying, and construction work. It has a fairly low vapor pressure, but still produces sufficient vapors in enclosed work areas to cause headaches. By itself, nitroglycerine is extremely sensitive to initiation by impact. The sensitivity is due to the presence of tiny vapor bubbles present in the liquid. These bubbles are heated by compression during an impact and raise the local temperature around them to a sufficiently high level to cause initiation of reaction in the adjacent liquid. Alfred Nobel discovered that when the nitroglycerine is absorbed in a spongelike material, the tiny bubbles do not form. This made the explosive far less sensitive. The absorbent material first used by Nobel was called kieselguhr, which is the German name for diatomaceous earth. We use this material extensively as a filter medium today. Woodmeal, a very fine sawdust, is generally used in modern dynamites

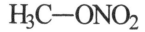

Figure 3.23. Methyl nitrate.

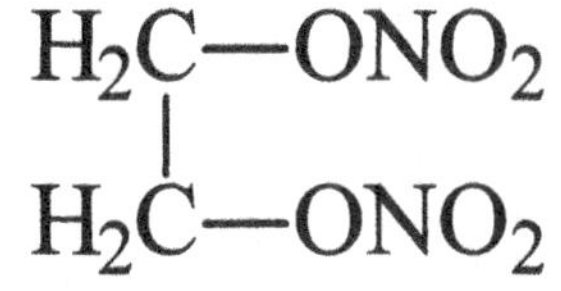

$$H_2C-ONO_2$$
$$H_2C-ONO_2$$

Figure 3.24. Nitroglycol.

as the absorbent. NG is made by direct nitration of dry glycerine with mixed acid.

Erythritol tetranitrate (Figure 3.26) is a solid. It forms colorless crystals and has a melting point around 61°C. It is made by dissolving erythritol in nitric acid and then precipitating out the explosive by adding sulfuric acid to the solution. It is not nearly as impact sensitive as nitroglycerine or the other liquids mentioned. It finds little use as a common explosive.

Like erythritol tetranitrate, mannitol hexanitrate (Figure 3.27) is prepared by dissolving mannitol in nitric acid and then precipitation following the addition of concentrated sulfuric acid. It is quite impact sensitive, about twice as sensitive as erythritol tetranitrate and is used as the primary explosive in some modern nonelectric blasting caps instead of lead azide.

Getting away from the straight-chain molecules, the explosive shown in Figure 3.28, a pentane isomer derivative, is one of the more important ones in present use. Although PETN is quite impact sensitive, it is less so than any of the straight-chain aliphatic nitrate ester explosives we have discussed so far. PETN is made by direct nitration of pentaerythritol with nitric acid (no sulfuric). It forms colorless crystals that have a melting point of around 141°C. As PETN is heated, it undergoes severe thermal decomposition long before the melting point is reached. Therefore, its useful temperature range is limited, normally not to exceed 70 to 75°C.

PETN is used extensively in explosive products such as MDF, prima cord,

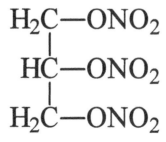

Figure 3.25. Nitroglycerine, NG.

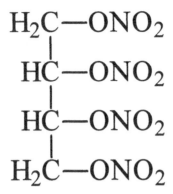

Figure 3.26. Erythritol tetranitrate.

blasting cap output charges, EBW detonator initial pressings, Detasheet, and is mixed with TNT in the castable explosive, pentolite.

By very carefully controlling the reaction of pentaerythritol in nitric acid, PETRIN (Figure 3.29) instead of PETN can be obtained. PETRIN is not a particularly desirable explosive, but because of the hydroxyl group left on the last of the outer carbons, this material has one particularly useful feature. The hydroxyl can be reacted to the acid group in acrylic acid to form a polymerizable material, PETRIN-acrylate. PETRIN-acrylate polymer, a plastic, is used as an energetic binder in some composite rocket propellants.

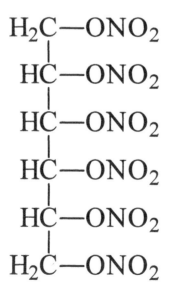

Figure 3.27. Mannitol hexanitrate.

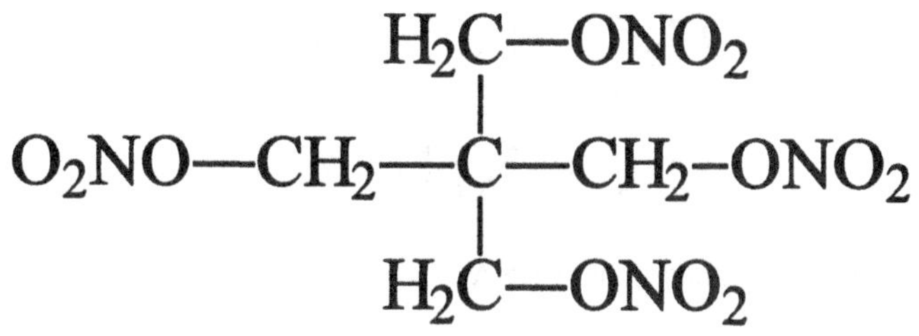

$$H_2C-ONO_2$$
$$O_2NO-CH_2-\overset{|}{\underset{|}{C}}-CH_2\text{-}ONO_2$$
$$H_2C-ONO_2$$

Figure 3.28. Pentaerythritol tetranitrate, PETN.

3.3.2 Nitramines

All the preceding aliphatic explosives utilized the nitrate ester substituent as the oxidizer. The other most common oxidizer substituent in aliphatic explosives is nitramine, which is in the following pure explosives shown in Figures 3.30 through 3.35.

EDNA (Figure 3.30) is the amino analog of ethylene dinitrate. It is not a liquid, however; it forms colorless crystals and melts at 176°C with considerable thermal decomposition as it approaches that temperature. It is made by reacting ethylene-urea with mixed acid. Dinitroethyleneurea is then formed and rapidly decarboxylizes, losing CO_2 and leaving EDNA. The major use of EDNA is in a castable mixture with TNT, called EDNATOL.

One of the crystal forms of NQ (Figure 3.31) is fiber- or featherlike; this enables the crystals to mechanically interlock with large void spaces left between them. This property enables NQ to maintain fairly decent mechanical properties (it does not flake or fall apart) at low pressing densities. The ability to maintain uniform low density makes NQ a useful laboratory explosive where some experiments require controllable low detonation velocity and pressure. NQ is made by dehydration of guanidine nitrate, which in turn was made by the reaction of ammonium nitrate with dicyanodiamide. NQ is also used as a major ingredient in triple base gun propellants.

Similar to NQ, nitro urea (Figure 3.32) is also made by dehydration (with

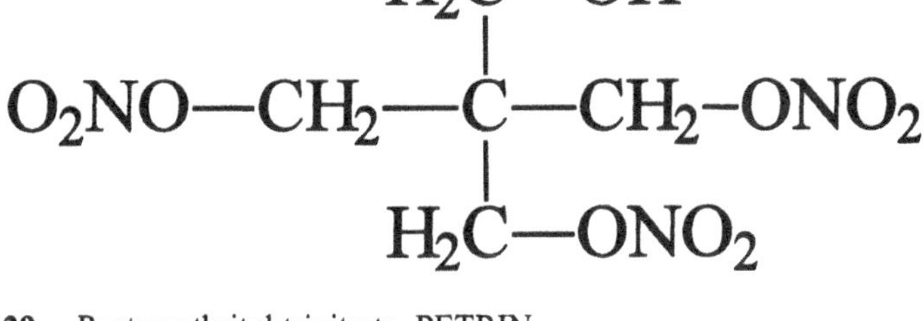

$$H_2C-OH$$
$$O_2NO-CH_2-\overset{|}{\underset{|}{C}}-CH_2\text{-}ONO_2$$
$$H_2C-ONO_2$$

Figure 3.29. Pentaerythritol trinitrate, PETRIN.

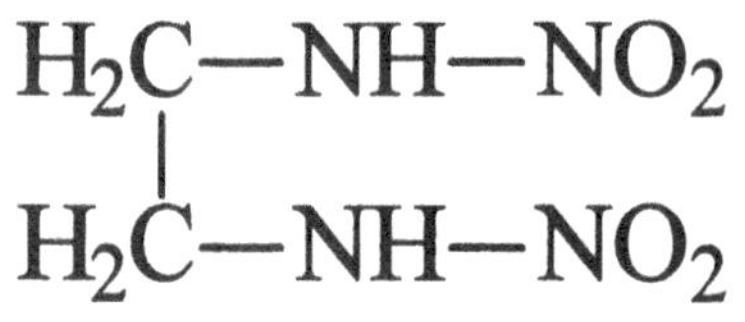

$$H_2C-NH-NO_2$$
$$|$$
$$H_2C-NH-NO_2$$

Figure 3.30. Ethylenedinitramine, EDNA.

sulfuric acid). The starting material in this case is urea nitrate. Unlike NQ, it does not have the feather-type crystals, and therefore is not useful in that application.

The next few explosives we will examine are also nitramines, but they differ from the previously discussed aliphatics in that they are cycloaliphatics.

The explosive shown in Figure 3.33 is also called cyclonite or hexogen. In the United States the name RDX is used most frequently. RDX is one of the widest used of the high-output explosives. It finds use in many bulk HE formulations as well as being used alone in detonator and blasting cap outputs. It can be found in special prima cords, MDF, shaped charges, etc.

It is made by any of several processes, none of which is as immediately obvious as the straightforward nitrations we have mentioned so far. The Henning, Schnurr, Knoffler, Apel, and Bachmann processes all use an aliphatic ring compound, hexamethylenetetramine, as the basic starting material, which is reacted with concentrated nitric acid under various conditions and with other various additives. In these processes, three of the methylene groups and one of the amine groups are split off the parent molecule and lost. In the Eble and Wolfram processes, the ring is built up from short radicals that bond to each other. For the Eble process, the reactants are paraformaldehyde and ammonium nitrate mixed in acetic anhydride.

HMX (Figure 3.34) is also called octogen. It has a somewhat higher melting point than RDX as well as a higher detonation velocity. HMX originally was obtained as a byproduct from the Bachmann process for RDX. It can also be solely synthesized from other processes. HMX is used in moderately high-temperature applications and where its greater output properties are required. Like

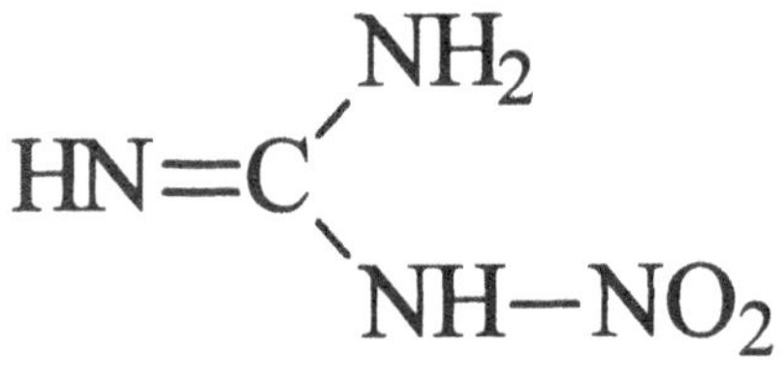

$$HN=C \overset{\displaystyle NH_2}{\underset{\displaystyle NH-NO_2}{}}$$

Figure 3.31. Nitroguanidine, NQ.

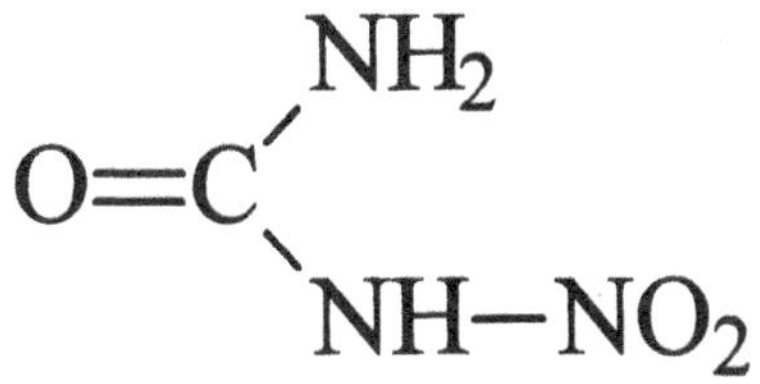

$$O=C \begin{array}{c} NH_2 \\ \\ NH-NO_2 \end{array}$$

Figure 3.32. Nitro urea.

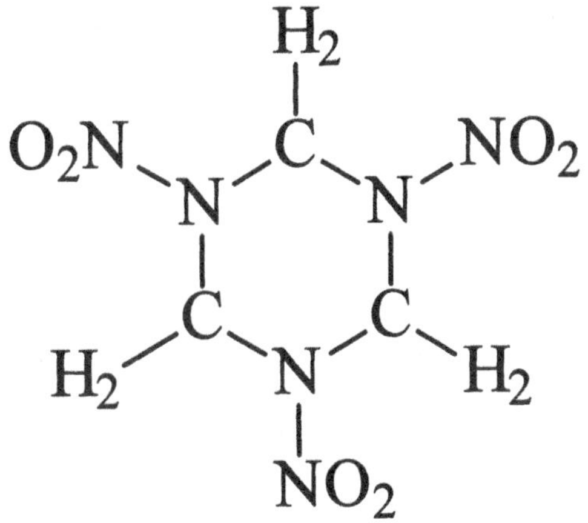

Figure 3.33. Cyclo-1,3,5-trimethylene-2,4,6-trinitamine, RDX.

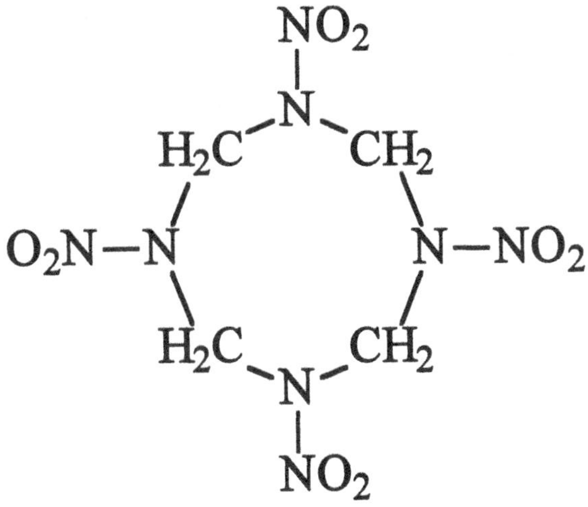

Figure 3.34. Cyclotetramethylenetetranitramine, HMX.

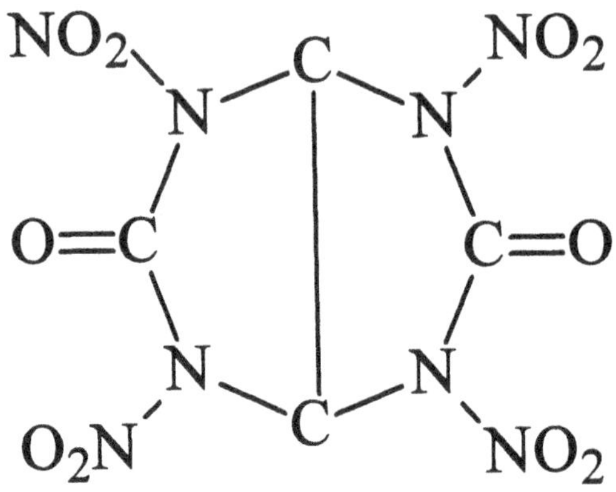

Figure 3.35. Tetranitroglycolurile, Sorguyl.

RDX, it is fairly insensitive to either friction or impact, and is very stable in storage.

The very interesting compound shown in Figure 3.35 is somewhat similar to HMX, except for the cross link between the opposite (top and bottom) carbons, and the ketone linkages at the side carbons. Although not yet in any regular use, this explosive has some interesting potential.

It is denser than HMX and has a higher detonation velocity, and hence a much higher detonation pressure. On the other side, however, it is far more sensitive to friction and impact (about twice as sensitive as PETN), and cannot be in formulations with TNT, where it decomposes on contact.

3.4 Inorganic Explosives

Inorganic compounds do not have hydrocarbon backbones forming the basis of the molecules. Generally, inorganic compounds are ionic acids or bases, or salts.

Table 3.4 Physical Properties of Some Inorganic Explosives

Name	TMD (g/cm^3)	Melting Point $(°C)$	Detonation Velocity at Density $(mm/\mu ms)$ @ (g/cm^3)	
Mercury fulminate	4.43	d^a	4.25	3.00
Lead azide	4.8	d^a	4.63	3.00
Silver azide	5.1	251 w/d	4.0^b	4.00
Ammonium nitrate	1.72	169.6	5.27 (confined)	1.30

References 3, 4, 5.
[a] d, decomposition.
[b] Calculated from thermodynamic data.

Figure 3.36. Mercury fulminate.

Some inorganic compounds contain covalent bonds as well as ionic bonds, however, and some inorganic explosives fall into that group. The inorganic explosives are usually primary explosives; that is, they are very sensitive to impact and friction, and also are easily ignited and grow to detonation from hot spots such as are caused by sparks, flame, and other sources of heat and high temperature. Some of these explosives are shown in Figures 3.36 through 3.39, and Table 3.4 presents their physical properties.

Mercury fulminate (Figure 3.36) was one of the earliest of the initiating explosives. It was used extensively in blasting caps and primers. It has since been displaced by lead azide in modern initiators. Mercury fulminate is made by dissolving mercury in nitric acid, forming a mercury nitrate solution. This solution is then poured into alcohol, forming mercury fulminate, which then precipitates out of the solution.

Another fulminate, that of silver, is even more sensitive and can detonate while the crystals are still suspended in the solution from which they formed. Silver fulminate is sometimes formed accidently during the process of silvering a glass vessel if alcohol cleaning agents are still present in the vessel. Many accidents from detonation of silver fulminate formed in this manner have occurred.

Lead azide (Figure 3.37) is by far the most common of the initiating explosives. It is used in most blasting caps as well as in the vast majority of hot-wire detonators. It has excellent thermal stability up to 250°C. It is made by precipitating the lead compound out of a reaction mixture of acidic lead nitrate and sodium azide. Materials such as dextrin or polyvinyl alcohol are added to the reaction mix to prevent the formation of very large crystals during the precipitation. The larger crystals are undesirable because they can spontaneously detonate upon breaking.

Figure 3.37. Lead azide.

$$Ag\text{-}N_3$$

Figure 3.38. Silver azide.

$$NH_4^+ NO_3^-$$

Figure 3.39. Ammonium nitrate.

Silver azide (Figure 3.38) is similar to lead azide. It is also sensitive to light and will decompose with the possibility of spontaneous detonation upon exposure to bright light sources. Because of their sensitivity, bulk azides are stored and shipped wet and in small quantities per package. The wet packing is simply a liquid, usually water or a water-alcohol mixture covering the azide powder in a small jar. Stored wet azides must be regularly inspected to ensure that the liquid has not evaporated down to the level of the powder.

Ammonium nitrate (AN) (Figure 3.39) is not a particularly good explosive by itself. It is extremely difficult to initiate, and will propagate a detonation only in very large diameters (above a half-meter). It is used commercially as a fertilizer. When mixed with a fuel, however, it becomes a very powerful explosive. Explosives utilizing AN mixtures will be discussed later.

4

Use Forms of Explosives

Although the pure explosive compounds previously described are used in their pure form as liquids, pressed powders, or in some cases such as with TNT, as castings, the majority of uses for explosives require mechanical properties that the pure materials do not have. In order to change the mechanical properties, as well as some of the thermal, output, or sensitivity properties, the pure explosives are often blended with other explosives and other inert materials. The resulting mixtures can then be worked in various ways to form specific explosive products. A list of these types of products follow:

- Pressings
- Castings
- Plastic bonded, machined
- Putties
- Rubberized
- Extrudable
- Binary
- Blasting agents
- Slurries and gels
- Dynamites

4.1 Pressings

Many of the explosive materials have crystal forms that are not amenable to pressing operations. The pressed pellets do not hold together well and can easily

Table 4.1 Pellet Density Versus Loading Pressure[a]

| | Loading Pressure (kpsi) | | | | | | |
Explosive	3	5	10	12	15	20	TMD[b]
Ammonium picrate	1.33	1.41	1.47	1.49	1.61	1.64	1.72
RDX	1.46	1.52	1.60	1.63	1.65	1.68	1.82
EDNA		1.39	1.46		1.51	1.55	1.71
Lead azide	2.46	2.69	2.98	3.05	3.16	3.28	4.68
Lead styphnate	2.12	2.23	2.43	2.47	2.57	2.63	3.10
PETN		1.48	1.61				1.76
Picric acid	1.40	1.50	1.57	1.59	1.61	1.64	1.76
Tetryl	1.40	1.47	1.57	1.60	1.63	1.67	1.73
TNT	1.34	1.40	1.47	1.49	1.52	1.55	1.65

[a] Densities in g/cm^3; Ref. 6.
[b] Theoretical maximum density.

flake apart. Some powders are extremely sensitive to electrostatic buildup or friction during pressing; others will not flow well enough into pressing dies or molds. To alleviate these problems, various additives are blended with the explosives. Molding lubricants are generally either graphite or stearates. Phlegmatizing agents such as petroleum jelly and mineral oil are used. Colorants and taggants, as well as antistatic agents, are also added to some pressing mixes.

Pressings can be of two types: (1) canned or cartridged, and (2) free standing. Pressings directly into a can, cup, or cartridge may often be made with a pure explosive with no additives. Sometimes pressing lubricants or antistats are added. Pressings of free-standing pellets are most often of explosives with a binder added. Some explosives, such as TNT and tetryl, can form pellets of fairly good mechanical properties without the addition of binders, but these are exceptions.

The density of the final pressed pellet, whether free standing or in a cup, depends upon both the material itself and upon the pressure at which it is pressed. Table 4.1 shows pellet density versus loading pressure for several common pressed explosives. In general, pressed pellets have an aspect ratio (length divided by diameter) of less than one because the powders tend to bridge. This, along with wall-to-powder friction forces, causes a density gradient through the pellet.

4.2 Castables

All of the modern castable formulations are based on mixtures of relatively higher-melting crystalline explosives and molten TNT. Since TNT has a negative oxygen balance, oxidizers, such as nitrates and positive oxygen balance explosives, are often added to it. Table 4.2 shows the weight percent of constituents of many of the TNT-based castables in use. Many of the castable explosives are also machineable. Their machineability, however, is not as good as that of the PBXs.

4.3 Plastic Bonded (PBX)

The PBXs are powdered explosives to which plastic binders have been added. The binder is usually precipitated out of solution in the preparation process such that it coats the explosive crystals. Agglomerates of these coated crystals form pressing "beads." The beads are then either die pressed or isostatically pressed at temperatures as high as 120°C. Pressures from 1 to 20 kpsi then produce pellets, or billets, with densities as high as 97% of the theoretical maximum density (TMD). The billets thus produced have good mechanical strength and can be machined to very close tolerances. Table 4.3 lists various PBXs in common use in the U.S. Department of Energy's (DOE) weapons laboratories. Table 4.4 defines some of the more common plastic binders used in PBXs.

4.4 PUTTIES

Putties are mixtures of finely powdered RDX and plasticizers. The mixture is puttylike and can be molded by hand to any desirable shape. Like modeling clay, it retains its shape unsupported after molding. Although many different putty compositions have been made, only one is now prevalent in the United States. That is Composition C-4. It has the following formulation:

Component	Percent (weight)
RDX	91.0
Di(2-ethylhexyl) sebacate	5.3
Polyisobutylene	2.1
20-weight motor oil	1.6

The British military have a similar mix called PE-4. This mix has 88% RDX and 12% plasticizer.

4.5 Rubberized

Mixtures of RDX or PETN mixed with rubber-type polymers and plasticizers can be rolled into rubbery gasketlike sheets. The sheets maintain their dimen-

Table 4.2 TNT-Based Castable Explosives[a]

Name	TNT	Aluminum	Ammonium Nitrate	Ammonium Picrate	Barium Nitrate	Boric Acid	Calcium Chloride	EDNA	HMX	Lead Nitrate	PETN	Sodium Picrate	RDX	Tetryl	Wax
Amatex	x		x												
Amatol	60		40												
Amatol	50		50												
Amatol	20		80												
Ammonal	67	11	22												
Baronal	35	15			50										
Baratol	33				67										
Baratol	24				76										
Boracitol	40					60									
Comp B	36												63		1
Comp B-2	40												55		5
Comp B-3	40												60		
Cyclotol	50												50		
Cyclotol	35												65		
Cyclotol	30												70		
Cyclotol	25												75		
DBX (Minex)	40	18	21										21		
Ednatol	45							55							

H-6	30	20		0.5			45	5
HBX-1	38	17		0.5			40	5
HBX-3	29	35		0.5			31	5
HTA-3	29	22			49			
Minol-2	40	20	40					
Octol	23				77			
Octol	25				75			
Octol	30				70			
Pentolite	50					50		
Pentolite	90					10		
Picratol	48		52					
Plumbatol	30				70			
PTX-1	20						30	50
PTX-2	31					27	42	
Sodatol	50					50		
Tetrytol	20							80
Tetrytol	25							75
Tetrytol	30							70
Tetrytol	35							65
Tritonal	80	20						
Torpex	40.5	18					40.5	1

References 4 and 5.

Table 4.3 Plastic Bonded Explosives

Name	Explosive Ingredients	Binder Ingredients	Color
LX-14-0	HMX, 95.5%	Estane & 5702-Fl, 4.5%	Violet spots on white
LX-10-0	HMX, 95%	Viton A, 5%	Blue-green spots on white
LX-10-1	HMX, 94.5%	Viton A, 5.5%	Blue-green spots on white
PBX-9501	HMX, 95%	Estane, 2.5%; BDNPA-F, 2.5%	White
PBX-9404	HMX, 94%	NC, 3%; CEF, 3%	White or blue
LX-09-1	HMX, 93.3%	BDNPA, 4.4%; FEFO, 2.3%	Purple
LX-09-0	HMX, 93%	BDNPA, 4.6%; FEFO, 2.4%	Purple
LX-07-2	HMX, 90%	Viton A, 10%	Orange
PBX-9011	HMX, 90%	Estane and 5703-Fl, 10%	Off-white
LX-04-1	HMX, 85%	Viton A, 15%	Yellow
LX-11-0	HMX, 80%	Viton A, 20%	White
LX-15	HNS-I, 95%	Kel-F 800, 5%	Beige
LX-16	PETN, 96%	FPC 461, 4%	White
PBX-9604	RDX, 96%	Kel-F 800, 4%	
PBX-9407	RDX, 94%	FPC461, 6%	Black or white
PBX-9205	RDX, 92%	Polystyrene, 6%, DOP, 2%	White
PBX-9007	RDX, 90%	Polystyrene, 9.1%; DOP, 0.5%; rosin, 0.4%	White or mettled gray
PBX-9010	RDX, 90%	Kel-F 3700, 10%	White
PBX-9502	TATB 95%	Kel-F 800, 5%	Yellow
LX-17-0	TATB, 92.5%	Kel-F 800, 7.5%	Yellow
PBX-9503	TATB, 80%; HMX, 15%	Kel-F 800, 5%	Purple

Reference 4.

Table 4.4 Binders Used in Plastic Bonded Explosives

Material	Description
Estane 5702-F1	Polyurethane solution system
Viton A	Vinylidine fluoride/hexafluoropropylene copolymer, 60/40 wt %
BDNPA-F	Bis(2,2-dinitropropyl)acetal/bis(2,2-dinitropropyl) formal, 50/50 wt %
CEF	Tris-β-chlorethylphosphate
Kel-F 800	Chlorotrifluoroethylene/vinylidine fluoride copolymer, 3:1
FPC 461	Vinyl chloride/chlorotrifluoroethylene copolymer, 1.5:1
Polystyrene	Styrene polymer, 100%
DOP	Di(2-ethylhexyl)phthalate
FEFO	Bis(2-fluoro-2,2-dinitroethyl)formal

[a] Reference 4.

Table 4.5 Thickness and Weight of DuPont Detasheets

Designation	Thickness (in.)	Weight (g/sq. in)
C-1	0.04	1
C-2	0.08	2
C-3	0.12	3
C-4	0.16	4
C-6	0.24	6
C-10	0.40	10

sional stability, and are very easy to handle. They can be cut to specified shapes and glued to a desired surface. One of the commercial products is manufactured by duPont under the trade name Detasheet™ These olive-drab colored sheets come in varying thicknesses, shown in Table 4.5. Although several types (varying percentages of HE) were formerly made, only the "Detasheet C" is currently available from DuPont. It consists of 63% PETN, 8% nitrocellulose (12.34 N), and 29% ATBC (an organic plastsizer).

Another version of Detasheet is manufactured for the military, which DuPont calls "Deta Flex." It contains the same plasticizers, but has approximately 70% RDX or PETN. It comes only in one-quarter-inch thickness and is also olive colored. A DOE version of this explosive is called LX-02-1. This material is colored blue and contains the following:

PETN	73.5%
Butyl rubber	17.6%
ATBC	6.9%
Cab-o-sil	2.0%

NAX (North American Explosives), a subsidiary of Ensign-Bickford Co., in partnership with Royal Ordnance of England, manufactures a sheet product called Primasheet 1000. This mixture contains 65% PETN, 8% NC, and 27% plasticizer. It is olive colored and comes in various thicknesses that include 1, 1.5, 2, 3, 4, 5, 6, and 8 millimeters. Another sheet from this company, Primasheet 2000, is their commercial version of the British military sheet explosive SX2. This material is a deep, almost Kelly green color. It contains 88.2% RDX and the balance is plasticizer. It is available in thickness of 2–7 mm in 1-mm increments.

4.6 Extrudables

PETN mixed with uncured Sylgard 182™ silicone rubber and curing agent at 80% PETN and 20% rubber, forms a thick viscous material that can be extruded under moderate pressures (less than 100 psi). After extrusion in holes, molds, or channels, the temperature can be raised to polymerize and cure the Sylgard, leaving a tough rubberlike material. Three DOE laboratory specifications exist for this type of explosive: LX-13 (green), XTX-8003 (white), and XTX-8004 (white). All three are 80:20 mixes of the above; they differ in the particle size of the PETN used, but XTX-8004 uses RDX instead of PETN.

If not used immediately after mixing, the uncured mix is frozen and stored. The freezing temperatures (below 0°F) slow the curing rate of the Sylgard rubber sufficiently to allow storage for up to 1 year.

A different extrudable explosive is available from NAX that is packaged in plastic mastic tubes (each containing 0.5 kg of explosive) to fit a standard caulking gun. This explosive is called DEMEX-400 and contains RDX and nonsetting plasticizer.

4.7 Binary

In this product form, the explosive consists of two materials stored and shipped in separate packages. Each of the materials is considered to be nonexplosive until mixed. The exact materials are proprietary; however, they are probably as shown in Table 4.6. After mixing, the various products form either a liquid, slurry, or wetted powder. They are all cap sensitive (can be initiated directly with a standard commercial blasting cap); however, some show two distinct detonation-velocity regimes, a very low velocity (1 mm/μs) from cap initiation and a high velocity (6-8 mm/μs) when initiated from a high-explosive booster.

4.8 Blasting Agents

"A blasting agent is defined as any material or mixture, consisting of a fuel and an oxidizer, intended for blasting, not otherwise classified as an explosive and in which none of the ingredients is classified as an explosive, provided that the finished product, as mixed and packaged for use or shipment cannot be detonated by means of a No. 8 blasting cap." (Ref. 7). Blasting agents, therefore, must be initiated by means of a booster. The size of the booster required depends not only on the particular blasting agent, but also on the amount of confinement in which it will be fired.

Blasting agents are primarily mixtures of ammonium nitrate (AN) and a fuel. The fuel is usually fuel oil (FO), added at about 6% fuel oil, by weight. Other fuels and various additives such as colorants are also used. The various blasting agents differ from each other not only in the particular mix used, but also in the particle sizes of the ingredients and in the packaging. Most blasting agents, when properly boosted, have a detonation velocity in the range of 2.4 to 4.8 mm/μs. Table 4.7 lists some commercial blasting agents.

Table 4.6 Commercial Binary Explosives

Name	Manufacturer	Component A	Component B	Mixed form	Detonation Velocity (mm/μs)	Detonation Pressure (kbar)
Kinestik	Atlas Powder Co.	Ammonium nitrate	Nitromethane	Wet powder	5.5	90
Kinepouch	Atlas Powder Co.	Ammonium nitrate	Nitromethane	Wet powder	5.5	90
Kinepak	Kinetics Intl. Corp.	Ammonium nitrate	Nitromethane	Wet powder	5.5	90
Boulder-Busters	Kinetics Intl. Corp.	Ammonium nitrate	Nitromethane	Wet powder	5.5	90
Marine Pac-Liquid	XPLO Corp.	Nitroparaffin or nitroso compound	Mixed aliphatic amines and aclicyclic polyamides	Liquid	6.2	
CAJUN Marine Pac	XPLO Corp.	Hydrazone or osazone	Nitrogen-containing inorganic salts	Liquid	7.9	
ASTRO-PAK (astrolite)	Explosives Corp. of America (EXCOA)	Ammonium nitrate plus other oxidizing inorganics like perchlorates	Ammonia and/or hydrazine or similar nitrogen-containing liquid	Liquid	8.0	
				Liquid		

Reference 7.

Table 4.7 Commercial, Prepackaged, Blasting Agents

Manufacturer	Name	Density (g/cm^3)	Detonation Velocity (mm/μs)	Detonation Pressure (kbar)
Apache Powder Co.	Hi-Density Carbonite	1.08	3.8	39
Apache Powder Co.	Carbonite P & PB	0 80		
Apache Powder Co.	Carbaglo			
Apache Powder Co.	Carbomal		0.9	
Atlas Powder Co.	Pellite		4.0	16
Atlas Powder Co.	Pellite-CR		4.2	27
Atlas Powder Co.	Pellite-HD		4.3	34
Atlas Powder Co.	Petron-A		3.4	43
E.I. duPont de Nemours & Co., Inc.	Nitramon A, HH, S, S-EL, WW, WW-EL			
	Nitramite 2, WW, WW-EL			
	Nilite ND, 303			
	Aluvite 1, 2, 3		3.6-4.8	
	Tovite		3.6-4.8	
	ANFO-P, -HD		2.7-4.8	
Austin Powder Co.	Austimon S		3.9-4.3	
	Austinite 15, 30		3.9-4.3	
Gulf Oil Chemicals Company	Gulf N-D-N 200, 500, 750, 51-B		4.1-4.5	
Energy Sciences and Consultants, Inc.	Thermoprimer D-30, D-35, D-40, D-45, D-50, D-55 (D-, indicates % by weight of metal particles in mix)			
	Temprel 3, 6, 9, 12, 15			
Hercules, Inc.	Vibronite S		3.1	
	Vibronite S-1		4.3	
	Dynatex B		3.8	
	Dynatex B-WR		3.0	
	Tritex 2		3.0	
	Hercomix 1, 2		3.8	
Trojan Division	Hydratol-5, -SA		2.6-2.9	
IMC Chemical Group, Inc.	Anoil-FR, HD, LD,		2.9-4.2	

Reference 7.

4.9 Slurries and Aqueous Gels

Of the explosives used in largest volume, slurries and aqueous gels are relatively new on the scene. They were introduced in the late 1950s and early 1960s. This mainstay of commercial blasting is basically a thickened supersaturated solution of AN in water with a fuel and sensitizer added. As shown earlier, AN,

(NH_4NO_3), has a very high positive oxygen balance ($+20$). The most common fuel used in slurries is aluminum powder. However, many other fuels are also used; among them are water-soluble fuels such as glycols and alcohols. The slurries are extremely insensitive to initiation; therefore, many of them are sensitized by addition of other powdered explosives such as PETN and TNT. They are also sensitized by the addition of glass microballoons. Slurries are thickened with gelatins and guar gums, as well as with water-soluble polymerizable plastics. Sensitized, heavily gelled slurries are packaged in cartridges like dynamites. Slurry ingredients can also be brought unmixed to the use site, where they are pumped from tanker trucks into mixing valves and from there directly into the blast pattern boreholes. This method uses most of the tonnage of slurries. The amount of slurries used in 1980 in the United States alone exceeded 4 billion pounds. For comparison, the military was using just about a half-billion pounds of explosives annually at the height of the war in Viet Nam.

Some slurries or aqueous gels are classified as blasting agents because they are not cap sensitive and do not contain any explosive ingredients. Some slurries are not classified as blasting agents because they do contain an explosive ingredient, usually as a sensitizer, but are not cap sensitive. Other slurries are cap sensitive. Some of the slurried explosives are also classified as permissable by the U.S. Bureau of Mines. This means that they may be used in coal mines with potentially explosive atmospheres. Table 4.8 lists a number of explosive slurries and aqueous gels.

4.10 Dynamites

Dynamites, for the purpose of this text are defined as prepared, cap-sensitive, explosive charges that contain nitroglycerin (NG). NG, an oily liquid, is extremely sensitive to initiation by shock. The sensitivity is due to the presence of bubbles of NG vapor, which are usually present throughout the liquid. Alfred Nobel discovered and patented the technique of absorbing NG in a porous medium that suppressed the effects of the bubbles and allowed the mixture to be safely handled. Originally, Nobel used kieselguhr (diatomaceous earth) as the absorbent. Today a wide variety of absorbents are used, along with any number of other additives. Some of the other additives are used to provide water resistance and to adjust oxygen balance and density. Sometimes halogen salts are added in order to suppress the formation of a large and hot fireball from the detonation. This is done to lower the chance of the dynamite igniting flammable and explosive gases present in some mines. Such dynamites are called "permitted." The following are some of the materials added to NG in modern dynamites:

Absorbents: wood pulp (sawdust), wood flour (very fine sawdust), apricot pits, ground coarse or fine, corn flour (very fine corn meal), bagasse (sugar cane pitch), oat hulls, "white" starch (usually corn or milo), chalk, zinc oxide, ground coal, charcoal, sulfur, and lampblack.

Table 4.8 Some Commercial Slurry or Aqueous Gel Explosives

Manufacturer	Product Name	Cap Sensitive	Classified as "Blasting Agents"	Detonation Velocity (mm/μs)
Bulk-site mixed				
E.I. duPont de Nemours & Co., Inc.	Tovan-Extra	No	No	
ERECO Chemicals	Iregel	No	Yes	3.0-4.6
Prepackaged				
Apache Powder Co.	arbagel, 5,10,15	No	Yes	3.5-4.0
	Dynagel, B	Yes	No	4.4
Atlas Powder Co.	Aquagel 70,80,270,280	Yes	No	3.9-4.4
	Aquanal, SS	No	Yes	5.0-5.5
	Aquaram	No	Yes	5.6
	Aquaflo	No	Yes	5.8
Austin Powder Co.	Slurimite 40 NCN	No	Yes	4.9
E.I. duPont de Nemours & Co., Inc.	Tovex 100,200,700,800, P,T-1,S,300,310,320	Yes	No	3.4-4.8
	Tovex 500, 650	Marginal	No	4.3-4.5
	Tovex Extra	No	No	5.7
	Pourvex Extra	No	No	4.9
	Drivex	Marginal	No	5.3
Energy Sciences and Consultants, Inc.	Thermoprimer W-30,40	No	Yes	
	MS-80-10, -80-15, -80-20, -80-25	No	Yes	
	Dellek 3,6	No	Yes	
Gulf Explosives, Gulf Oil Chemicals Company	Slurran 800, 805, 815	No	Yes	4.6-5.2
	Slurran 915, 915-W	Yes	No	4.6-4.8
	Detagel	Yes	No	4.9
Hercules Powder Co.	Gel-Power 0-1,-2,-3,4	No	Yes	4.3
	Gel-Strip	No	Yes	5.2
	Gel-Power A-1,-2,-3,4	Yes	No	4.0
IRECO Chemicals	Iregel-335,355,375,405,455	No	Yes	3.9-4.3
	Iretol DBA-1	Marginal	No	5.0
	Iretol 55T25	Marginal	No	5.3
	Iremex-F	Marginal	No	5.9
	Iremite	Yes	No	
	Ireprime	Yes	No	
IMC-Trojan Div.	Trojel EZ pour	No	No	5.0
	Trojan Trojel	No	No	6.1+

Reference 7.

Table 4.9 U.S. Commercial Dynamites

Manufacturer Product Name	Type Straight	Gelatin	Extra	Extra Gelatin	Permissible (Y or N)	Range of strength (%)	Detonation Velocity (mm/ms) Confined	Unconfined
Apache Powder Co.								
Straight	x				No	15–60		2.5–5.5
Standard			x		No	15–60		2.2–3.9
Special			x		No	20–55		1.8–3.1
Gelatin		x			No	20–100		3.5–7.2
Special Gelatin				x	No	30–90		4.0–6.0
Ammogel				x	No	5–70		1.1–1.9
Quarry			x		No	60		
Seismograph Hvg.		x			No	60		
Seismograph Spec. Gel			x	No	60			
Seismograph Ammogel			x	No	60			
Seismograph Pattern			x		No	60		
Atlas Powder Co.								
Ammodyte			x		No	40–60	3.6–4.0	2.6–3.1
Power Primer			x		No	75	5.5	5.3
Giant Gelatin				x	No	30–60	3.2–5.5	2.1–4.9
Petrogel		x			No	60		6.1
Gelodyn		x			No	65	3.7–4.6	3.0–3.7
Powerdyn		x			No	75	4.3	3.0
Coalite	x				Yes		2.0–3.6	1.8–2.7
Gelcoalite		x			Yes		4.05–5.2	2.9–4.3
Kleen-Kut					No	13–52	2.7–4.3	2.1–3.0
Floridyn			x		No	50	4.0	3.1
Austin Powder Co.								
NG & Ditching	x				No	30–60	3.0–4.7	3.4–5.7
Extra			x		No	20–60		
Apcogel B–1		x			No	69	5.3	
Extra Gelatin				x	No	30–80	4.2–6.2	

(Continued)

Table 4.9 U.S. Commercial Dynamites *(Continued)*

Manufacturer Product Name	Type				Permissible (Y or N)	Range of strength (%)	Detonation Velocity (mm/ms)	
	Straight	Gelatin	Extra	Extra Gelatin			Confined	Unconfined
Austin Powder Co. *(cont.)*								
Seismograph HV		x			No	60	5.3	
40 Extra Gelatin				x	No	40	5.3	
Red–E–Split B					No	40	5.3	
Red–E–Split D					No	58	3.0	
Red Diamond 2 thru 12	x				Yes	50–65		1.8–3.5
Red Diamond Gelatin B		x			Yes			4.9
E. I. duPont de Demeurs & Co., Inc.								
Red Arrow					No	70		4.0
Hi–Drive		x			No	85		6.9
Straight	x				No	30–50		3.5–5.5
Red Cross Extra			x		No	20–60		2.4–3.7
Red Cross Blasting FR			x		No	25–60		1.2–1.6
Toval 2			x		No	40		3.8
Stripkolex					No			
DuPont Gelatin		x			No	20–90		3.2–6.8
HV Gelatin		x			No	40–90		6.1–6.7
DuPont Extra HVA–HVH			x		No	20–55		2.7–3.3
DuPont Extra LVA–LVG			x		No	25–55		2.0–2.5
Hi–Cap			x		No	25–50		2.1–2.7
Gypsum A					No	30		2.1
Special Gelatin				x	No	25–80		4.0–5.2
Toval		x			No			4.0
Gelex		x			No	30–60		3.4–4.0
Loggers' & Loggers' 2			x		No	20–60		2.4–3.7
Seismograph 60% HV Gel		x			No	60		
Seismogel A		x			No			
Seismex, PW			x		No			
Hi–Cap S			x		No			
Sausage Powder					No	A,B,C	2.1–3.1	1.8–2.7
Trimtex		x			No			

Product						Grade	Density
Trimtex-Z			x		No	A thru D	2.4–2.8
Duobel A thru D	x				Yes	AA thru E	1.8–2.7
Monobel AA, A thru E	x				Yes	C, CC	1.8,1.7
Lump Coal C, CC	x				Yes	AA, C	5.0,3.7
Gelobel AA, C		x					
Hercules, Inc.							
NG Dynamite	x		x		No	50	5.3
Extra			x		No	40–60	3.0–3.8
Hercol			x		No	65–68	1.6–3.2
Hercon 2, 3			x		No	65	3.5–3.3
40% Gelatin		x			No	40	5.5
Hi-Pressure Gelatin		x			No	60	6.0
Gelatin Extra				x	No	40–75	5.5–7.0
Hercosplit WR					No		4.4
Unigel					No	66	3.0–4.6
Gelamite		x			No	62–70	6.0
Bivrogel 3		x			No		1.6
Bivrocol 1			x		No	65	1.6–3.2
Red H–A thru H–L	x		x		Yes	A–L	3.0
Collier–C	x				Yes		5.2
Hercogel–A		x			Yes		
Trojan (U.S. Powder) IMC							
Ditching	x		x		No	50	5.2
Special				x	No	20–60	2.4–4.0
Special Gelatin		x			No	40–60	4.8–5.2
Prima–Mite		x	x		No	75	5.8
Mine–Gel 1A, 1B, 2 thru 5					No	30–60	3.0–3.8
Super, Grade B thru H		s			No	20–45	2.1–2.9
Tru-Cut, Grade A thru F	x	x			No	15–50	2.7–4.6
Seismic Gelatin-2, 3, 60	x	x			No	60	5.0–5.5
Super Permissibles 2 thru 25		x			Yes	2 thru 25	2.0–2.9
Super Gel A thru I		x			Yes	A thru I	3.6–5.0

Reference 7.

Oxidizers: ammonium nitrate, sodium nitrate, and potassium nitrate (saltpeter).

Others: calcium carbonate and sodium chloride for flame suppression, and paraffin and guncotton (nitrocellulose) for water resistance and gelling.

NG-based dynamites have been replaced to a large extent by AN-based mixtures, which are much safer to use and are also less expensive. The number of manufacturers and the varieties of dynamites available is declining rapidly. Table 4.9 shows many of the dynamites that were still available through the late 1980s (it is estimated that by the turn of the century, NG dynamites will no longer be manufactured). Each type listed was sold in a variety of packaging types and strengths. The range of detonation velocities given is for the range of strengths offered for that type in that particular packaging configuration.

The types in the table are: straight, those having NG as the only energy-contributing constituent; gelatin, NG gelled with nitrocellulose or similar energy-contributing gelatinizing constituent; extra, AN and/or other oxidizing constituent added to ungelled NG; and extra/gelatin, AN and/or other oxidizers added to gelled NG.

U.S. Army military dynamite does not contain any NG. Instead, it is made of a mixture containing TNT, RDX, cornstarch, and nondetergent engine oil. It is equivalent to 60% straight dynamite and has an unconfined detonation velocity of approximately 6.1 mm/μs. It has no problems with freezing or oozing and extruding because it contains no NG. There are three designations for military dynamite: M1, M2, and M3. They differ only in the size of the cartridge, not in the ingredients.

M1 = 1.25 in. diam. x 8 in. long, $\frac{1}{2}$ lb. each,
M2 = 1.5 in. diam. x 8 in. long, $\frac{3}{4}$ lb. each,
M3 = 1.5 in. diam x 12 in long, 1 lb. each.

Military dynamite cartridges are paraffin-impregnated paper with a paraffin coating. Although the cartridges are not absolutely waterproof, military dynamite remains usable after 24 h submersion in water.

5

Estimating Properties of Explosives

The references listed at the end of this section are excellent sources for properties data for many explosives. However, if those references are not available, or if properties data for new or proposed explosive compounds are required, then some means of reasonably estimating the values must be used. In this chapter we will explore some surprisingly accurate methods of estimating the theoretical maximum density (TMD) of an explosive as well as its detonation velocity at TMD, knowing only the structural formula for the explosive compound. Further, we will be able to estimate the detonation velocity at densities other than at TMD, once that value is known; and also, once having found the detonation velocity and density, we can estimate the detonation (or Chapman-Jouguet, CJ) pressure from those values.

5.1 Estimation of Theoretical Maximum Density

The following methodology and data were generated by L. T. Eremenko, of the U.S.S.R. (Refs. 8 and 9). Eremenko found that TMD could be estimated to less than 2–3% error by a simple linear relationship between the density and the hydrogen content of substituted organic molecules.

$$\rho(\text{TMD}) = a_i - k_i H$$

where $\rho(\text{TMD})$ is the theoretical maximum density; a_i, a constant; k_i, a constant; and H, percent by weight of hydrogen in the explosive molecule. The method depends upon H being greater than zero and less than around 6% ($0 < H < 6\%$).

The values of a_i and k_i depend upon the structural group, or homolog, to which the explosive molecule belongs. The definitions of the groups and the values for a_i and k_i are as follows:

Group I Liquid aliphatics with relatively symmetrical substitutions around the main chain. Figure 5.1 shows an example of this in the compound 1,3-dinitro-propane. The nitro groups are at each end of the molecule, and the molecule is said to be symmetrical. Another example, given in Figure 5.2, is 2,2-dinitropro-pane. Here also, the substituents are symmetrically placed on the main chain. Also included in this group are those molecules that have two or more kinds of substituents, even if they are not placed exactly symmetrically.

For this group of liquid aliphatic explosive compounds, the TMD is found from:

$$\rho(\text{TMD}) = 1.780 - 0.096H$$

TMD is at 20°C, and is in the units g/cm^3. For the dinitropropane example above, both the 1,3-, or the 2,2-homologs are symmetrical, $C_3H_6N_2O_4$:

$$H = 100 \times \text{weight of hydrogen/molecular weight}$$

$$H = (100)(6)(1.008)/[(3)(12.01)+(6)(1.008)+(2)(14.01)$$

$$+(4)(16.00)] = 4.51$$

$$\rho(\text{TMD}) = 1.780 - (0.096)(4.51) = 1.347 \text{ g/cm}^3$$

Reference data show that the measured TMD of 1,3-dinitropropane at 20°C is 1.354. Therefore, the error encountered is $100(1.347 - 1.354)/(1.354) = -0.52\%$.

Group II Liquid aliphatics in which the substituents are definitely not symmet-rically placed around the main chain. An example is 1,1-dintropropane, shown in Figure 5.3. Elementally, this is the same as the previous example, $C_3H_6N_2O_4$, but the nitro groups are bunched at one end. For these unsymmetrical aliphatic liquids, the TMD is found from:

$$\rho(\text{TMD}) = 1.584 - 0.067H$$

Evaluating our example, where $H = 4.51$,

$$\rho(\text{TMD}) = 1.584 - (0.067)(4.51) = 1.282 \text{ g/cm}^3 .$$

Reference data show this compound to have a TMD at 20°C of 1.258 g/cm^3. This time our error is $(100)(1.282 - 1.258)/(1.258) = 1.91\%$.

$$O_2N-CH_2-CH_2-CH_2-NO_2$$

Figure 5.1.   1,3-Dinitropropane compound.

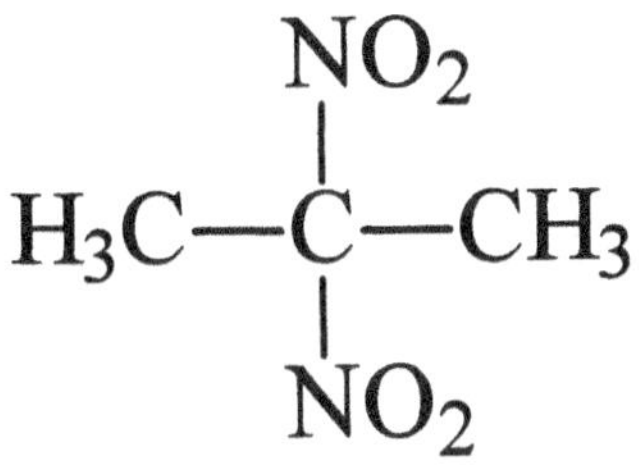

$$NO_2$$

$$H_3C-\overset{\overset{\displaystyle NO_2}{|}}{\underset{\underset{\displaystyle NO_2}{|}}{C}}-CH_3$$

Figure 5.2. 2,2-Dinitropropane compound.

Group III Solid, noncyclic, aliphatic compounds containing only nitro and (or) nitrate substituents:

$$\rho(TMD) = 2.114 - 0.169H$$

Group IV Solid, noncyclic, aliphatic secondary polynitroalkylamines and polynitroalkylamides:

$$\rho(TMD) = 2.114 - 0.151H$$

The alkylamine and alkylamide compounds contain the structural groups shown in Figure 5.4.

Group V Solid, noncyclic, aliphatic secondary nitramines containing terminated ethylenenitrate and (or) 2,2-dinitropropyl groups:

$$\rho(TMD) = 2.114 - 0.134H$$

The secondary nitramines are those inside a chain as shown in Figure 5.5. Eremenko defines terminated ethylenenitrate as shown in Figure 5.6. Figure 5.7 shows the 2,2-dinitropropyl groups.

Group VI Solid, noncyclic, aliphatic primary nitramines and nitrates of primary amines:

$$\rho(TMD) = 2.118 - 0.103H$$

Primary nitramines are at the terminus, or end, of the aliphatic chain.

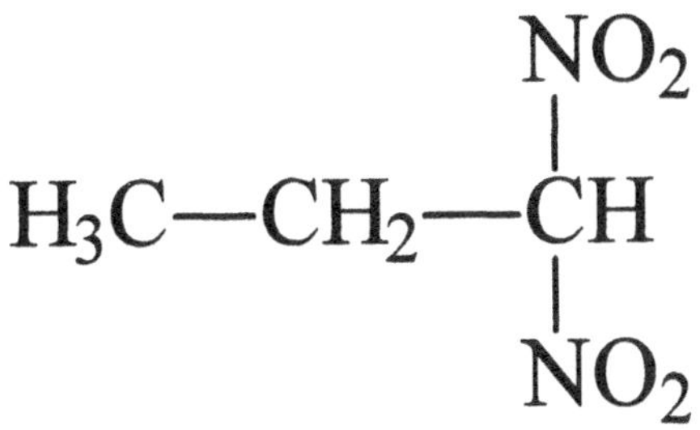

$$NO_2$$

$$H_3C-CH_2-\overset{\overset{\displaystyle NO_2}{|}}{\underset{\underset{\displaystyle NO_2}{|}}{CH}}$$

Figure 5.3. 1,1-Dintropropane compound.

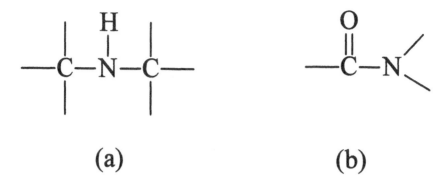

Figure 5.4.   (a) Alkylamine; (b) alkylamide.

Group VII Solid, nitro- and (or) nitroxycyclanes and -oxacyclanes:

$$\rho(\text{TMD}) = 2.085 - 0.143H$$

These are aliphatic cyclo compounds that contain oxygen as one of the members of the ring.

Group VIII Solid, nitrazacyclanes and nitrazaoxacyclanes:

$$\rho(\text{TMD}) = 2.086 - 0.093H$$

These are aliphatic rings that contain either nitrogen, or nitrogen and oxygen as members of the ring. RDX, HMX, and Sorguyl are in this group. The TMD of RDX is calculated as an example and shown in Figure 5.8. The elemental formula is $C_3H_6N_6O_6$.

$$H = (100)(6)(1.008)/[(3)(12.01)+(6)(1.008)+(6)(14.01)$$
$$+(6)(16)] = 2.7226$$
$$\rho(\text{TMD}) = 2.086 - (0.093)(2.7226) = 1.833 \ \text{g/cm}^3$$

Reference data show for RDX that $\rho(\text{TMD})$ at 20°C is 1.816 g/cm³. The error encountered is $(100)(1.833 - 1.816)/(1.816) = 0.94\%$.

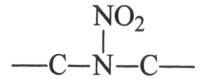

Figure 5.5.   Secondary nitramines.

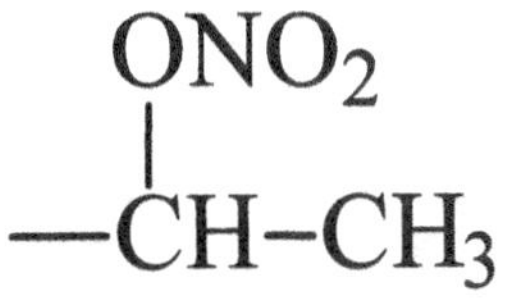

$$\overset{\displaystyle ONO_2}{\underset{\displaystyle -CH-CH_3}{|}}$$

Figure 5.6. Terminated ethylenenitrate.

Group IX Solid, normal (straight-chain) nitrazaalkanes:

$$\rho(TMD) = 2.114 - 0.114H$$

These preceeding nine groups constitute the estimation of TMD for aliphatic explosives. The following four groups are for estimation of TMD of aromatic explosives.

Group X Compounds containing nonhydrogen substituents in an aromatic ring or rings of condensed or jointed aromatic systems. The bridges that join the aromatic rings may be different by nature, but atoms in the chain must be only any one or a combination of carbon, hydrogen, nitrogen, oxygen, chlorine, or fluorine.

$$\rho(TMD) = 1.948 - 0.141H$$

An example of this group is shown in Figure 5.9. The elemental formula for HNS is $C_{14}H_6N_6O_{12}$. Therefore, the value of H is:

$$H = (100)(6)(1.008)/[(14)(12.01)+(6)(1008)$$

$$+(6)(14.01)+(12)(16)] = 1.343 \text{ g/cm}^3$$

$$\rho(TMD) = 1.948 - (0.141)(1.343) = 1.759 \text{ g/cm}^3$$

The reference data show for HNS at 20°C that $\rho(TMD) = 1.740$, and the error $= (100)(1.759 - 1.740)/(1.740) = 1.1\%$.

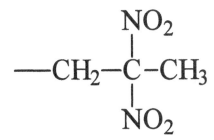

Figure 5.7. 2,2-Dinitropropyl groups.

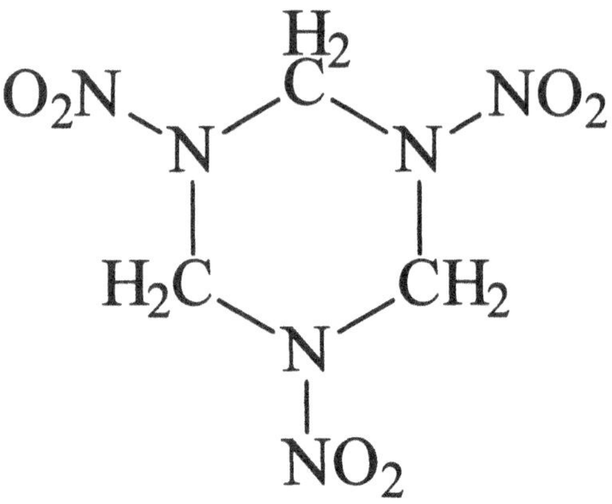

Figure 5.8. RDX, hexahydro-1,3,5-trinitro-1,3,5-triazine.

Group XI Aromatic hydrocarbons, heteroatomic aromatic compounds as well as compounds of Group X, which have additional hydrogen-containing substituents or functions.

$$\rho(\text{TMD}) = 1.954 - 0.130H$$

Examples of this are TNT (the —CH$_3$ group is the additional hydrogen-bearing substituent) and trinitrophenol (picric acid), in which the —OH group is the additional hydrogen bearer.

Group XII Anilines substituted on the ring or related heterocompounds having no more than one substituent or function on an amino nitrogen:

$$\rho(\text{TMD}) = 1.984 - 0.124H$$

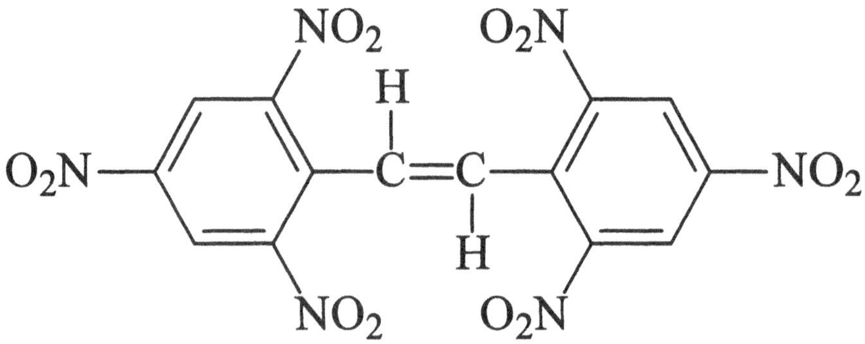

Figure 5.9. HNS, 2,2',4,4',6,6'-hexanitrostilbene.

Trinitroaniline (TNA) is an example of this group. Tetryl is not because both a nitro- and a methyl group are substituted on the amino nitrogen.

Group XIII Compounds having substituents that form strong one- or two-dimensional intramolecular hydrogen bonds:

$$\rho(TMD) = 2.094 - 0.132H$$

By fitting a compound, structurally, into any one of the above thirteen groups, we can closely estimate its TMD. The values in the equations yield TMD at 20°C. When there are differing crystal forms for the same compound, this method tends to give the denser one.

5.2 Estimation of Detonation Velocity AT TMD

This method, like the previous one for TMD, is based upon empirical calculation of values based on molecular structure. This method, developed by L.R. Rothstein at the U.S. Naval Weapons Station in Yorktown, PA (Refs. 10 and 11) gives the detonation velocity, D', for an explosive at its TMD. Rothstein's data show that this method is good to within $\pm3.5\%$ error for 99% of the 80 explosives tested, and within $\pm2.4\%$ for 95% of those tested. Two functions must be calculated to find the detonation velocity, D', at TMD. The first is the chemical structural function, which yields the factor, F. The second is a simple linear equation relating F to D':

$$F = 100\left(\frac{\Phi + \Psi}{MW}\right) - G$$

$$\Phi = n(O) + n(N) + n(F) - \left(\frac{n(H) - n(HF)}{2n(O)}\right)$$

$$\Psi = \frac{A}{3} - \frac{n(B\,/\,F)}{1.75} - \frac{n(C)}{2.5} - \frac{n(D)}{4} - \frac{n(E)}{5}$$

where $A = 1$ if the compound is aromatic, otherwise $A = 0$; $G = 0.4$ for liquid explosives, and $G = 0$ for solids; $n(O)$ is the number of oxygen atoms in molecule; $n(N)$ the number of nitrogen atoms in molecule; $n(H)$ the number of hydrogen atoms in molecule; $n(F)$ the number of fluorine atoms in molecule; $n(HF)$ the number of hydrogen fluoride molecules that can possibly form from available hydrogen; $n(B/F)$ the number of oxygen atoms in excess of those available to form CO_2 and H_2O and/or the number of fluorine atoms in excess of those available to form HF; $n(C)$ the number of oxygen atoms doubly bonded directly to carbon (as in a ketone or ester); $n(D)$ the number of oxygen atoms singly bonded directly to carbon (as in C—O—R, and where R can be —H, —NH₄ , —C, etc.); and $n(E)$ the number of nitrato groups existing either as a nitrate ester or as a nitric acid salt such as hydrazine mononitrate. If $n(O) = 0$, or if $n(HF) > n(H)$, the term in large parentheses in the equation for $\Phi = 0$.

Having obtained the value of factor F, the detonation velocity D' at TMD is found from:

$$D' = \frac{F - 0.26}{0.55}$$

Let us examine two different examples, as shown in Figures 5.10 and 5.11.
The elemental formula of Figure 5.10 is $C_3H_5N_3O_9$.

$A = 0$, since NG is not aromatic,

$G = 0.4$, since nitroglycerine is a liquid,

$n(O) = 9$, since the number of oxygen atoms in the NG molecule is 9,

$n(N) = 3$, since the number of nitrogen atoms in NG is 3,

$n(H) = 5$, since the number of hydrogen atoms in NG is 5,

$n(F) = 0$, since there is no fluorine in NG,

$n(HF) = 0$, since there is no fluorine in NG, no hydrogen fluoride can be formed,

$n(B/F) = 0.5$, 9 oxygen atoms are available. Two and a half of these are required to form 2.5 moles of water from the five hydrogen and six of the oxygens and needed to form 3 moles of carbon dioxide from the three carbons. This is a total of 8.5 oxygen needed to form CO_2 and H_2O, leaving 0.5 atoms left over,

$n(C) = 0$, since no oxygen atoms are double bonded to carbon in the NG molecule,

$n(D) = 0$, since no oxygen atoms are bonded singly to both carbon and another group other than in the form of a nitrate ester (which is covered in the next variable),

$n(E) = 3$, since there are three nitrate ester groups,

$MW = 227.1$, the molecular weight of NG is $(3)(12.01) + (5)(1.008) + (3)(14.01) + (9)(16)$.

Armed with these variables now quantified, we can calculate the value of F:

$$F = 100 \left(\frac{9 + 3 + 0 - \dfrac{5 - 0}{2 \times 9} + \dfrac{0}{3} - \dfrac{0.5}{1.75} - \dfrac{0}{2.5} - \dfrac{0}{4} - \dfrac{3}{5}}{227.1} \right)$$

$$- 0.04 = 4.372$$

$$D' = \frac{4.372 - 0.26}{0.55} = 7.48 \text{ km/s}$$

From the references, the detonation velocity of NG is found to be 7.60 mm/μs. The error of the estimation in this example is $100(7.48 - 7.60)/7.60 = -1.6\%$.

The example shown in Figure 5.11 is a fluorinated solid explosive called PF. PF is 1-fluoro-2,4,6-trinitrobenzene (picryl fluoride). The elemental formula is

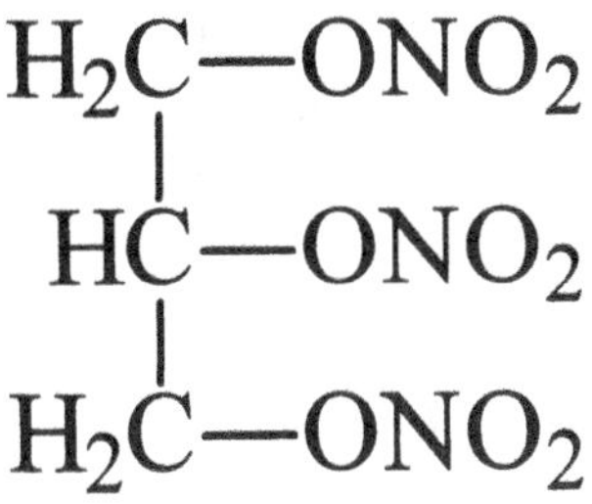

$$H_2C-ONO_2$$
$$HC-ONO_2$$
$$H_2C-ONO_2$$

Figure 5.10.   A liquid, nitroglycerine.

$C_6H_2N_3O_6F$. The molecular weight $=$ $(6)(12.01)+(2)(1.008)+(3)(14.01)$ $+(6)(16)+(18.998) = 231.1$.

> $A = 1$ (this molecule is aromatic),
> $G = 0$ (this is a solid),
> $n(O) = 6$,
> $n(N) = 3$,
> $n(H) = 2$,
> $n(F) = 1$,
> $n(HF) = 1$,
> $n(B/F) = 0$ (this molecule is obviously underoxidized and underfluorinated, so there is no extra O or F),
> $n(C) = 0$,
> $n(D) = 0$,
> $n(E) = 0$.

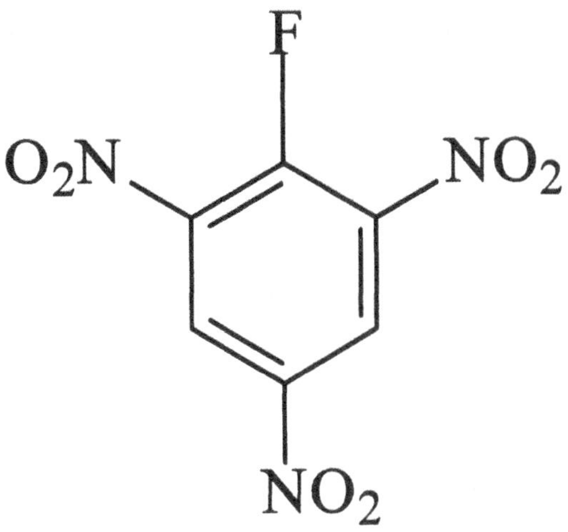

F

O_2N NO_2

NO_2

Figure 5.11.   PF (fluorinated solid explosive).

$$F = 100 \left(\frac{6 + 3 + 1 - \dfrac{2 - 1}{2 \times 6} + \dfrac{1}{3} - \dfrac{0}{1.75} - \dfrac{0}{2.5} - \dfrac{0}{4} - \dfrac{0}{5}}{231.1} \right)$$

$$- 0 = 4.435$$

$$D' = \frac{4.435 - 0.26}{0.55} = 7.59 \text{ km/s}$$

Data in the reference gives $D' = 7.50$ mm/μs. The error is $(100)(7.59 - 7.50)/(7.50) = +1.2\%$.

We can now estimate both the TMD and D at TMD for an explosive, based only upon its structural formula. Of course we can seldom, if ever, utilize an explosive at its TMD. Pressings, at best, can approach TMD perhaps within several percent. More often, we work at densities considerably lower. It is necessary, therefore, to be able to correct D' to its value at some lower density.

5.3 Detonation Velocity as a Function of Density

The density of an explosive affects the detonation velocity. For most explosives, the relationship between detonation velocity D and density of the unreacted explosive is close to linear over reasonable ranges of density. This means it takes the form:

$$D = a + b\rho$$

where a and b are empirical constants specific to each particular explosive. If D is known at one density, then it can be estimated at another by only needing to know the value of b.

$$(D_1 = a + b\rho_1) - (D_2 = a + b\rho_3) \rightarrow (D_1 - D_2) = b(\rho_1 - \rho_2)$$

For most purposes, since ρ is usually changed only over a small range, it can be assumed that $b = 3$. This approximation is normally good within a 10 to 15% range of change in density. The approximation is based on the average b from data for a number of explosives found in the references:

$$D_1 = D_2 + 3(\rho_1 - \rho_2)$$

If the reference or starting detonation velocity was for density at TMD, and we are seeking the detonation rate at some lower density, then:

$$D = D' - 3(\text{TMD} - \rho)$$

As an example, we can find that at TMD (1.77 g/cm^3), PETN has a detonation velocity of $D' = 8.29$ km/s. In order to find the velocity at a lower density, say 1.67 g/cm^3:

$$D = (8.29) - (3)(1.77 - 1.67) = 7.99 \text{ km/s}$$

Note that this estimate is probably only good down to about 0.85 TMD. Reference experimental data show that at 1.67 g/cm³, D for PETN is 7.98 km/s. This is an error of $(100)(7.99 - 7.98)/(7.98) = 0.13\%$.

An in-depth look at the effects of density on detonation is given in Chapter 21.

5.4 Estimating Detonation Velocity of Mixtures

We have just seen how to estimate ρ(TMD) and D' (D at TMD) from the chemical structure of an explosive molecule. That is a lot of estimating power. Another tool to add to this kit is estimation of D for mixtures of explosives. In the 1940s, Manny Urizer at Los Alamos Scientific Laboratory found that by adding the detonation velocity on a partial volume basis one could arrive at the D for a mixture (Ref. 4).

$$D_{mix} = \Sigma D_i V_i$$

where D_{mix} is the detonation velocity of the mixture, D_i the detonation velocity at TMD of individual explosive component or characteristic velocity of the non-explosive component, and V_i the volume fraction of the component.

Not only does this work for mixtures of explosives, but also for mixtures of explosives with additional oxidizers, inert fillers, and even void space. This is achieved by assigning an "equivalent characteristic velocity" to the nonexplosive ingredients. These values, used in like manner in the equation above, yield excellent estimates of the measured D_{mix}. Table 5.1 gives the characteristic velocity of various fillers and oxidizers as well as of air. A note on using this: Urizer assumes air or void as one of the ingredients, and so D' must be used for the explosive components of the mixture.

This method also gives us an additional and even more accurate method (as compared to that in Section 5.3) to estimate detonation velocities of pure explosives at densities other than TMD if we know the velocity at TMD. Notice that void space is listed with a characteristic velocity of 1.5 km/s. If we use the Urizer method where the mix consists only of an explosive at TMD and voids, then we can derive

$$D = D' \left(\frac{\rho}{\rho_{TMD}} \right) + 1.5 \left(1 - \frac{\rho}{\rho_{TMD}} \right) \text{ or}$$

$$D = 1.5 + \rho \left(\frac{D' - 1.5}{\rho_{TMD}} \right)$$

Whereas the slope of the rule given in Section 5.3 was only good for ranges of 10 to 15% change in densities, this estimate applies well over the entire range of densities achievable. An example of this is shown in Figure 5.12, where we see experimentally derived values of detonation velocity of PETN plotted versus

Table 5.1   Characteristic Velocities D_i

Material	Density ρ (g/cm^3)	Characteristic Velocity D_i (km/s)
Polymers and plasticizers		
Adiprene L	1.15	5.69
AFNOL	1.48	6.35
Beeswax	0.92	6.50
BDNPA-F (50/50 wt % eutectic)	1.39	6.31
BDNPF	1.42	6.50
CEF	1.45	5.15
DNPA	1.47	6.10
EDNP	1.28	6.30
Estane 5740-X2	1.2	5.52
Exon-400 XR61	1.7	5.47
Exon-454 (85/15 wt % PVC/PVA)	1.35	4.90
FEFO (as constituent to ~35%)	1.60	7.20
Fluoronitroso rubber	1.92	6.09
Halowax 1014	1.78	4.22
Kel-F wax		5.62
Kel-F elastomer	1.85	5.38
Kel-F 800/827	2.00	5.83
Kel-F 800	2.02	5.50
Neoprene CNA	1.23	5.02
NC	1.58	6.70
Paracril BJ (Buna-N nitrile rubber)	0.97	5.39
Polyethylene	0.93	5.55
Polystyrene	1.05	5.28
Saran F-242		5.55
Silastic 160		5.72
Sylgard 182	1.05	5.10
Teflon	2.15	5.33
Viton A	1.82	5.39
Inorganic additives		
Air or void		1.5
Al	2.70	6.85
Ba(NO$_3$)$_2$	3.24	3.80
KClO$_4$	2.52	5.47
LiClO$_4$	2.43	6.32
LiF	2.64	6.07
Mg	1.74	7.2
Mg/Al alloy (61.5/38.5 wt %)	2.02	6.9
NH$_4$ClO$_4$	1.95	6.25
Si02 (Cab-O-Sil)	2.21	4.0

Table 5.1 Characteristic Velocities D_i *(Continued)*

Material	Density ρ (g/cm^3)	Characteristic Velocity D_i (km/s)
	Pure explosives at TMD	
DATB	1.84	7.52
FEFO (invalid when <35% present)	1.61	7.50
HMX	1.90	9.15
NQ	1.81	8.74
PETN	1.78	8.59
RDX	1.81	8.80
TATB	1.94	8.00
TNT	1.654	6.97

Reference 4.

the initial density over a broad range that even includes fine particles of PETN dispersed in air.

5.5 Estimating Detonation Pressure

From the preceding material we are able to estimate the detonation velocity of an explosive at any particular density. These two parameters, D, the detonation velocity, and ρ, the density of the unreacted explosive, can be used to estimate the detonation or Chapman-Jouguet (CJ) pressure, P_{CJ}. It can be shown (but not within the context of this module) that the CJ pressure is:

$$P_{CJ} = \frac{\rho D^2}{\gamma + 1}$$

where P_{CJ} is the CJ, or detonation pressure, in gigapascals (GPa); ρ is the density of the unreacted explosive, in g/cm^3; γ is the ratio of specific heats of the detonation product gases; and D is the detonation velocity, in km/s.

In general, the detonation product gases are molecules such as H_2O, CO, CO_2, N_2, etc. The particular composition or molar ratio of the product gases is, as we have seen, a function of the composition of the explosive. However, for most explosives, the product composition is fairly similar and for the mixture, at the high temperatures and pressures encountered in detonations, is also similar. In the range of explosive densities from around 1 to 1.8, γ is approximately equal to 3.

If we insert that value into the above relationship, we find:

$$P_{CJ} = \rho D^2/4$$

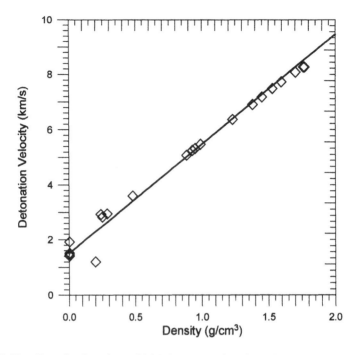

Figure 5.12.   Data for D and ρ_o of PETN compared to the Urizer estimate.

As an example, let us estimate the CJ pressure of RDX at an initial density of 1.767 g/cm^3. At this density, the detonation velocity is 8.70 km/s; so the CJ pressure is:

$$P_{CJ} = (1.767)(8.70)^2/4 = 33.4 \text{ GPa}$$

Experimental data in the referenced literature show that P_{CJ} for RDX at 1.767 g/cm^3 is equal to 33.8 GPa. The error for our estimate is, therefore, $(100)(33.4 - 33.8)/(33.8) = -1.2\%$.

6

Decomposition

We have been discussing fast oxidation reactions such as burning and detonation and have seen that in those reactions, the products formed were at the maximum oxidation state possible for the particular fuel-to-oxidizer ratio, or oxygen balance, of the explosive being examined. Another reaction of importance when dealing with explosives is the decomposition reaction that occurs at a very slow rate and at low temperatures, relative to burning and detonation.

6.1 Decomposition Reactions

When an explosive slowly decomposes, the products may not follow the previously described hierarchy or be at the maximum oxidation states. The nitro, nitrate, nitramines, acids, etc., in an explosive molecule can break down slowly. This is due to low-temperature kinetics as well as the influence of light, infrared, and ultraviolet radiation, and any other mechanism that feeds energy into the molecule. Upon decomposition, products such as NO, NO_2, H_2O, N_2, acids, aldehydes, ketones, etc., are formed. Large radicals of the parent explosive molecule are left, and these react with their neighbors. As long as the explosive is at a temperature above absolute zero, decomposition occurs. At lower temperatures the rate of decomposition is infinitesimally small. As the temperature increases, the decomposition rate increases. Although we do not always, and in fact seldom do, know the exact chemical mechanism, we do know that most explosives, in the use range of temperatures, decompose with a zero-order reaction rate. This means that the rate of decomposition is usually independent of the composition of or the presence of the reaction products. The rate depends only upon temperature. This process is described by the Arrhenius equation, shown below. Energy transfer mechanisms such as conduction, convection, shock, impact or friction heating, and even nuclear radiation provide the temperature increase that drives the decomposition reaction.

$$\frac{d(A'/A)}{dt} = k$$

$$k = Ze^{-E_a/RT}$$

where A' is the the amount of explosive that has not yet decomposed, A the amount of explosive with which we started, t the time, k the reaction rate constant, Z the reaction rate frequency factor, E_a the activation energy for the decomposition reaction, R the universal gas constant, and T the absolute temperature.

To determine the values of the various parameters in the reaction-rate equation, extensive experiments must be run. Unfortunately, these parameters have not been evaluated for very many explosives. Table 6.1 lists lumped parameter values for a few very common explosives.

An example of the use of the decomposition-rate equation is as follows: Suppose that RDX will be held at a high temperature (175°F) for one year. How much, or what percent, of the RDX will have decomposed in that time?

(a) 175°F = 352.4 K,
(b) 1 yr = 3.1536×10^7 s,
(c) from the rate equation,

$$\frac{d(A'/A)}{dt} = Ze^{-E_a/RT}$$

and the values for Z and E_a/R for RDX from Table 6.1, we get

$$\frac{d(A'/A)}{dt} = (3.5575 \times 10^{18})(e^{-24,027/352.4}) = 8.72 \times 10^{-12} \ s^{-1}$$

(d) Since we now have the rate of decomposition at our storage temperature, we can find the change in weight for the time period at that temperature.

Table 6.1 Decomposition-Rate Equation Constants[a]

Explosive	Z (s^{-1})	E_a/R (K)
TNT	6.3529×10^{14}	22105
RDX	3.5575×10^{18}	24027
PETN	4.05565×10^{16}	20933
Nitrocellulose	9.8916×10^{13}	17056

[a] Calculated from values in Ref. 12.

$d(A'/A) = 8.72 \times 10^{-12} \, dt$

and dt was 3.1536×10^7 s, then

$d(A'/A) = 2.75 \times 10^{-4}$, or 0.0275%

Unfortunately, even when data such as in Table 6.1 are available, their use is rather limited. The rates are based on experiments using very pure materials. The impurities present often catalyze or even alter the decomposition reaction. Therefore, to use rate data realistically, the rates must be for the particular lot or lot purity of the material that is actually going to be used in the configuration of a design.

Often, accelerated aging tests are conducted with samples drawn from a production lot. These tests look at aging at elevated temperatures over short periods of time, using the Arrhenius relationship to allow trading temperature for time.

It is also important to note that most explosives do not decompose with zero-order kinetics at higher temperatures. The reaction order varies with the kind of explosive and can vary with the extent of decomposition. Decomposition at higher temperatures can be autocatalytic, which means that the presence of decomposition products makes the reaction proceed even faster.

More often we are not interested so much in the exact rate of decomposition as we are in how a particular explosive compares to others in its storage life. For that purpose, much simpler tests are run, such as the thermal stability tests.

6.2 Thermal Stability Tests

In general, thermal stability testing runs the gamut from visual inspection of samples that have been heated for some period of time to measurement of the quantity and composition of the gases evolved.

6.2.1 Vacuum Stability Test

A sample, the weight of which depends upon the particular standard test being run, is placed in a vessel, and a vacuum is drawn to remove air from the system. The vessel is attached to a manometer apparatus and then is placed in a constant-temperature bath and held at temperature for a given period of time. The amount of gas produced can be calculated from pressure readings and a knowledge of the vessel's volume and temperature.

The standard test for military explosives uses 5 g for the sample size (1 g for initiating explosives). The samples are heated for 40 h at various temperatures (90, 100, 120, 135, and 150°C). The results of the tests are reported as standard cubic centimeters of gas produced per 40 h at each of the specified temperatures. A similar test conducted by Lawrence Livermore National Laboratory (LLNL) uses 1-g samples heated at 120°C for 48 h.

In the LLNL chemical reactivity test (CRT), a sample is heated at 120°C for

22 h, and the gases produced are separated by a gas chromatograph into individual volumes of N_2, NO, CO, N_2O, and CO_2. The data from the LLNL CRT, however, are still reported as total gases evolved per 0.25 g of sample. Table 6.2 gives some typical vacuum stability and CRT data for several explosives.

6.2.2 Explosion Temperature (Henkin) Test

This test bridges the gap in the growth from thermal decomposition reaction to explosion and eventually involves fast oxidation reactions. A small sample of explosive is pressed into a blasting cap cup made of gilding metal. The cup is then inserted into a molten Wood's Metal bath. The time it takes from insertion in the bath until some noticeable reaction takes place (usually a mild explosion) is noted. The test is repeated at several different bath temperatures. See Table 6.3. A smooth curve is drawn through the data points (time to explosion versus bath temperature), and the temperatures that cause reaction in 1, 5, and 10 s are interpolated from the graph.

An additional test is done with this same apparatus, the ''0.1 second test.'' In this test, a pinch (yep, that's right, a pinch!) of high explosive is thrown onto the surface of the melted Wood's Metal. If it does not react immediately, then the bath temperature is raised, and the test is repeated until the temperature is found that causes the thrown sample to ''pop'' on contact with the bath surface. This is reported as the ''explosion temperature at 0.1 s.''

Very slow heating rates can mask an explosive material's ability to react violently. This is often observed in DTA tests.

6.2.3 Differential Thermal Analysis

The following is quoted from Ref. 4.

> In the usual DTA [differential thermal analysis] analysis, identical containers are set up (one containing the sample and the other containing a standard reference substance) in identical thermal geometries with temperature sensors arranged to give both the temperature of each container and the difference in temperatures between containers. The data are displayed as DTA thermograms; the temperature difference is plotted against the temperature of the sample. The standard reference material chosen is one whose thermal behavior does not change rapidly (over the temperature range of interest). Such a plot is nearly a straight line if the sample also has no rapidly changing thermal behavior (or if it is very similar to the standard material). Excursions above and below a background line (baseline) result from endo- or exothermic (heat-absorbing or heat-releasing) changes. The DTA analyses permit interpretation for phase changes, decomposition and kinetic information, melting points, and thermal stability. Sample sizes are less than 40 mg.

Figure 6.1, a DTA thermogram of pure (99.9%) HMX, is a good example. The pronounced endotherm (dip) is due to a change in crystal form (a phase transformation) at the temperature shown ($\sim 200°C$). Soon after that, decomposition becomes quite evident, and at around 250°C the sharp break in the curve

Table 6.2 Vacuum Stability Test Data

| Explosive | Picatinny Test (5 g/40 h) | | | | LLNL Vacuum Stability Test (1 g/48 h) 120°C | LLNL Chemical Reactivity Test (CRT) (0.25 g/ 22 h) 120°C |
	100°C	120°C	135°C	150°C		
HMX	0.37	0.45		0.62	0.07	<0.01
HNS						0.01
Lead azide	1.0	0.07			<0.4	
Lead styphnate	0.4	0.3			<0.4	
NC (12.0% N)	1.0	11+(16 h)			5.0	1.0–1.2
NQ	0.37	0.44				0.02–0.05
PETN	0.5	11+				0.10–0.14
RDX	0.7	0.9		2.5	0.12–0	0.02–0.025
Tetryl	0.3	1.0		11+		0.036
TNT	0.1	0.23	0.44	0.65	0.005	0–0.012

References 4 and 5.

Table 6.3 Temperature for Explosion (°C)

Explosive	0.1 s	1 s	5 s	10 s
Ammonium nitrate			465	
Baratol			385	
Black Powder	510	490	427	356
Composition B	526	368	278	255
Composition C-4			290	
DNT				310
EDNA	265	216	189	178
HMX	380		327	306
Lead azide	396	356	340	335
Lead styphnate			282	276
Mercury fulminate	263	239	210	199
Nitroglycerine			222	
Nitroguanidine			275	
Octol (75/25)			350	
Pentolite (50/50)	290	266	220	204
PETN	272	244	225	211
RDX	405	316	260	240
Silver azide	310		290	
Tetryl	340	314	257	238
TNT	570	520	475	465

Reference 5.

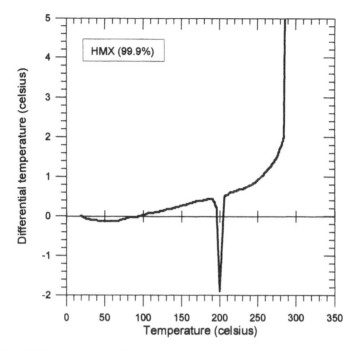

Figure 6.1. DTA thermogram of HMX.

indicates a change in the decomposition mechanism to a more rapid reaction. Following that, the HMX decomposes faster and faster as it approaches its melting point (285°C). Once melting takes place, the sample will ignite spontaneously because the liquid-phase decomposition kinetics are much faster than the solid-phase decomposition kinetics.

A similar test is performed on the differential scanning calorimeter (DSC) apparatus. The thermograms are similar but allow for quantitative interpretation of the results. Currently, the DSC is the more common apparatus used. In addition to its enabling quantitative interpretation of results, the DSC is a more sensitive instrument and thus allows smaller samples to be used. A 40-mg sample, upon violent deflagration, can result in substantial damage to the instrument.

6.2.4 Thermogravimetric Analysis

In the TGA test, a small sample, approximately 2 to 10 mg. is placed in an apparatus that continuously weighs it while it is being heated. The heating rate is held as constant as possible, usually at around 10°C per minute.

The data produced are displayed as a plot of weight remaining versus temperature, or if temperature is held constant, the plot is weight versus time. The data are useful for determining thermal stability and chemical reaction. Figure

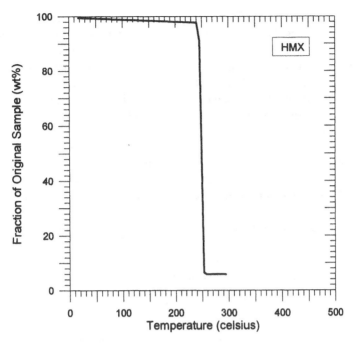

Figure 6.2.　TGA for pure HMX.

6.2, the TGA for pure HMX, verifies what we saw previously in the DTA thermogram; that noticeable decomposition begins at about 240°C and rapidly increases in rate at around 250°C.

6.3 Chemical Compatibility

It was alluded to earlier that decomposition reaction rate and mechanism were affected by the presence of other substances. The chemical reactions between explosives and other substances are called chemical incompatibilities. These are very important in explosive systems that must be stored for long periods of time. Compatibility studies are directed not only to address the problem of degradation of explosives by other materials in a system, but also to address the degradation of other parts of a system caused by the explosive. Examples of the latter are: (1) a case where a substance in the explosive compound caused severe corrosion of the metal bridge wire in a hot wire initiator, and (2) where NO_2 vapors given off by the slow decomposition of an explosive in one part of a weapon system caused severe corrosion and subsequent failure of a printed circuit in another area of the weapon.

The approach to studying compatibility is a logical, step-by-step method. It requires anticipation of a problem and an analysis of the occurrence of changes in a given system. For organic explosives, this change is usually the evolution of gaseous products. For metals, the changes are manifested as corrosion of the surfaces; for other materials, a combination of these two. First, visual inspection is used to see if any noticeable changes occur in color, texture, or other surface features. Next, suspect materials, such as adhesives, etc., are mixed with the explosive sample, and these are then subjected to DTA, DSC, and TGA analyses to see if the presence of the material has any effect on the thermogram of the explosive. These are compared to the thermograms of the individual material components in question. An incompatibility is usually indicated by a shift to a lower temperature of the explosive decomposition thermogram. Gross incompatibilities can usually be screened by these methods. CRTs are then run with both single materials and a mixture, using gas and liquid chromatography to analyze the samples. Results from the mixture are compared to results from the single materials to determine whether any reaction has occurred in the mixture. Additional data are gathered from "coupon tests," where sandwiches of explosive and other materials are assembled and exposed to elevated temperatures for a given time. A match in contact surface areas is attained with the coupon tests. This is important since a given design has surface contact of the potentially incompatible materials, not mixtures of them. After exposure to high temperatures, the coupons are disassembled and studied with the aid of both optical and scanning electron microscopes to detect changes in the surface structures. This is then, in many cases, followed by long-term aging in the use design configuration and subsequent complete chemical and microscopic analyses. Aging of the use design configuration is the final proof of a compatibility test because it incorporates the actual materials, surfaces, and processes that can subtly hide a compatibility problem.

References

1. *Handbook of Chemistry and Physics*, 53rd Ed., PPC-1 thru C-52, "Definitive Rules for Nomenclature of Organic Chemistry," Chem Rubber Publ., Cleveland, 1973.

2. Stillman, J. C., *Picric Acid Warning*, Information Bulletin No. 6, IABTI, February 1983.

3. Meyer, R., *Explosives*, 2nd Ed., Verlag Chemie, New York, 1981.

4. Dobratz, B. M., Ed., *LLNL Handbook of Explosives*, UCRL-52997, Lawrence Livermore National Laboratory, Livermore, California, 1981.

5. AMCP 706-177, Engineering Design Handbook, Explosive Series, *Properties of Explosives of Military Interest*, U.S. Army Material Command, Washington D.C., 1971.

6. AMCP 706-179, Engineering Design Handbook, Explosive Series, *Explosive Trains*, U.S. Army Material Command, Washington, D.C., 1974.

7. Hermann, S., *Explosives Data Guide*, Explosives Research Institute, Arizona, 1977.

8. Eremenko, L. E., *Interrelationship between Density and Structure in an Explosive*, in Proceedings of the 11th Symposium on Explosives and Pyrotechnics, Philadelphia, Pennsylvania, 1981.

9. Eremenko, L. T., *Density Calculation of Aromatic Crystals Using Their Structure*, in Proceedings of the 8th International Pyrotechnics Symposium, Colorado, 1982.

10. Rothstein, L. R., and Petersen, R., *Predicting High Explosives Detonation Velocities from Their Composition and Structure*, Propellants and Explosives, Vol. 4, No. 4, 1979.

11. Rothstein, L. R., *Predicting High Explosives Detonation Velocities from Their Composition and Structure*, Propellants and Explosives, Vol. 6, No. 4, 1981.

12. Popolato, A., *High Explosive Materials*, Proceedings of Behavior and Utilization of Explosives in Engineering Design, 12th Annual Symposium, Albuquerque, New Mexico, 1972.

Bibliography

Cook, M. A., *The Science of High Explosives*, Chapman & Hall, London, 1958 (reprint 1971).

Taylor, W., *Modern Explosives*, Royal Institute of Chemistry, London, 1959.

Dutton, W. S., *One Thousand Years of Explosives*, Holt Rinehart & Winston, New York, 1960.

Urbanski, T., *Chemistry and Technology of Explosives* (4 volumes), Pergamon Press, Oxford, 1964–1967.

Fortnam, S., *High Explosives and Propellants*, Pergamon Press, Oxford, 1966.

Davis, T. L., *The Chemistry of Powder and Explosives*, Wiley, New York, 1966.

Ellern, H., *Military and Civilian Pyrotechnics*, Chem. Publ. Co., New York, 1968.

TM9-1300k-214/T0llA-1-34, *Military Explosives*, Depts. of the Army and Air Force, Washington, D.C., 1970.

Johansson, C. H., and, Persson, P. A., *Detonics of High Explosives*, Academic Press, London, 1970.

Newhouser, C. R., *Introduction to Explosives*, National Bomb Data Center, Gaithersburg, Maryland, 1973.

Rossi, B., and Pozdnyadov, Z., *Commercial Explosives and Initiators* (translated from Russian), NTIS, U.S. Dept. of Commerce, 1973.

Cook, M. A., *The Science of Industrial Explosives*, IRECO Chemicals, Inc., Salt Lake City, Utah, 1974.

Fair, H. D., and Walker, R. F., *Energetic Materials*, Vols. I & II, Plenum Press, New York, 1977.

Fickett, W., and Davis, W. C., *Detonation*, University of California Press, Berkeley, California, 1979.

McLain, J. H., *Pyrotechnics*, Franklin Institute Press, Philadelphia, Pennsylvania, 1980.

Encyclopedia of Explosives and Related Items, Vols. 1–10, Picatinny Arsenal (U.S. AARADCOM), Dover, New Jersey, 1960–1982.

SECTION

II

ENERGETICS OF EXPLOSIVES

Introduction

We will review the basic quantities of thermodynamics: energy, temperature, heat, work, and the ideal gas law. These quantities will be used to explain the principles of thermophysics and thermochemistry, which will be applied to the specific reactions of combustion and detonation. Using the thermochemical data of heats of detonation or explosion, we will see how to calculate adiabatic reaction temperatures. These data in turn will be used to analyze or predict pressures of explosions in closed vessels. We shall also see how, using thermochemical data, to predict detonation velocities and detonation pressures.

7

Basic Terms of Thermodynamics

We shall begin with the basics of thermodynamics and see how these basics are used to develop engineering calculations and estimates dealing with explosives. Hougan, Watson, and Ragatz, in *Chemical Process Principles* (Ref. 1) give an excellent basic description of energy, temperature, and heat. The following paragraphs are quoted, with permission, directly from that text.

7.1 Energy

The properties of a moving ball, a swinging pendulum, or a rotating flywheel are different from those of the same objects at rest. The differences lie in the motions of the bodies and in the ability of the moving objects to perform work, which is defined as the action of a force moving under restraint through a distance. Likewise, the properties of a red-hot metal bar are different from those of the same metal bar when cold. The red-hot bar produces effects on the eye and the touch that are very different from those of the cold bar.

Under the classification of potential energy are included all forms not associated with motion but resulting from the position and arrangement of matter. The energy possessed by an elevated weight, a compressed spring, a charged storage battery, a tank of gasoline, or a lump of coal is potential energy. Similarly, potential energy is stored within an atom as the result of forces of attraction among its subatomic parts. Thus potential energy can be further classified as external potential energy, which is inherent in matter as a result of its position relative to the earth, or as internal potential energy, which resides within the structure of matter.

In contrast, energy associated with motion is referred to as kinetic energy. The energy represented by the flow of a river, the flight of a bullet, or the rotation of a flywheel is kinetic energy. Also, individual molecules possess kinetic energy by virtue of their translational, rotational, and vibrational motions. Like potential energy, kinetic energy is subclassified as internal kinetic energy, such as that associated with molecular and atomic structure, and as external kinetic energy, such as that associated with external motion.

In addition to the forms of energy associated with composition, position, or motion of matter, energy exists in the forms of electricity, magnetism, and radiation, which are associated with electronic phenomena.

The science pertaining to the transformation of one form of energy to another is termed *thermodynamics*. Early studies of the transformation of energy led to the realization that, although energy can be transformed from one form to another, it can never be destroyed, and that the total energy of the universe is constant. This principle of the conservation of energy is referred to as the first law of thermodynamics. Many experimental verifications have served to establish the validity of this law.

7.2 Temperature and Heat

Energy may be transferred not only from one form to another but also from one aggregation of matter to another without change of form. The transformation of energy from one form to another or the transfer of energy from one body to another always requires the influence of some driving force. As an example, if a hot metal bar is placed in contact with a cold one, the former will be cooled and the latter warmed. The sense of ''hotness'' is an indication of the internal energy of matter. The driving force which produces a transfer of internal energy is termed temperature and that form of energy which is transferred from one body to another as a result of a difference in temperature is termed heat.

7.3 Internal Energy

The general concepts of energy, temperature, and heat were introduced under the broad classification of potential and kinetic energies. Both these forms were subclassified into external forms determined by the position and motion of a mass of matter relative to the earth or other masses of matter and into internal forms determined by the inherent composition, structure, and state of matter itself, independent of its external position or motion as a whole.

The internal energy of a substance is defined as the total quantity of energy that it possesses by virtue of the presence, relative positions, and movements of its component molecules, atoms, and subatomic units. A part of this energy is contributed by the translational motion of the separate molecules and is particularly significant in gases where translational motion is nearly unrestricted in contrast to the situation in liquids and solids. Internal energy also includes the rotational motion of molecules and of groups of atoms that are free to rotate within the molecules. It includes the energy of vibration between the atoms of a molecule and the motion of electrons within the atoms. These kinetic portions of the total internal energy are determined by the temperature of the substance and by its molecular structure. The remainder of the internal energy is present as potential energy resulting from the attractive and repulsive forces acting between molecules, atoms, electrons, and nuclei. This portion of the internal energy is determined by molecular and atomic structures and by the proximity of the molecules and atoms to one another. At the absolute zero of temperature all translational energy disappears, but a great reservoir of potential energy and a small amount of vibrational energy remain.

The total internal energy of a substance is unknown, but the amount relative to some selected temperature and state can be accurately determined. The crystalline state and hypothetical gaseous state at absolute zero temperature are commonly used as references for scientific studies, whereas engineering calculations are based on a variety of reference conditions arbitrarily selected.

7.4 Energy in Transition: Heat and Work

In reviewing the several forms of energy previously referred to, it will be noted that some are capable of storage, unchanged in form. Thus, the potential energy of an elevated weight or the kinetic energy of a rotating flywheel is stored as such until by some transformation they are converted, in part at least, to other forms.

Heat represents energy in transition under the influence of a temperature difference. When heat flows from a hot metal bar to a cold one, the internal energy stored in the cold bar is increased at the expense of that of the hot bar, and the amount of heat energy in transition may be expressed in terms of the change in internal energy of the source or of the receiver. Under the influence of a temperature gradient, heat flows also by the bodily convection and mixing of hot and cold fluids and by the emission of radiant energy from a hotter to a colder body without the aid of any tangible intermediary.

It is inexact to speak of the storage of heat. The energy stored within a body is internal energy, and when heat flows into the body, it becomes internal energy and is stored as such. Zemansky writes as follows:

The phrase ''the heat in a body'' has absolutely no meaning. Perhaps an analogy will clinch the matter. Consider a fresh water lake. During a shower, a certain amount of rain enters the lake. After the rain has stopped, there is no rain in the lake. There is water in the lake. Rain is a word used to denote water that is entering the lake from the air above. Once it is in the lake, it is no longer rain. Another form of energy in transition of paramount interest is work, which is defined as the energy that is transferred by the action of a mechanical force moving under restraint through a tangible distance. It is evident that work cannot be stored as such but is a manifestation of the transformation of one form of energy to another. Thus, when a winch driven by a gasoline engine is used to lift a weight, the internal energy of the gasoline is transformed in part to the potential energy of the elevated weight, and the work done is the energy transferred from one state to the other.

7.5 Energy Units

The basic concept that mechanical work is equal to force times the distance through which the force acts leads to definitions of units of mechanical energy. The common units are as follows: (1) The erg is the amount of work done (energy expended) when a force of one dyne acts through a distance of one centimeter. (2) The joule. Since one erg is an inconveniently small unit, the joule, equal to 10^7 ergs, is more commonly used. (3) The newton-meter is the work done where a force of one newton acts over a distance of one meter: 1 newton-meter = 1 joule. (4) The foot-pound of mechanical energy is expended when a force of one pound acts through a distance of 1 foot.

The common units of energy in the field of electrical engineering are the watt-second and the kilowatt-hour. These energy units are usually thought of as being inherently electrical in nature, yet they are, in reality, mechanical energy units. The watt-second, for example, is defined as being equal to 1 (joule) = 10^7 (erg) = 10^7 (dyne) (cm).

In problems dealing with the production, generation, and transfer of heat, it is customary to use special units of energy called heat units. For many years, these units of thermal energy were defined in terms of the heat capacity of water. A variety of units developed because of the fact that various masses of water and various temperature scales were selected to define the units. Furthermore, it was soon recognized that the heat capacity of water varies with temperature, and, accordingly, a temperature specification was included in the definitions of the units of thermal energy. The units of thermal energy bear no derivable relation to mechanical energy units, and it was therefore necessary to determine, by experiment, the ''mechanical equivalent of heat,'' which related the two independent sets of units.

The common units of thermal energy as formerly defined were as follows:

1. The gram-calorie: the energy required to heat one gram of water through a temperature range of one degree centigrade. Because of the variable heat capacity of water it was customary either to specify the temperature of the water or to take a mean value over a specified temperature range. The 15-degree gram-calorie was defined as the energy required to heat one gram of water from 14.50 to 15.50° (C), at a pressure of one atmosphere. The mean gram-calorie was defined as 1/100 of the energy required to heat one gram of water from 0 to 100° (C) at a pressure of one atmosphere.
2. The kilogram-calorie. Because the gram-calorie is a rather small unit, it frequently is more convenient to use a unit 1000 times as great, the kilogram-calorie. The 15-degree kilogram-calorie and the mean kilogram-calorie were formerly defined in a manner similar to the way in which the corresponding gram-calories were defined, except that one kilogram of water was involved in the definition.
3. The British thermal unit (Btu): the energy required to heat one pound of water through a temperature range of 1° (F). Because of the variable heat capacity of water, it was necessary with this unit, just as with the gram-calorie, either to specify the temperature of the water or to use a mean value. The 60-degree Btu and the mean Btu between 32 and 212° (F) were in common use. In both instances, a constant pressure of 1 atmosphere was included in the definition.

While thermal energy units were defined for many years as indicated above, it is now customary to define them arbitrarily in terms of mechanical units, with no reference to the heating of water. At the present time, there are two ''defined'' gram-calories in wide use, the United States National Bureau of Standards thermochemical gram-calorie and the so-called steam gram-calorie (also called the I.T. gram-calorie because it was defined in 1929 by the International Steam Table Conference). . . .

Although the basic definition of the gram-calorie is no longer associated with the heat capacity of water, it is still true that, as a very close approximation, the National Bureau of Standards thermochemical calorie represents the energy required to heat one gram of water from 14.50 to 15.50° (C) (the old 15 calorie). Also, the I.T. gram-calorie still represents, as a very close approximation, 1/100 of the energy required to raise one gram of water from 0 to 100° (C) (the old mean calorie). From a practical engineering standpoint, it is therefore still legitimate to think of the gram-calorie as being the energy required to heat one gram of water one centigrade degree and the Btu as being the energy required to heat one pound of water one Fahrenheit degree.

7.6 Enthalpy

In the preceding quoted material, the concept of internal energy of a substance was introduced. In this section Q denotes internal energy per unit mass of a substance and m denotes the mass of the substance. Initially, let us consider a closed system (no material either enters or leaves). Heat, however, can enter or leave the system. The net heat added to the system (heat added minus heat extracted) we will call q. If this system has rigid boundaries (i.e., the volume of the system remains constant), then all the net heat entering the system goes to increase the internal energy of the material in the system.

$$q_v = mQ_2 - mQ_1, \text{ or } q_v = m \, \Delta Q \tag{7.1}$$

where q_v is the heat entering a constant-volume process, m the amount of material in the system (either in weight units or moles), Q_1 the internal energy per unit amount of material at the start of the process, and Q_2 the internal energy per unit amount of material at end of process.

If the system does not have rigid boundaries, that is, it expands as heat is added, and we allow the expansion to occur such that the pressure in the system remains constant, then the system is doing work in expansion against this pressure. This work is equal to the pressure times the change in volume and is equivalent to an energy output of the system. Therefore, in order that all energy entering the process must equal all energy leaving the process:

$$q_p = mQ_2 - mQ_1 + w \tag{7.2}$$

where q_p is the heat entering a constant-pressure process, w the work done in expansion, and $w = mPV_2 - mPV_1$, and where P is the pressure of the system, V_1 the starting volume per unit amount of material, and V_2 the ending volume per unit amount of material.

So, for a constant-pressure process:

$$q_p = mQ_2 - mQ_1 + mPV_2 - mPV_1, \text{ or} \tag{7.4}$$

$$q_p = m(Q_2 + PV_2) - m(Q_1 + PV_1) \tag{7.5}$$

The quantity, $U + PV$, is defined as enthalpy, and H is the enthalpy per unit amount of material (either weight or moles).

$H = U + PV$

Therefore, Eq. (7.5) becomes:

$$q_p = mH_2 - mH_1, \text{ or}$$

$$q_p = m \, \Delta H$$

(7.6)

Reiterating the above, for a constant-volume process, $q_v = m \, \Delta Q$, and for a constant-pressure process, $q_p = m \, \Delta H$.

CHAPTER

8

Thermophysics

Now that we defined heat, temperature, energy, work, and enthalpy, let us apply these definitions to the behavior of matter when its internal energy is changed. We shall examine in this chapter quantitatively how a solid is heated to its melting point and then melts. We will follow the liquid melt to the boiling point, through vaporization and into the gas phase. We shall be able to calculate the temperatures and amounts of energy required to change a given material from any given thermal state to any other state.

8.1 Heat Capacity of Gases

The heat capacity, C, of a substance is defined as the quantity of energy a given amount of that substance must absorb in order to raise its temperature one degree.

$$mC = \frac{dq}{dT} \tag{8.1}$$

where m is the amount of material (weight or moles), C the heat capacity, dq the change in heat, and dT the change in temperature.

Usually we use calories per gram per Kelvin, or BTU per pound per degree Rankine as units of heat capacity. Numerically, the two groups of units are identical. We also use heat capacity on a molar basis, in which case the units are per gram mole or per pound mole, and again these are numerically identical.

$$1\frac{(cal)}{(g)(K)} = 1\frac{(BTU)}{(lb)(^0 R)}, \text{ or}$$

$$1\frac{(cal)}{(g \text{ mole})(K)} = 1\frac{(BTU)}{(lb \text{ mole})(^0 R)}$$

(8.2)

If we heat a substance and restrict its volume such that the volume remains constant, then all the heat goes into increasing the substance's internal energy. Therefore, the heat capacity at constant volume is:

$$C_v = \frac{dQ}{dT}$$

(8.3)

If we allow the substance to expand as it is heated, then as we saw earlier, some of the heat added increases the internal energy, and some is used in the work involved in expansion. So the heat capacity at constant pressure is:

$$C_p = \frac{dH}{dT}$$

(8.4)

From this it is easy to see that the quantity C_p is greater than C_v.

Let us examine this for an ideal gas. Keep in mind we are looking at a constant-pressure process.

$$C_p = \frac{dH}{dT}$$

(8.5)

$$H = Q + PV$$

(8.6)

$$dH = dQ + P\,dV, \text{ and therefore}$$

(8.7)

$$C_p = \frac{dQ + P\,dV}{dT} = \frac{dQ}{dT} + \frac{P\,dV}{dT}$$

(8.8)

For an ideal gas, $PV = nRT$, where n is the number of moles and R is the universal gas constant. Since we are looking at a constant pressure process with a constant amount of material:

$$P\,dV = nR\,dT, \text{ and } P\frac{dV}{dT} = nR$$

(8.9)

If we are dealing on a unit molar basis, then $n = 1$, and

$$P\frac{dV}{dT} = R$$

(8.10)

When we defined heat capacity, we said that $C_v = dQ/dT$, so replacing this in Eq. (8.8) along with Eq. (8.10) we have:

$$C_p = C_v + R$$

(8.11)

For monatomic gases such as helium, argon, or xenon, etc., the molar heat capacity at constant volume is approximately 3(cal)/(g mole)(K). In the same system of units R is about 2; so the heat capacity at constant pressure for an ideal, monatomic gas is around 5(cal)/(g mole)(K).

For multiatomic gases, much of the energy absorbed goes into rotational and vibrational energy in the bonds; so the heat capacity must be higher than that for the monatomic gases.

The heat capacity of all substances increases with increasing temperature. Table 8.1 lists the molar heat capacities of several common gases at various temperatures.

Table 8.1 Molal heat capacities of gases at constant pressure (P=0); units: (g cal) / (g mole)(K)

TK	H_2	N_2	CO	Air	O_2	NO	H_2O	CO_2
300	6.896	6.961	6.965	6.973	7.019	7.134	8.026	8.894
400	6.974	6.991	7.013	7.034	7.194	7.162	8.185	9.871
500	6.993	7.070	7.120	7.145	7.429	7.289	8.415	10.662
600	7.008	7.197	7.276	7.282	7.670	7.468	8.677	11.311
700	7.035	7.351	7.451	7.463	7.885	7.657	8.959	11.849
900	7.139	7.671	7.787	7.785	8.212	7.990	9.559	12.678
1000	7.217	7.816	7.932	7.928	8.335	8.126	9.861	12.995
1100	7.308	7.947	8.058	8.050	8.440	8.243	10.145	13.26
1200	7.404	8.063	8.168	8.161	8.530	8.343	10.413	13.49
1300	7.505	8.165	8.265	8.258	8.608	8.426	10.668	13.68
1400	7.610	8.253	8.349	8.342	8.676	8.498	10.909	13.85
1500	7.713	8.330	8.419	8.416	8.739	8.560	11.134	13.99
1600	7.814	8.399	8.481	8.483	8.801	8.614	11.34	14.10
1700	7.911	8.459	8.536	8.543	8.859	8.660	11.53	14.2
1800	8.004	8.512	8.585	8.597	8.917	8.702	11.71	14.3
1900	8.092	8.560	8.627	8.647	8.974	8.738	11.87	14.4
2000	8.175	8.602	8.665	8.692	9.030	8.771	12.01	14.5
2100	8.254	8.640	8.699	8.734	9.085	8.801	12.14	14.6
2200	8.328	8.674	8.730	8.771	9.140	8.828	12.26	14.6
2300	8.398	8.705	8.758	8.808	9.195	8.852	12.37	14.7
2400	8.464	8.733	8.784	8.841	9.249	8.874	12.47	14.8
2500	8.526	8.759	8.806	8.873	9.302	8.895	12.56	14.8
2750	8.667	8.815	8.856	8.945	9.431	8.941	12.8	14.9
3000	8.791	8.861	8.898	9.006	9.552	8.891	12.9	15.0
3250	8.899	8.900	8.933	9.060	9.663	9.017	13.1	15.1
3500	8.993	8.934	8.963	9.108	9.763	9.049	13.2	15.2
3750	9.076	8.963	8.990	9.150	9.853	9.079	13.2	15.3
4000	9.151	8.989	9.015	9.187	9.933	9.107	13.3	15.3
4250	9.220	9.013	9.038	9.221	10.003	9.133	13.4	15.4
4500	9.282	9.035	9.059	9.251	10.063	9.158	13.4	15.5
4750	9.338	9.056	9.078	9.276	10.115	9.183	13.5	15.5
5000	9.389	9.076	9.096	9.308	10.157	9.208	13.5	15.6

Reference 1.

Table 8.2 Empirical constants for molal heat capacities of gases at constant pressure $(p = 0)$ $C_p = a + bT + cT^2$, where T is in degrees Kelvin; (g cal) / (g mole) (K); temperature range 300 to 1500 K

Gas	a	$b(10^3)$	$c(10^6)$
H_2	6.946	−0.196	+0.4757
N_2	6.457	1.389	−0.069
O_2	6.117	3.167	−1.005
CO	6.350	1.811	−0.2675
NO	6.440	2.069	−0.4206
H_2O	7.136	2.640	+0.0459
CO_2	6.339	10.14	−3.415
SO_2	6.945	10.01	−3.794
SO_3	7.454	19.13	−6.628
HCl	6.734	0.431	+0.3613
C_2H_6	2.322	38.04	−10.97
CH_4	3.204	18.41	−4.48
C_2H_4	3.019	28.21	−8.537
Cl_2	7.653	2.221	−0.8733
Air	6.386	1.762	−0.2656
NH_3^a	5.92	8.963	−1.764

Reference 1.

Fortunately the plots of molar heat capacity of gases versus absolute temperature are very smooth and fit the form of a simple quadratic equation over the temperature range in which we normally use them (from room temperature to several thousand degrees kelvin).

$$C_p = a + bT + cT^2$$

The constants a, b, and c vary with each particular gas. Table 8.2 gives the values for several common gases.

Example 8.1 Using the values from Table 8.2, calculate the molar heat capacity of carbon dioxide (CO_2) at 600 K (327°C).

1. The table gives the values of the empirical constants as: $a = 6.339$, $b = 10.14 \times 10^{-3}$, and $c = -3.415 \times 10^{-6}$.
2. $C_p = a + bT + cT^2 = 6.339 + (10.14 \times 10^{-3})(600) + (-3.415 \times 10^{-6})(600)^2 = 11.194$ cal/(g mole)(K)

The experimental value given in Table 8.1 is 11.311. The agreement between them is 1.04%. Heating calculations using a fixed value of C_p (or C_v) are good only over a relatively short temperature interval.

Example 8.2 How much thermal energy (heat) is required to bring one mole of CO_2 gas from 590 to 650 K?

1. $C_p = \dfrac{dq}{dT}$

$dq = C_p \, dT$

2. C_p (at 600 K) = 11.311 (cal)/(g mole)(K)

$$dT = 650 - 590 = 60 \text{ K}$$

3. $dq = (11.311)(60) = 678.7 \text{ cal}$

Over broad temperature ranges, the value of C_p (or C_v) changes sufficiently to make it necessary to take the temperature effect into account. To do this we merely integrate the quadratic approximation.

Example 8.3   How much heat is required to bring one mole of CO_2 gas from room temperature, 300 K (27°C, 80.6°F) to 1000 K?

1. $dq = C_p \, dT$

$$= (a + bT + cT^2) \, dT$$

$$= a \, dT + bT \, dT + cT^2 \, dT$$

2. $q = \displaystyle\int dq = \int_{T_o}^{T_1} (a \, dT + bT \, dT + cT^2 \, dT)$

$$= a \int_{T_o}^{T_1} dT + b \int_{T_o}^{T_1} T \, dT + c \int_{T_o}^{T_1} T^2 \, dT$$

$$= a(T_1 - T_o) + \frac{b}{2}(T_1^2 - T_o^2) + \frac{c}{3}(T_1^3 - T_o^3)$$

and "plugging in" the values:

$$q = (6.339)(1000 - 300) + \frac{(10.4 \times 10^{-3})}{2}(1000^2 - 300^2)$$
$$+ \frac{(-3.415 \times 10^{-6})}{3}(1000^3 - 300^3) = 7943 \text{ cal}$$

The products of a detonation or burning reaction are mixtures of different reaction products, usually gases. It would be very inconvenient to treat each one separately. The mixed gases can be treated as a single gas with an average heat capacity value. The heat capacity of the mixture is equal to the sum of the products of the mole fraction of each gas component times its heat capacity.

$$\overline{C}_p = \sum_i n_i C_{pi}$$

where $\overline{C}_p$ is the molar heat capacity of the mixture, n_i the mole fraction of component i, and C_{pi} the molar heat capacity of component i.

As an example, let us look at the detonation of PETN.

1. $C_5H_8N_4O_{12} \rightarrow 2N_2 + 4H_2O + 2CO + 3CO_2$

2. For each mole of PETN there are eleven moles of product gases. The mole fraction of each component, therefore, is:

$$n(N_2) = 2/11 = 0.1818$$

$$n(H_2O) = 4/11 = 0.3636$$

$$n(CO) = 2/11 = 0.1818$$

$$N(CO_2) = 3/11 = 0.2727$$

3.

From Table 8.2, the molar heat capacities of these gases are:

$$C_p(N_2) = 6.457 + 1.389 \times 10^{-3}T - 0.069 \times 10^{-6}T^2$$

$$C_p(H_2O) = 7.136 + 2.640 \times 10^{-3}T + 0.0459 \times 10^{-6}T^2$$

$$C_p(CO) = 6.350 + 1.811 \times 10^{-3}T - 0.2675 \times 10^{-6}T^2$$

$$C_p(CO_2) = 6.339 + 10.14 \times 10^{-3}T - 3.415 \times 10^{-6}T2$$

and the average molar heat capacity for the mixture, $\overline{C}_p$, is:

$$\overline{C}_p = n(N_2)C_p(N_2) + n(H_2O)C_p(H_2O) + \ldots$$

$$= 6.652 + 4.307 \times 10^{-3}T - 0.9758 \times 10^{-6}T^2$$

8.2 Heat Capacity of Liquids

The heat capacity of a liquid is generally higher than that of either the solid or gas phase of the same material. As with gases (and solids) the heat capacity increases with increasing temperature. The relationship between the value of heat capacity and temperature for liquids is fairly linear over modest temperature ranges, and therefore is easier to deal with than in gases.

$$C_p = C_{p0} + aT$$

where C_{p0} and a are constants and T is the temperature. Tables 8.3 and 8.4 give heat capacities for some inorganic and organic liquids, respectively.

When the experimental or measured value of C_p for a particular liquid is not available, a fair approximation for C_p near room temperature can be obtained by Kopp's Rule, which states that the heat capacity of a liquid is approximately equal to the sum of the atomic heat capacities of its individual atoms. For the purpose of Kopp's Rule, these individual atomic heat capacities (molar basis) are given in Table 8.5.

Example 8.4   Using Kopp's Rule for liquids, estimate the C_p for sulfuric acid at room temperature.

1. H_2SO_4, elements are 2H, 1S, and 4O .
   $C_p = (2)(4.3) + (7.4) + (4)(6.0) = 40$ (cal)/(g mole)(K)

Table 8.3 Heat capacities of inorganic liquids C_p is the heat capacity, (g cal) / (g)(°C) at T(°C); a, the temperature coefficient in the equation $C_p = C_{p0} + aT$ over the indicated temperature range

Liquid	Formula	t(°C)	C_p	C_{p0}	a	Temperature Range (°C)
Ammonia	NH_3	−40	1.051			
		0	1.098			
		60	1.215			
		100	1.479			
Mercury	Hg	0	0.0335			
		60	0.0330			
		100	0.0329			
		200	0.0329			
		280	0.0332			
Nitric acid	HNO_3	25	0.417			
Silicon tetrachloride	$SiCl_4$	25	0.204			
Sodium nitrate	$NaNO_2$	350	0.430			
Sulfuric acid	H_2SO_4			0.339	0.00038	10 to 45°C
	H_2SO_4	25	0.369			
Sulfuryl chloride	SO_2Cl_2	25	0.234			
Sulfur dioxide	SO_2	−20	0.3130	0.318	0.00028	10 to 140°C
Water	H_2O	0	1.008			
		15	1.000			
		100	1.006			
		2200	1.061			
		300	1.155			

Reference 1.

2. Measured values in Table 8.3 are listed in (cal)/(g)(K); so we need the molecular weight of H_2SO_4, which is 98.

   40 (cal)/(g mole)(K) / 98 (g)/(g mole) = 0.408 (cal)/(g)(K)
3. Table 8.3 gives C_p for H_2SO_4 at 25°C as 0.369; so our value estimated from Kopp's Rule is off by +10%.

8.3 Heat Capacity of Solids

The heat capacity of most of the solid crystalline elements is approximately 6 (cal)/(g mole)(K) at or near room temperature. The heat capacities of solid compounds are higher. Tables 8.6 and 8.7 give heat capacities for various solid compounds.

When data are not available for the heat capacity of a solid, it can also be estimated by Kopp's Rule; however, the values of heat capacity for each element are different from those used for liquids. Table 8.8 gives heat capacity of the elements for use in Kopp's Rule for solids. Kopp's Rule for solids is usually

Table 8.4 Heat Capacities of Organic Liquids; data from International Critical Tables unless otherwise indicated; C_p is the heat capacity, calories per gram per °C at t °C C; a the temperature coefficient in $C_p = C_{p0} + at$, applying over the indicated temperature range

Liquid	Formula	t (°C)	C_p	C_{p0}	a	Temperature Range (°C)
Carbon tetrachloride	CCl_4	20	0.201	0.198	0.000031	0 to 70
Carbon disulfide	CS_2			90.235	0.000246	−100 to +150
Chloroform	$CHCl_3$	15	0.226	0.221	0.000330	−30 to +60
Formic acid	CH_2O_2	0	0.496	0.496	0.000709	40 to 140
Methyl alcohol	CH_4O	0	0.566			
		40	0.616			
Acetic acid	$C_2H_4O_2$			0.468	0.000929	0 to 80
Ethyl alcohol	C_2H_6O	−50	0.473			
		0	0.535			
		25	0.580			
		50	0.652			
		100	0.825			
		150	1.053			
Glycol	$C_2H_6O_2$	0	0.544	0.544	0.001194	−20 to +200
Allyl alcohol	C_3H_6O	0	0.3860			
		21 to 96	0.665			
Acetone	C_3H_6O			0.506	0.000764	−30 to +60
Propane	C_3H_5	0	0.576	0.576	0.001505	−30 to +20
Propyl alcohol	C_3H_5O	−50	0.456			
		0	0.525			
		+50	0.654			
Glycerol	$C_3H_5O_3$	−50	0.485			
		0	0.540			
		+50	0.598			
		+100	0.668			
Ethyl acetate	$C_4H_8O_2$	20	0.478			
n-Butane	C_4H_{10}	0	0.550	0.550	0.00191	−15 to 20
Ether	$C_4H_{10}O$	0	0.529			
		30	0.548			
		120	0.802			
Isopentane	C_5H_{12}	0	0.5??			
		8	0.526			
Nitrobenzene	$C_6H_5NO_2$	10	0.358			
		50	0.329			
		120	0.393			
Benzene	C_6H_6	5	0.389			
		20	0.406			
		60	0.444			
		90	0.473			
Aniline	C_6H_7N	0	0.478			
		50	0.521			
		100	0.547			

Table 8.4   *(Continued)*

Liquid	Formula	t (°C)	C_p	C_{p0}	a	Temperature Range (°C)
n-Hexane	C_6H_{14}	20 to 100	0.600			
Toluene	C_7H_8	0	0.386			
		50	0.421			
		100	0.470			
n-Heptane	C_7H_{16}	0 to 50	0.507			
		30	0.518	0.476	0.00142	30 to 80
Decane (BP 172°)	$C_{10}H_{22}$	0 to 50	0.502			
n-Hexadecane	$C_{16}H_{34}$	0. to 50	0.496			
Steric acid	$C_{18}H_{36}O_2$	75 to 137	0.550			
Diphenyl	$C_{12}H_{10}$			0.300	0.00120	80 to 300

With permission from Reference 1.

more accurate than for liquids. An extensive listing of heat capacities of solid explosives is given in Ref. 2.

8.4 Latent Heat of Fusion

When a solid material melts (changes to a liquid), energy is absorbed. This energy increases the kinetic energy of the molecules or atoms sufficiently to overcome the attractive forces that bind the molecules or atoms together into their crystal form. This increase in enthalpy, which accompanies melting, is called the *latent heat of fusion*. Although some volume change is involved, $P\,dV$ is very small compared to dQ. There are no general relationships that allow easy estimation of heats of fusion; however, the quantity λ_f / T_f (where λ_f is the latent heat of fusion, (cal)/(g mole), and T_f is the absolute melting or fusion temperature K) falls into three ranges depending upon the type of material involved. For

Table 8.5   Heat Capacity of Elements at 20°C for Kopp's Rule for Liquids

Element	C_p, (cal)/(g mole)(K)
Carbon (C)	2.8
Hydrogen (H)	4.3
Boron (B)	4.7
Silicon (Si)	5.8
Oxygen (O)	6.0
Flourine (F)	7.0
Phosphorous (P)	7.4
Sulfur (S)	7.4
All others	8.0

Table 8.6 Heat Capacities of Solid Inorganic Compounds; C_p = cal gram (°C)

Compound	Formula	t (°C)	C_p
Aluminum sulfate	$Al_2(SO_4)_3$	50	0.184
	$Al_2(SO_4)_3 \cdot 18H_2O$	34	0.353
Ammonium chloride	NH_4Cl	0	0.357
Antimony	Sb	25	0.05
Antimony trisulfide	Sb_2S_3	0	0.0830
		100	0.0884
Arsenic	As	25	0.0796
Arsenic oxide	As_2O_2	0	0.117
Barium carbonate	$BaCO_3$	100	0.110
		400	0.123
		800	0.130
Barium chloride	$BaCl_2$	0	0.0853
		100	0.0945
Barium sulfate	$BaSO_4$	0	0.1112
		1000	0.1448
Bismuth	Bi	25	0.0292
Bismuth trioxide	Bi_2O_3	25	0.0584
Boron	B	25	0.264
Boron oxide	B_2O_2	25	0.2138
Cadmium	Cd	25	0.0551
Cadmium sulfate	$CdSO_4 \cdot 8H_2O$	0	0.1950
Cadmium sulfide	CdS	0	0.0881
		100	0.0924
Calcium chloride	$CaCl_2$	61	0.164
	$CaCl_2 \cdot 6H_2O$	0	0.321
Calcium fluoride	CaF_2	0	0.204
		40	0.212
		80	0.216
Calcium sulfate	$CaSO_4 \cdot 2H_2O$	0	0.2650
		50	0.198
Calcium sulfide	CaS	25	0.157
Carbon (diamond)	C	25	0.147
Cesium	Cs	25	0.0558
Chromium	Cr	25	0.147
Chromium oxide	Cr_2O_2	0	0.168
		50	0.188
Cobalt	Co	25	0.1045
Cupric oxide	CuO	25	0.133
Copper sulfate	$CuSO_4$	0	0.148
	$CuSO_4 \cdot H_2O$	0	0.1717
	$CuSO_4 3 \cdot H_2O$	0	0.2280
	$CuSO_4 5 \cdot H_2O$	0	0.2560
Ferrous carbonate	$FeCO_3$	54	0.193
Ferrous sulfate	$FeSO_4$	45	0.167
Gold	Au	25	0.0306
Iodine	I	25	0.0518
Lead carbonate	$PbCO_3$	32	0.080
Lead chloride	$PbCl_2$	0	0.0649
		200	0.0704
		400	0.0800
Lead nitrate	$Pb(NO_3)_2$	45	0.1150
Lead sulfate	$PbSO_4$	45	0.0838
Lithium	Li	25	0.815
Magnesium chloride	$MgCl_2$	48	0.193

Table 8.6 *(Continued)*

Compound	Formula	t (°C)	C_p
Magnesium sulfate	$MgSO_4$	61	0.222
	$MgSO_4 \cdot H2O$	9	0.239
	$MgSO_4 \cdot 6H2O$	9	0.349
	$MgSO_4 \cdot 7H2O$	12	0.361
Manganese dioxide	MnO_2	0	0.152
Manganese oxide	MnO	58	0.158
Manganic oxide	Mn_2O_3	58	0.162
Mercuric chloride	$HgCl_2$	0	0.0640
Mercuric sulfide	HgS	0	0.0506
Mercurous chloride	$HgCl$	0	0.0499
Molybdenum	Mo	25	0.0585
Nickel sulfide	NiS	0	0.116
		100	0.128
		200	0.138
Palladium	Pd	25	0.059
Platinum	Pt	25	0.0326
Potassium chloride	KCl	0	0.1625
		200	0.1725
		400	0.1790
Potassium chlorate	$KClO_2$	0	0.1910
		200	0.2960
Potassium chromate	K_2CrO_4	46	0.1864
Potassium dichromate	$K_2Cr_2O_7$	0	0.178
		400	0.236
Potassium nitrate	KNO_3	0	0.214
		200	0.267
		500	0.292
Potassium perchlorate	$KClO_4$	25	0.190
Potassium sulfate	K_2SO_4	0	0.0848
Selenium	Se	25	0.0755
Silver chloride	$AgCl$	0	0.0848
		200	0.0974
		500	0.101
Silver nitrate	$AgNO_2$	50	0.146
Sodium borate	$Na_2B_4O_7$	45	0.234
(Borax)	$Na_2B_4O_7 \cdot 10H_2O$	35	0.385
Sodium carbonate	Na_2CO_3	45	0.256
Sodium chloride	$NaCl$	0	0.204
		100	0.217
		400	0.229
		600	0.236
Sodium nitrate	$NaNO_3$	0	0.2478
		100	0.294
		250	0.358
Sodium sulfate	Na_2SO_4	0	0.202
		100	0.220
Sulfur (monoclinic)	S	25	0.1765
Titanium	Ti	25	0.1255
Titanium dioxide	TiO_2	25	0.165
Tungsten	W	25	0.0325
Water (ice)	H_2O	−40	0.435
		0	0.492

Reference 1.

Table 8.7 Heat Capacities of Miscellaneous Materials C_p = cal/(gram)(°C)

Substance	C_p	Temperature Range (°C)
Alundum	0.186	100
Asbestos	0.25	
Asphalt	0.22	
Bakelite	0.3 to 0.4	
Brickwork	0.2 (approx.)	
Carbon (gas retort)	0.204	
Cellulose	0.32	
Cement	0.186	
Charcoal (wood)	0.242	
Chrome brick	0.17	
Clay	0.224	
Coal	0.26 to 0.37	
Coal tar	0.35	40
	0.45	200
Coke	0.265	21–400
	0.359	21–800
	0.403	21–1300
Concrete	0.156	70–312
	0.219	72–1472
Cryloite	0.253	16–55
Fireclay brick	0.198	100
	0.298	1500
Fluorspar	0.21	30
Glass (crown)	0.16 to 0.20	
(flint)	0.117	
(Pyrex)	0.20	
(silicate)	0.188 to 0.204	0–100
	0.24 to 0.26	0–700
(Wool)	0.157	
Granite	0.20	20–100
Magnesite brick	0.222	100
	0.195	1500
Pyrites (copper)	0.131	19–50
(iron)	0.136	15–98
Sand	0.191	
Steel	0.12	

Reference 6.

most elements, $2 < \lambda_f / T_f < 3$; for most solid inorganic compounds, $5 < \lambda_f / T_f < 7$; and for most solid organic compounds, $9 < \lambda_f / T_f < 11$. These ranges are very broad, and there are many exceptions. Table 8.9 lists the molar or gram-atom latent heats of fusion for some solid materials.

8.5 Heat of Vaporization

When a liquid vaporizes, that is, changes from a liquid to a gas, the energy absorbed increases the kinetic energy of the molecules or atoms sufficiently for

Table 8.8 Heat Capacity of Elements For Use In Kopp's Rule For Solids At 20°C

Element	C_p, (cal)/(g mole)(K)
Carbon (C)	1.8
Hydrogen (H)	2.3
Boron (B)	2.7
Silicon (Si)	3.8
Oxygen (O)	4.0
Fluorine (F)	5.0
Phosphorous (P)	5.4
All others	6.2

Table 8.9 Heats of Fusion; λ_f, heat of fusion, g cal per g atom or g mole; t_f, melting point, °C (T_f, melting point, K) (to convert heats of fusion to Btu per pound mole, multiply table values by 1.8)

Material	λ_f	t_f (°C)	λ_f/T_f
Elements			
Ag	2,700	961	2.19
Al	2,600	660	2.8
Cu	3,110	1083	2.29
Fe	3,660	1535	2.0
Na	629	98	1.7
Ni	4,200	1455	2.4
Pb	1,220	327	2.03
S (rhombic)	300	115	0.8
Sn	1,690	232	3.38
Zn	1,595	419	2.303
Compounds			
H_2O	1,436.3	0.0	3.13
Sb_2S_2	11,200	547	5.40
CO_2	1,999	−56.2	4.97
$CaCl_2$	6,780	782	4.90
NaOH	1,700	318	2.01
NaCl	6,800	808	5.71
Carbon tetrachloride	600	22.9	1.47
Methyl alcohol	757	−98	1.45
Acetic acid	2,800	16.6	5.65
Ethyl alcohol	1,200	−114.6	3.33
Benzene	2,370	5.4	5.10
Aniline	1,950	−7.0	4.30
Naphthalene	4,550	80	8.45
Diphenyl	4,020	71	7.59
Stearic acid	13,500	64	25.4

Reference 1.

them to overcome the attractive and surface forces holding them in the liquid state. Very large volume changes are involved in vaporization, and therefore $P\, dV$ is quite large. Therefore, the enthalpy changes in vaporization are much greater than those in fusion (melting), where small volume changes are the rule. With vaporization, the ratio λ_b/T_b is very useful in predicting heats of vaporization. For nonpolar liquids, λ_b/T_b is very close to 21 (cal)/(g mole)(K). This is known as "Trouton's Rule." Trouton's ratio increases somewhat with increasing temperature. Kistyakowski corrected for this by means of his equation for nonpolar liquids:

$$(\lambda_b/T_b) = 8.75 + 4.571 \log 10 T_b$$

where λ_b is the heat of vaporization in (cal)/(g mole) at the normal boiling point and T_b is the normal boiling point in degrees kelvin. Trouton's ratio for polar liquids is usually much higher than 21. Table 8.10 gives heats of vaporization for a number of both polar and nonpolar liquids.

8.6 Heat of Transition

In a solid, the change from one crystal structure to another is also a state change and involves a change in enthalpy. The energy absorbed increases the kinetic energy of the molecules or atoms just enough to cause them to shift the crystal structure to a different geometry. This does not occur in all solids, but where it does, it requires heat of transition. This is shown in Table 8.11 for five elements that undergo crystal structural changes (lattice shifts) at specific temperatures.

8.7 Summary

In this chapter we covered the definitions of the various heating terms for each state and change of state of materials. These are:

C_v, the heat capacity at constant volume;
C_p, the heat capacity at constant pressure;
λ_f, the heat of fusion;
λ_b, the heat of vaporization; and
λ_t, the heat of transition.

Along with tables of the above quantities for a limited number of elements and compounds, we also had several methods for estimating the value of these quantities when experimental data are unavailable. The following example should serve as a review involved in the processes of heating.

Example 8.5 One mole of water, H_2O, in the solid state (ice) is heated at constant pressure (1 atm) from its initial temperature of 233 K through both its melting and vaporization points. Heating is stopped when the temperature of the vapor reaches 700

Table 8.10 Heats of Vaporization; λ the heat of vaporization at t_b (°C), g cal per g mole; t, temperature, °C; t_b, normal boiling point, °C (to convert heats of vaporization to Btu per pound mole, multiply table values by 1.8)

Substance	λ	t_b (°C)
Ammonia	5,581	−33.4
Argon	1,590	−185.8
Bromine	7.340	25.0
Carbon dioxide	6,030	−78.4
Carbon disulfide	6,400	46.25
Carbon monoxide	1,444	−191.5
Carbon oxyfulfide	4,423	−50.2
Carbon tetrachloride	7,170	76.7
Chlorine	4,878	−34.1
Dichlorodifluoromethane	4,850	−30.5
Dichloromonofluoromethane (Freon 21)	6,400	8.9
Helium	22	−268.9
Hydrogen	216	−252.7
Hydrogen bromide	4,210	−66.7
Hydrogen chloride	3,860	−85.0
Hydrogen cyanide	6,027	25.7
Hydrogen fluoride	1,800	19.9
Hydrogen iodide	4,724	−35.35
Hydrogen sulfide	4,724	−60.3
Mercury	13,890	356.6
Nitric oxide	3,950	−151.7
Nitrogen	1,336	−195.8
Nitrous oxide	3,950	−88.5
Oxygen	1,620	−183.0
Silicon tetrafluoride	6,150	−95.5
Sulfur	25,000	444.6
Sulfur dioxide	5,950	−10.0
Sulfur trioxide	9,990	43.3
Trichloromonofluoromethane (Freon 11)	5,960	23.6
Water	9,717	100.0

K. Calculate the total heat required to accomplish this. Plot a graph of total heat absorbed versus absolute temperature for the entire process.

The approach taken here will break the heating process up into five distinct steps. The first step will be to heat the ice in the solid phase up to its normal melting point. The second step will be to melt the ice, changing it to liquid water at the normal melting point. The third step will be to heat the liquid water from its melting point up to its boiling point. The fourth step will be to vaporize the water, changing it to a vapor (steam) at its normal boiling point. The fifth step will be to heat the steam from its boiling point to the final temperature.

Table 8.11 Heats of Transition; λ_f the heat absorbed in transition, g cal per g atom; to convert to Btu per pound mole, multiply by 1.8; t_f, the temperature of transformation, °C

Transition	λ_f	t_f (°C)
Sulfur		
Rhombic → monoclinic	7.0	114–151
Iron (electrolytic)		
$\alpha \rightarrow \beta$	363	770
$\beta \rightarrow \gamma$	313	910
$\gamma \rightarrow \delta$	106	1400
Manganese		
$\alpha \rightarrow \beta$	1325	1070–1130
Nickel		
$\alpha \rightarrow \beta$	78	320–330
Tin		
White → gray	530	0

Source: International Critical Tables (1929).

1. Heating ice from 233 to 273 K (mp),

$$q = n\int_{T_o}^{T_i} C_p \, dT = nC_p(T_1 - T_o)$$

Table 8.6 gives C_p of water (ice) as 0.435 (cal)/(g)(°C) at −40°C (233 K), and 0.492 (cal)/(g)(°C) at 0°C (273 K). If we assume that C_p versus T is linear over this fairly narrow range, then the average C_p is $(0.435 + 0.492)/2 = 0.4635$ (cal)/(g)(°C). The molecular weight of water is 18.016, and so the molar heat capacity is $0.4635 \times 18.016 = 8.35$ (cal)/(g mole)(K).

$$q = (1)(8.35)(273 - 233) = 334 \text{ (cal)}$$

2. The ice is now at its normal melting point. In order to melt it

$$q = n\lambda_f$$

From Table 8.9, $\lambda_f = 1436.3$ (cal)/(g mole), n is one mole; so

$$q = (1)(1436.3) = 1436.3 \text{ (cal)}$$

3. The water is now in the liquid state at 0°C (273 K). In order to heat it to 100°C (373 K), the normal boiling point

$$q = n\int_{T_1}^{T_2} C_p \, dT = nC_p(T_2 - T_1)$$

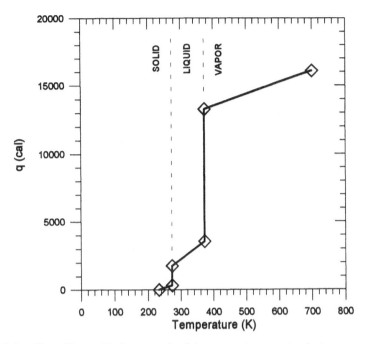

Figure 8.1.   Plot of heat added versus absolute temperature, water, 1 atm pressure.

The heat capacity of liquid water is 18.016 (cal)/(g mole)(K).

$$q = (1)(18.016)(373 - 273) = 1801.6 \text{ (cal)}$$

4. The water is now at the normal boiling point. In order to vaporize it

$$q = n\lambda_b$$

from Table 8.10, $\lambda_b = 9717$ (cal)/(g mole),

$$q = (1)(9717) = 9717 \text{ (cal)}$$

5. The water is now vapor (steam) at 100°C (373 K), and the last step is to heat this vapor to 700 K.

$$q = n\int_{T_2}^{T_3} C_p \, dT = n\int_{T_2}^{T_3} (a + bT + cT^2) \, dT$$

$$q = \left(a(T_3 - T_2) + \frac{b}{2}(T_3^2 - T_2^2) + \frac{c}{3}(T_3^3 - T_2^3) \right)$$

From Table 8.2: $a = 7.136$, $b = 2.64 \times 10^{-3}$, and $c = 0.0459 \times 10^{-6}$, which yields C_p as (cal/(g mole)(K), $T_3 = 700$ K, and $T_2 = 373$ K.

6. Adding the above steps:

T(K)	Process	q(cal)	Accumulated q (cal)
$T_o = 233$	Start		
	1. Heating ice from $T_o \rightarrow T_1$	334	334
$T_1 = 273$	2. Melting ice	1436	1770
	3. Heating water from $T_1 \rightarrow T_2$	1802	3572
$T_2 = 373$	4. Vaporizing water	9717	13289
	5. Heating steam from $T_2 \rightarrow T_3$	2801	16090
$T_3 = 700$	End		

The whole process is shown in Figure 8.1.

9

Thermochemistry

In the previous chapter we saw that every change in physical state, such as in temperature and phase, involves a change in internal energy (U) and enthalpy (H). We referred to the calculations, or quantification of these changes in energy due to changes in physical state, as *thermophysics*. Changes in the chemical state, that is, in the makeup or composition of molecules, also involves a change in internal energy and enthalpy, and we call the quantification of these changes *thermochemistry*.

9.1 Heat of Reaction

A chemical reaction is a process where one or more chemical species change their molecular configuration to one that is different. As an example, let us consider the simple burning of gaseous hydrogen molecules with gaseous oxygen molecules to form steam (water in the gaseous state).

$$2H_2 + O_2 \rightarrow 2H_2O$$

We started with three separate molecules, two of hydrogen and one of oxygen, and at the end of the process we had changed these into two molecules of water. The hydrogen-hydrogen (H—H) bonds were destroyed, as were the oxygen-oxygen (O—O) bonds, and hydrogen-oxygen (H—O) bonds were created. Internal energy is stored in these bonds in the various modes of flexing, vibration, and rotation. The internal energy stored in the H—O bonds is different from that stored in the H—H and O—O bonds. So we have a difference in internal energy

(U) between the starting and ending chemical states due to changes in the bonds. The number of molecules per unit reaction changed from three to two, resulting in a change in volume if the process were carried out at constant pressure. Therefore, there was work done equal to $P \Delta V$. Remembering that enthalpy, H, is equal to $U + PV$, we see that the process created a change in enthalpy. This change in enthalpy between the starting and ending chemical states is called the *heat of reaction.*

Earlier, it was stated that absolute enthalpy could not be determined for a substance, and therefore we can deal only with changes or differences in this quantity. To simplify calculations of heat of reaction and to make them consistent, we must therefore arbitrarily define a standard state to which we reference all changes in enthalpy for chemical reactions. The standard state used for most engineering calculations is defined as 25°C (298 K) and 1 atm pressure.

9.2 Heat of Formation

The calculational device used to tie or reference heats of reaction to the standard state is the heat of formation. The heat of formation can be considered a special case of the heat of reaction. It is the heat of reaction, or enthalpy change, involved in making a particular compound, or molecule, from its elements where both the elements and the final compound are at standard-state conditions. It is also required that both the elements and compound are in their normal state of aggregation at the standard-state conditions (i.e., gaseous, liquid, solid, etc.). The heat of formation at the standard state is designated as ΔH_f^0.

As an example, let us consider the compound n-hexane, C_6H_{14}, which is made up of 6 atoms of the element carbon and 14 atoms of the element hydrogen.

$$6C(s) + 7H_2(g) \rightarrow C_6H_{14}(l)$$

At the standard state (25°C and 1.0 atm), carbon is a solid and is designated with the notation (s) in the above reaction equation. Likewise, hydrogen at the standard state exists as gaseous (g) hydrogen molecules, and n-hexane at this state is a liquid (l). As stated above, the heat of reaction is the difference between the absolute enthalpy of a system at the starting and ending states; so for this special case, the heat of formation:

$$\Delta H_f^0(C_6H_{14}, l) = H^0(C_6H_{14}, l) - 6H^0(C, s) - 7H^0(H_2, g)$$

where $H^0(i)$ is the absolute enthalpy of component i at the standard state.

The heat of formation is always given on a per mole basis, and its units are (cal)/(g mole), (kcal)/(g mole), or (Btu)/(lb mole). By convention, the heat of formation of all the elements in their normal state of aggregation in the standard state is zero. Table 9.1 lists the standard heats of formation of a number of inorganic compounds. Table 9.2 lists the standard heats of formation of a number of explosive compounds and common explosive reaction products.

Table 9.1 Heats of Formation of Inorganic Compounds; Reference conditions: 25°C (298.16 K), 1 atm pressure, gaseous substances in ideal state. ΔH_f^0 = standard heat of formation, kcal per g mole; Multiply values by 1800 to obtain Btu per pound mole.

Compound	Formula	State[a]	ΔH_f^0
Aluminum chloride	$AlCl_3$	c	−166.2
Aluminum hydroxide	$Al(OH)_3$	amorph	−304.2
Aluminum oxide	Al_2O_3	c	−399.09
Aluminum silicate	Al_2SiO_5	c	−648.9
Aluminum sulfate	$Al_2(SO_4)_3$	c	−820.98
Ammonia	NH_3	g	−11.04
Ammonia	NH_3	l	−16.06
Ammonium carbonate	$(NH_4)_2CO_3$	dil	−225.11
Ammonium bicarbonate	$(NH_4)HCO_3$	c	−203.7
Ammonium chloride	NH_4Cl	c	−75.38
Ammonium hydroxide	NH_4OH	aqueous	−87.64
Ammonium nitrate	NH_4NO_3	c	−87.27
Ammonium oxalate	$(NH_4)_2C_2O_4$	c	−268.72
Ammonium sulfate	$(NH_4)_2SO_4$	c	−281.86
Ammonium acid sulfate	$(NH_4)HSO_4$	c	−244.83
Antimony trioxide	Sb_2O_3	c	−168.4
Antimony pentoxide	Sb_2O_5	c	−234.4
Antimony sulfide	Sb_2S_3	c	−43.5
Arsenic acid	H_3AsO_4	c	−215.2
Arsenic trioxide	As_2O_3	c	−156.4
Arsenic pentoxide	As_2O_5	c	−218.6
Arsine	AsH_3	g	41.0
Barium carbonate	$BaCO_3$	c	−291.3
Barium chlorate	$Ba(ClO_3)_2$	c	−181.7
Barium chloride	$BaCl_2$	c	−205.56
Barium chloride	$BaCl_2 \cdot 2H_2O$	c	−349.35
Barium hydroxide	$Ba(OH)_2$	c	−226.2
Barium oxide	BaO	c	−133.4
Barium peroxide	BaO_2	c	−150.5
Barium silicate	$BaSiO_3$	c	−359.5
Barium sulfate	$BaSO_4$	c	−350.2
Barium sulfide	BaS	c	−106.0
Bismuth oxide	Bi_2O_3	c	−137.9
Boric acid	H_3BO_3	c	−260.2
Boron oxide	B_2O_3	c	−302.0
Bromine chloride	$BrCl$	g	+3.51
Cadmium chloride	$CdCl_2$	c	−93.0
Cadmium oxide	CdO	c	−60.86
Cadmium sulfate	$CdSO_4$	c	−221.36
Cadmium sulfide	CdS	c	−34.5
Calcium carbide	CaC_2	c	−15.0
Calcium carbonate	$CaCO_3$	c	−288.45
Calcium chloride	$CaCl_2$	c	−190.0
Calcium chloride	$CaCl_2 \cdot 6H_2O$	c	−623.15
Calcium fluoride	CaF_2	c	−290.3
Calcium hydroxide	$Ca(OH)_2$	c	−235.80

Table 9.1   *(Continued)*

Compound	Formula	State[a]	ΔH_f^0
Calcium nitrate	$Ca(NO_3)_2$	c	-224.0
Calcium oxalate	$CaC_2O_4 \cdot H_2O$	c	-399.1
Calcium oxide	CaO	c	-151.9
Calcium phosphate	$Ca_3(PO_4)_2$	c	-986.2
Calcium silicate	$CaSiO_3$	c	-378.6
Calcium silicate	Ca_2SiO_4	c	-538.0
Calcium sulfate	$CaSO_4$	c	-342.42
Calcium sulfide	CaS	c	-115.3
Carbon graphite	C	c	0
Diamond	C	c	$+0.4532$
Amorphous (in coke)	C	amorph	$+2.6$
Carbon monoxide	CO	g	-26.4157
Carbon dioxide	CO_2	g	-94.0518
Carbon disulfide	CS_2	g	$+27.55$
Carbon disulfide	CS_2	l	$+21.0$
Carbon tetrachloride	CCl_4	g	-25.50
Carbon tetrachloride	CCl_4	l	-33.34
Chloric acid	$HClO_3$	dil	-23.50
Chromium chloride	$CrCl_3$	c	-134.6
Chromium chloride	$CrCl_2$	c	-94.56
Chromium oxide	CrO_2O_3	c	-269.7
Chromium trioxide	CrO_3	c	-138.4
Cobalt oxide	CoO	c	-57.2
Cobalt oxide	Co_3O_4	c	-210
Cobalt chloride	$CoCl_2$	c	-77.8
Cobalt sulfide	CoS	ppt	-21.4
Copper carbonate	$CuCO_3$	c	-142.2
Copper chloride	$CuCl_2$	c	-52.3
Copper chloride	$CuCl$	c	-32.5
Copper nitrate	$Cu(NO_3)_2$	c	-73.4
Copper oxide	CuO	c	-37.1
Copper oxide	Cu_2O	c	-39.84
Copper sulfate	$CuSO_4$	c	-184.00
Copper sulfide	CuS	c	-11.6
Copper sulfide	Cu_2S	c	-19.0
Cyanogen	C_2N_2	g	$+73.60$
Hydrobromic acid	HBr	g	-8.66
Hydrochloric acid	HCl	g	-22.063
Hydrogen cyanide	HCN	g	$+31.2$
Hydrofluoric acid	HF	g	-63.2
Hydriodic acid	HI	g	$+6.20$
Hydrogen oxide	H_2O	g	-57.7979
Hydrogen oxide	H_2O	l	-68.3174
Deuterium oxide	D_2O	g	-59.5628
Deuterium oxide	D_2O	l	-70.4133
Hydrogen peroxide	H_2O_2	l	-44.84
Hydrogen sulfide	H_2S	g	-4.815
Iron carbide	Fe_3C	c	$+5.0$
Iron carbonate	$FeCO_3$	c	-178.70

Table 9.1 *(Continued)*

Compound	Formula	State[a]	ΔH_f^0
Iron chloride	$FeCl_2$	c	−81.5
Iron chloride	$FeCl_3$	c	−96.8
Iron hydroxide	$Fe(OH)_2$	c	−135.8
Iron hydroxide	$Fe(OH)_3$	c	−197.0
Iron nitride	Fe_4N	c	−2.55
Iron oxide	FeO	c	−64.3
Iron oxide	$Fe_{0.96}O$	c	−63.7
Iron oxide	Fe_3O_4	c	−267.0
Iron oxide	Fe_2O_3	c	−196.5
Iron silicate	$FeO \cdot SiO_2$	c	−276
Iron silicate	$2FeO \cdot SiO_2$	c	−343.7
Iron sulfate	$Fe_2(SO_4)_3$	aqueous	−653.62
Iron sulfate	$FeSO_4$	c	−220.5
Iron sulfide	FeS	c	−22.72
Iron sulfide	FeS_2	c	−42.52
Lead carbonate	$PbCO_3$	c	−167.3
Lead chloride	$PbCl_2$	c	−85.85
Lead nitrate	$PbNO_3$	c	−107.35
Lead oxide (yellow)	PbO	c	−52.07
Lead peroxide	PbO_2	c	−66.12
Lead suboxide	Pb_2O	c	−51.2
Lead sesquioxide	Pb_3O_4	c	−175.6
Lead sulfate	$PbSO_4$	c	−219.50
Lead sulfide	PbS	c	−22.54
Lithium chloride	$LiCl$	c	−97.70
Lithium hydroxide	$LiOH$	c	−116.45
Magnesium carbonate	$MgCO_3$	c	−266
Magnesium chloride	$MgCl_2$	c	−153.40
Magnesium hydroxide	$Mg(OH)_2$	c	−221.00
Magnesium oxide	MgO	c	−143.84
Magnesium silicate	$MgSiO_3$	c	−357.9
Magnesium sulfate	$MgSO_4$	c	−305
Manganese carbonate	$MnCO_3$	c	−213.9
Manganese carbide	Mn_3C	c	−1
Manganese chloride	$MnCl_2$	c	−115.3
Manganese oxide	MnO	c	−92.0
Manganese oxide	Mn_3O_4	c	−331.4
Manganese oxide	Mn_2O_3	c	−232.1
Manganese dioxide	MnO_2	c	−124.5
Manganese dioxide	MnO_2	amorph	−117.0
Manganese silicate	$MnO \cdot SiO_2$	c	−302.5
Manganese silicate	$MnO \cdot SiO_2$	glass	−294.0
Manganese sulfate	$MnSO_4$	c	−254.24
Manganese sulfide	MnS	c	−48.8
Mercury bromide	$HgBr_2$	c	−40.5
Mercury chloride	$HgCl_2$	c	−55.0
Mercury chloride	Hg_2Cl_2	c	−63.32
Mercury nitrate	$Hg(NO_3)_2$	dil	−58.0
Mercury nitrate	$Hg_2(NO_3)_2 \cdot 2H_2O$	c	−206.9
Mercury oxide	HgO	c	−21.56

Table 9.1 *(Continued)*

Compound	Formula	State[a]	ΔH_f^0
Mercury oxide	Hg_2O	c	-21.8
Mercury sulfate	$HgSO_4$	c	-168.3
Mercury sulfate	Hg_2SO_4	c	-177.34
Mercury sulfide	HgS	c	-13.90
Mercury thiocyanate	$Hg(CNS)_2$	c	$+48.0$
Molybdenum oxide	MoO_2	c	-130
Molybdenum oxide	MoO_3	c	-180.33
Molybdenum sulfide	MoS_2	c	-55.5
Nickel chloride	$NiCl_2$	c	-75.5
Nickel cyanide	$Ni(CN)_2$	c	$+27.1$
Nickel hydroxide	$Ni(OH)_3$	c	-162.1
Nickel hydroxide	$Ni(OH)_2$	c	-128.6
Nickel oxide	NiO	c	-58.4
Nickel sulfide	NiS	c	-17.5
Nickel sulfate	$NiSO_4$	c	-213.0
Nitrogen oxide	NO	g	$+21.600$
Nitrogen oxide	N_2O	g	$+19.49$
Nitrogen oxide	NO_2	g	$+8.091$
Nitrogen pentoxide	N_2O_5	g	$+3.6$
Nitrogen pentoxide	N_2O_5	c	-10.0
Nitrogen tetroxide	N_2O_4	g	$+2.309$
Nitrogen trioxide	N_2O_3	g	$+20.0$
Nitric acid	HNO_3	l	41.404
Oxalic acid	$H_2C_2O_4 \cdot 2H_3O$	c	-340.9
Oxalic acid	$H_2C_2O_4$	c	-197.6
Perchloric acid	$HClO_4$	l	-11.1
Phosphoric acid (meta)	HPO_3	c	-228.2
Phosphoric acid (ortho)	H_3PO_4	c	-306.2
Phosphoric acid (pyro)	$H_4P_2O_3$	c	-538.0
Phosphorous acid (hypo)	H_3PO_2	l	-143.2
Phosphorous acid (ortho)	H_3PO_3	l	-229.1
Phosphorus trichloride	PCl_3	g	-73.22
Phosphorus pentoxide	P_2O_5	c	-360.0
Platinum chloride	$PtCl_4$	c	-62.9
Platinum chloride	$PtCl$	c	-17.7
Potassium acetate	$KC_2H_3O_2$	c	-173.2
Potassium carbonate	K_2CO_3	c	-273.93
Potassium chlorate	$KClO_3$	c	-93.50
Potassium chloride	KCl	c	-104.175
Potassium chromate	K_2CrO_4	c	-330.49
Potassium cyanide	KCN	c	-26.90
Potassium dichromate	$K_2Cr_2O_7$	c	-485.90
Potassium fluoride	KF	c	-134.46
Potassium nitrate	KNO_3	c	-117.76
Potassium oxide	K_2O	c	-86.4
Potassium sulfate	K_2SO_4	c	-342.66
Potassium sulfide	K_2S	c	-100
Potassium sulfite	K_2SO_3	c	-266.9

Table 9.1 *(Continued)*

Compound	Formula	State[a]	ΔH_f^0
Potassium thiosulfate	$K_2S_2O_3$	aq	-274
Potassium hydroxide	KOH	c	-101.78
Potassium nitrate	KNO_3	c	-117.76
Potassium permanganate	$KMnO_4$	c	-194.4
Selenium oxide	SeO_2	c	-55.0
Silicon carbide	SiC	c	-26.7
Silicon tetrachloride	$SiCl_4$	l	-153.0
Silicon tetrachloride	$SiCl_4$	g	-145.7
Silicon dioxide	SiO_2	c	-205.4
Silver bromide	AgBr	c	-23.78
Silver chloride	AgCl	c	-30.362
Silver nitrate	$AgNO_3$	c	-29.43
Silver sulfate	Ag_2SO_4	c	-170.50
Silver sulfide	Ag_2S	c	-7.60
Sodium acetate	$NaC_2H_3O_2$	c	-169.8
Sodium arsenate	Na_3AsO_4	c	-365
Sodium tetraborate	$Na_2B_4O_7$	c	-777.7
Sodium borate	$Na_2B_4O_7 \cdot 10H_2O$	c	-1497.2
Sodium bromide	NaBr	c	-86.030
Sodium carbonate	Na_2CO_3	c	-270.3
Sodium carbonate	$Na_2CO_3 \cdot 10H_2O$	c	-975.6
Sodium bicarbonate	$NaHCO_3$	c	-226.5
Sodium chlorate	$NaClO_3$	c	-85.73
Sodium chloride	NaCl	c	-98.232
Sodium cyanide	NaCN	c	-21.46
Sodium fluoride	NaF	c	-136.0
Sodium hydroxide	NaOH	c	-101.99
Sodium iodide	NaI	c	-68.84
Sodium nitrate	$NaNO_3$	c	-101.54
Sodium oxalate	NaC_2O_4	c	-314.3
Sodium oxide	Na_2O	c	-99.4
Sodium triphosphate	Na_3PO_4	c	-460
Sodium diphosphate	Na_2HPO_4	c	-417.4
Sodium monophosphate	NaH_2PO_4	aqueous	-367.7
Sodium phosphite	Na_2HPO_3	c	-338
Sodium selenate	Na_2SeO_4	c	-258
Sodium selenide	Na_2Se	c	-63.0
Sodium sulfate	Na_2SO_4	c	-330.90
Sodium sulfate	$Na_2SO_4 \cdot 10H_2O$	c	-1033.48
Sodium bisulfate	$NaHSO_4$	c	-269.2
Sodium sulfide	Na_2S	c	-89.2
Sodium sulfide	$Na_2S \cdot 4\ 1/2\ H_2O$	c	-416.9
Sodium sulfite	Na_2SO_3	c	-260.6
Sodium bisulfite	$NaHSO_3$	dil	-206.6
Sodium silicate	Na_2SiO_3	glass	-360
Sodium silicofluoride	Na_3SiF_6	c	-677
Sulfur dioxide	SO_2	g	-70.96
Sulfur trioxide	SO_3	g	-94.45
Sulfuric acid	H_2SO_4	l	-193.91
Tellurium oxide	TeO_2	c	-77.69

Table 9.1   *(Continued)*

Compound	Formula	State[a]	ΔH_f^0
Tin chloride	SnCl$_4$	l	-130.31
Tin chloride	SnCl$_2$	c	-83.6
Tin oxide	SnO$_2$	c	-138.8
Tin oxide	SnO	c	-68.4
Titanium oxide	TiO$_2$	amorph	-207
Titanium oxide	TiO$_2$	c	-218.0
Tungsten oxide	WO$_2$	c	-136.3
Vanadium oxide	V$_2$O$_5$	c	-373
Zinc bromide	ZnBr$_2$	c	-78.17
Zinc carbonate	ZnCO$_3$	c	-194.2
Zinc chloride	ZnCl$_2$	c	-99.40
Zinc hydroxide	Zn(OH)$_2$	c	-153.5
Zinc iodide	ZnI$_2$	c	-49.98
Zinc oxide	ZnO	c	-83.17
Zinc sulfate	ZnSO$_4$	c	-233.88
Zinc sulfide	ZnS	c	-48.5
Zirconium oxide	ZrO$_2$	c	-258.2

[a]   Abbreviations: c, crystalline state; l, liquid state; g, gaseous state; dil, in dilute aqueous solution; □, infinite dilution; ppt, precipitated solid; amorph, amorphous state.

Source: Selected Values of Chemical Thermodynamic Properties, as of July 1, 1953, edited by D. D. Wagman, National Bureau of Standards.

9.3 Heats of Reaction from Heats of Formation

Knowing the standard heats of formation, we can readily calculate the enthalpy change involved in a chemical reaction at the standard state. The heat of reaction at the standard state, ΔH_r^0, is equal to the difference between the standard heats of formation of the reaction products and the standard heats of formation of the reactants.

$$\Delta H_r^0 = \Sigma\ \Delta H_f^0(\text{products}) - \Sigma\ \Delta H_f^0(\text{reactants})$$

Examining this in detail will show that the heat of reaction calculated in the above manner is indeed equal to the difference in the absolute enthalpies of the products and reactants. This will also show that the convention adopted that sets ΔH_f^0 of the elements equal to zero is valid in that the absolute enthalpies of the elements will mathematically cancel out of the equations.

Let us examine this by way of an example. Hydrochloric acid, HCl, will react with sodium hydroxide, NaOH, to form sodium chloride, NaCl, and water, H$_2$O.

$$\text{HCl} + \text{NaOH} \rightarrow \text{NaCl} + \text{H}_2\text{O}$$

Calculating the standard heat of reaction from the standard heats of formation, we have:

$$\Delta H_r^0 = [\Delta H_f^0(\text{NaCl}) + \Delta H_f^0(\text{H}_2\text{O})]_{\text{products}}$$

$$- [\Delta H_f^0(\text{HCl}) + \Delta H_f^0(\text{NaOH})]_{\text{reactants}} \quad (9.1)$$

Table 9.2 Heats of Formation for Some Pure Explosive Compounds, ΔH_f^0 is given in kcal/g mole, reference state is 25°C, 1 atm

Explosive Compound	Molecular Weight	ΔH_f^0, Heat of Formation
Ammonium nitrate, AN	80.05	−87.27
Ammonium perchlorate, AP	117.5	−70.58
BTF	252.1	+144.5
DATB	243.1	−23.6
DEGN	196	−99.4
DIPAM	454.1	−6.8
DNPA	204.1	−110
EDNP	220.2	−140
Explosive D	246	−94
FEFO	320.1	−177.5
HMX	296.2	+17.93
HNAB	452.2	+67.9
HNS	450.3	+18.7
Lead azide	291	+112
Lead styphnate	468.3	−200.0
NC (12.0% N)	263.9	−173.7
NC (13.35% N)	283.9	−163.0
NC (14.14% N)	297.1	−156.0
NG	227.1	−88.6
NM	61.0	−27.0
NQ	104.1	−22.1
PETN	316.2	−128.7
Picric acid	229.1	−51.3
RDX	222.1	+14.71
Tacot	388.2	+110.5
TATB	258.2	−36.85
Tetryl	287.0	+4.67
TNM	196.0	+13.0
TNT	227.1	−16.0

Data from Ref. 6.

Remember that the standard heat of formation is equal to the difference between the absolute enthalpy of a compound and the absolute enthalpy of the elements from which it is made (all at the standard state). Then:

$$\Delta H_f^0(NaCl) = H^0(NaCl) - H^0(Na) - \frac{1}{2} H^0(Cl_2)$$

$$\Delta H_f^0(H_2O) = H^0(H_2O) - H^0(H_2) - \frac{1}{2} H^0(O_2)$$

$$\Delta H_f^0(HCl) = H^0(HCl) - \frac{1}{2} H^0(H_2) - \frac{1}{2} H^0(Cl_2)$$

$$\Delta H_f^0(NaOH) = H^0(NaOH) - H^0(Na) - \frac{1}{2} H^0(O_2) - \frac{1}{2} H^0(H_2)$$

(9.2)

Substituting the above equations into Eq. (9.1), we get:

$$\Delta H_r^0 = H^0(\text{NaCl}) + H^0(\text{H}_2\text{O}) - H^0(\text{HCl}) - H^0(\text{NaOH})$$

$$+ [H^0(\text{Na}) - H^0(\text{Na})] + \left[\frac{1}{2} H^0(\text{Cl}_2) - \frac{1}{2} H^0(\text{Cl}_2)\right]$$

$$+ \left[H^0(\text{H}_2) - \frac{1}{2} H^0(\text{H}_2) - \frac{1}{2} H^0(\text{H}_2)\right] + \left[\frac{1}{2} H^0(\text{O}_2) - \frac{1}{2} H^0(\text{O}_2)\right]$$

So we see that by utilizing the standard heats of formation, we are indirectly taking the differences of the absolute enthalpies, and also that the convention of having the standard heat of formation of all the elements equal to zero is calculationally valid and consistent.

As a numerical example, let us calculate the standard heat of reaction of the acid-base reaction shown above. In this reaction, gaseous HCl reacts with solid sodium hydroxide, forming solid sodium chloride and liquid water.

$$\text{HCl(g)} + \text{NaOH(s)} \rightarrow \text{NaCL(s)} + \text{H}_2\text{O(l)}$$

$$\Delta H_r^0 = \Sigma \Delta H_f^0(\text{products}) - \Sigma \Delta H_f^0(\text{reactants}) \tag{9.3}$$

$$= \Delta H_f^0(\text{NaCl}) + \Delta H_f^0(\text{H}_2\text{O}) - \Delta H_f^0(\text{HCl})$$

$$- \Delta H_f^0(\text{NaOH})$$

From Table 9.1:

$$\Delta H_f^0(\text{NaCl, s}) = -98.232 \text{ (kcal)/(g mole)}$$

$$\Delta H_f^0(\text{H}_2\text{O, l}) = -68.3174 \text{ (kcal)/(g mole)} \tag{9.4}$$

$$\Delta H_f^0(\text{HCl, g}) = -22.063 \text{ (kcal)/(g mole)}$$

$$\Delta H_f^0(\text{NaOH, s}) = -101.99 \text{ (kcal)/(g mole)}$$

Applying these values to Eq. (9.3),

$$\Delta H_r^0 = -98.232 - 68.3174 - (-22.063 - 101.99)$$

$$= -42.5 \text{ kcal}$$

Note that ΔH_r^0 in this example is negative. A negative heat of reaction means that heat is liberated during the course of the reaction, making it an exothermic reaction. If ΔH_r^0 is positive, then heat must be supplied to the reaction, making it an endothermic reaction.

9.4 Heat of Combustion

The heat of combustion is a special case of the heat of reaction; it is the heat of reaction for burning a compound with molecular oxygen completely to its most oxidized state. It is useful because of the relative ease of obtaining experimental thermal data directly from calorimetric closed-bomb tests. The standard heat of combustion is designated as ΔH_c^0, and as with any heat of reaction is

$$\Delta H_c^0 = \Sigma \Delta H_f^0(\text{products}) - \Sigma \Delta H_f^0(\text{reactants})$$

An example is the standard heat of combustion of methane.

$$CH_4(g) + 2 O_2(g) \rightarrow CO_2(g) + 2 H_2O(l) \tag{9.5}$$

$$\Delta H_c^0(CH_4, g) = \Delta H_f^0(CO_2, g) + \Delta H_f^0(H_2O, l) - \Delta H_f^0(CH_4, g)$$

The heat of formation of O_2 is zero, and therefore is not shown in Eq. (9.5).

Since the heats of formation of combustion products (CO_2, H_2O, SO_2, etc.) are well known, the heat of formation of a compound can be easily calculated from the heat of combustion.

Thermochemical data for organic compounds are usually listed as heats of combustion, whereas thermochemical data for inorganic compounds are usually listed as heat of formation.

Compilations or tables of standard heats of combustion always list the final combustion products and their states to avoid confusion or ambiguity. Table 9.3 gives standard heats of combustion for some organic compounds.

As an example of using the heat of combustion to find the heat of formation of an organic compound, let us take the case of o-nitrophenol, $C_6H_5NO_3$. If o-nitrophenol were burned with oxygen completely to CO_2, water, and N_2, the reaction would be

$$C_6H_5NO_3 + 5.75 O_2 \rightarrow 6 CO_2 + 2.5 H_2O + 0.5 N_2$$

and the heat of combustion would be

$$\Delta H_c^0(C_6H_5NO_3, s) = 6\Delta H_f^0(CO_2, g) + \Delta H_f^0(H_2O, l)$$
$$- \Delta H_f^0(C_6H_5NO_3, s)$$

From Table 9.3, we have

$$\Delta H_c^0(C_6H_5NO_3, s) = 689 \text{ kcal/g mole}$$

From Table 9.1, we have

$$\Delta H_f^0(CO_2, g) = -94.05 \text{ kcal/g mole}$$

$$\Delta H_f^0(H_2O, l) = -68.32 \text{ kcal/g-mole}$$

Solving for the heat of formation, we get

$$\Delta H_f^0(C_6H_5NO_3, s) = 6\Delta H_f^0(CO2, g) + \Delta H_f^0(H_2O, l) - \Delta H_c^0(C_6H_5NO_3, s)$$
$$= (6)(-94.05)+(2.5)(-68.32)-(-689)$$
$$= -46.1 \text{ kcal/g mole}$$

Table 9.3 Standard Heats of Combustion; reference conditions: 25°C (298.16 K), 1 atm pressure, gaseous substances in ideal state; ΔH_c^0 is given in kcal/g mole, reference state is 25°C, 1 atm

Compound	Formula	State[a]	ΔH_c^0
Hydrocarbons			
Final products: $CO_2(g)$, $H_2O(l)$			
Carbon (graphite)	C	s	94.0518
Carbon monoxide	CO	g	76.6361
Hydrogen	H_2	g	58.3174
Methane	CH4	g	212.798
Ethyne (acetylene)	C_2H_2	g	310.615
Ethane (ethylene)	C_2H_4	g	337.234
Ethane	C_2H_6	g	372.820
Propyne (allylene, methylacetylene)	C_3H_4	g	463.109
Propene (propylene)	C_3H_6	g	491.987
Propane	C_3H_8	g	530.605
1,2-Butadiene	C_4H_6	g	620.71
2-Methylpropene (isobutylene, isobutene)	C_4H_8	g	646.134
2-Methylpropane (isobutane)	C_4H_{10}	g	686.342
n-Butane	C_4H_{10}	g	687.982
1-Pentene (amylene)	C_5H_{10}	g	806.85
Cyclopentane	C_5H_{10}	l	786.54
2,2-Diemthylpropane (neopentane)	C_5H_{12}	g	840.49
2-Methylbutane (isopentane)	C_5H_{12}	g	843.24
n-Pentane	C_5H_{12}	g	845.16
Benzene	C_6H_6	g	789.08
Benzene	C_6H_6	l	780.98
1-Hexene (hexylene)	C_6H_{12}	g	964.26
Cyclohexane	C_6H_{12}	l	936.88
n-Hexane	C_6H_{14}	l	995.01
Methylbenzene (toluene)	C_7H_8	g	943.58
Methylbenzene (toluene)	C_7H_8	l	934.50
Cycloheptane	C_7H_{14}	l	1086.9
n-Heptane	C_7H_{16}	l	1151.27
1,2-Demethylbenzene (o-xylene)	C_8H_{10}	g	1098.54
1,2-Dimethylbenzene (m-xylene)	C_8H_{10}	l	1088.16
1,3-Dimethylbenzene (m-xylene)	C_8H_{10}	g	1098.12
1,3-Dimethylbenzene (m-xylene)	C_8H_{10}	l	1087.92
1,4-Dimethylbenzene (p-xylene)	C_8H_{10}	g	1098.29
1,4-Dimethylbenzene (p-xylene)	C_8H_{10}	l	1088.16
n-Octane	C_8H_{18}	l	1307.53
1,3,5-Trimethylbenzene (mesitylene)	C_9H_{12}	l	1241.19
Naphthalene	C10H8	s	1231.6
n-Decane	$C_{10}H_{22}$	l	1620.06
Diphenyl	$C_{12}H_{10}$	s	1493.5
Anthracene	$C_{14}H_{10}$	s	1695
Phenanthrene	$C_{14}H_{10}$	s	1693
n-Hexadecane	$C_{16}H_{34}$	l	2557.64
Alcohols			
Final products: $CO_2(g)$, $H_2O(l)$			
Methyl alcohol	CH_4O	g	182.59
Methyl alcohol	CH_4O	l	173.65
Ethyl alcohol	C_2H_6O	g	336.82
Ethyl alcohol	C_2H_6O	l	326.70
Ethylene glycol	$C_2H_6O_2$	l	284.48

Table 9.3 *(Continued)*

Compound	Formula	State[a]	ΔH_c^0
Allyl alcohol	C_3H_6O	l	442.3
n-Propyl alcohol	C_3H_8O	g	494.26
n-Propyl alcohol	C_3H_8O	l	483.56
Isopropyl alcohol	C_3H_8O	g	493.02
Isopropyl alcohol	C_3H_8O	l	481.11
Glycerol	$C_3H_8O_3$	l	396.27
n-Butyl alcohol	$C_4H_{10}O$	g	649.98
n-Butyl alcohol	$C_4H_{10}O$	l	638.18
Amyl alcohol	$C_5H_{12}O$	l	786.7
Methyl-diethyl carbinol	$C_6H_{14}O$	l	926.9

Acids

Final products: $CO_2(g)$, $H_2O(l)$

Compound	Formula	State[a]	ΔH_c^0
Formic (monomolecular)	CH_2O_2	g	75.70
Formic	CH_2O_2	l	64.57
Oxalic	$C_2H_2O_4$	s	58.82
Acetic	$C_2H_4O_2$	g	219.82
Acetic	$C_2H_4O_2$	l	208.34
Acetic anhydride	$C_4H_6O_3$	g	432.34
Acetic anhydride	$C_4H_6O_3$	l	426.00
Glycolic	$C_2H_4O_3$	s	166.54
Propionic	$C_3H_6O_2$	g	378.36
Propionic	$C_3H_6O_2$	l	365.41
Lactic	$C_3H_6O_3$	s	325.8
d-Tartaric	$C_4H_6O_6$	s	274.9
n-Butyric	$C_4H_8O_2$	l	520
Citric (anhydrous)	$C_6H_8O_7$	s	474.3
Benzoic	$C_7H_6O_2$	s	771.5
o-Phthalic	$C_8H_6O_4$	s	770.8
Phthalic anhydride	$C_8H_4O_3$	s	781.4
o-toluic	$C_8H_8O_2$	s	928.6
Palmitic	$C_{16}H_{32}O_2$	s	2379
Stearolic	$C_{18}H_{32}O_2$	s	2628
Elaidic	$C_{18}H_{34}O_2$	s	2663
Oleic	$C_{18}H_{34}O_2$	l	2668
Stearic	$C_{18}H_{36}O_2$	s	2697

Carbohydrates, cellulose, starch, etc.

Final products: $CO_2(g)$, $H_2O(l)$

Compound	Formula	State[a]	ΔH_c^0
d-Glucose (dextrose)	$C_6H_{12}O_6$	s	673
l-Fructose	$C_6H_{12}O_6$	s	675
Lactose (anhydrous)	$C_{12}H_{22}O_{11}$	s	1350.1
Sucrose	$C_{12}H_{22}O_{11}$	s	1348.9

			g cal/gram
Starch			4177
Dextrin			4108
Cellulose			4179
Cellulose acetate			4495

Other CHO compounds:

Final products: $CO_2(g)$, $H_2O(l)$

Compound	Formula	State[a]	ΔH_c^0
Formaldehyde	CH_2O	g	134.67
Acetaldehyde	C_2H_4O	g	284.98
Acetone	C_3H_6O	g	435.32
Acetone	C_3H_6O	l	427.79

Table 9.3 *(Continued)*

Compound	Formula	State[a]	ΔH_c^0
Methyl acetate	$C_3H_6O_2$	g	397.5
Ethyl acetate	$C_4H_8O_2$	g	547.46
Ethyl acetate	$C_4H_8O_2$	l	538.76
Diethyl ether	$C_4H_{10}O$	l	652.59
Diethyl detone	$C_5H_{10}O$	l	738.05
Phenol	C_6H_6O	g	747.55
Phenol	C_6H_6O	l	731.46
Pyrogallol	$C_6H_6O_3$	s	639
Amyl acetate	$C_7H_{14}O_2$	l	1040
Camphor	$C_{10}H_{16}O$	s	1411

Nitrogen compounds

Final products: $CO_2(g)$, $N_2(g)$, $H_2O(l)$

Urea	CH_4N_2O	s	151.05
Cyanogen	C_2N_2	g	261.70
Trimethylamine	C_3H_9N	l	578.4
Pyridine	C_5H_5N	l	660
Trinitrobenzene (1,3,5)	$C_6H_3N_3O_6$	s	664.0
Trinitrophenol (2,4,6)	$C_6H_3N_3O_7$	s	620.0
o-Dinitrobenzene	$C_6H_4N_2O_4$	s	703.2
Nitrobenzene	$C_6H_5NO_2$	l	739
o-Nitrophenol	$C_6H_5NO_3$	s	689
o-Nitroaniline	$C_6H_6N_2O_2$	s	766
Aniline	C_6H_7N	l	812
Trinitrotoluene (2,4,6)	$C_7H_5N_3O_6$	s	821
Nicotine	$C_{10}N_{14}N_2$	l	1428

Halogen compounds

Final products: $CO_2(g)$, $H_2O(l)$, dil. soln. of HCl

Carbon tetrachloride	CCl_4	g	92.01
Carbon tetrachloride	CCl_4	l	84.17
Chloroform	$CHCl_3$	g	121.8
Chloroform	$CHCl_3$	l	114.3
Methyl chloride	CH_3Cl	g	182.81
Chloracetic acid	$C_2H_2ClO_2$	s	172.24
Ethylene dichloride	$C_2H_4Cl_2$	l	296.77
Ethyl chloride	C_2H_5Cl	g	339.66

Sulfur compounds

Final products: $CO_2(g)$, $SO_2(g)$, $H_2O(l)$

Carbonyl sulfide	COS	g	132.21
Carbon disulfide	CS_2	g	263.52
Carbon disulfide	CS_2	l	256.97
Methyl mercaptan	CH_4S	g	298.68
Dimethyl sulfide	C_2H_6S	g	457.12
Dimethyl sulfide	C_2H_6S	l	450.42
Ethyl mercaptan	C_2H_6S	l	448.0

References: Ref. 1 and:

1. *Selected Values of Physical and Thermodynamic Properties of Hydrocarbons and Related Compounds*, Am. Petroleum Inst. Research Prof. 44, ed. F. D. Rossini, Carnegie Institute of Technology (1952).

2. *International Critical Tables*, Vol. V (1929). The values taken from this source were converted to a reference temperature of 25°C.

3. John H. Perry, *Chemical Engineers Handbook*, 3rd ed., McGraw-Hill (1950).

[a] Abbreviations: s, solid; l, liquid; g, gaseous.

9.5 Heat of Detonation or Explosion

The energy, or heat, released from the chemical reaction that occurs during the burning of a propellant or detonation of an explosive is called the *heat of explosion* or *heat of detonation*. This is the heat of reaction for the reaction of the explosive itself going to the explosive products. It does not include any heat generated by secondary reactions of the explosive or its products with air. Usually the term *heat of explosion* is used for propellants and *heat of detonation* for explosives. These are designated as ΔH^0_{exp} and ΔH^0_d.

$$\Delta H^0_d = \Sigma \Delta H^0_f(\text{detonation products}) - \Delta H^0_f(\text{explosive})$$

$$\Delta H^0_{exp} = \Sigma \Delta H^0_f(\text{burning products}) - H^0_f(\text{propellant})$$

In a real detonation, the composition of the products is not always the same for the same explosive. Factors such as the initial density and temperature, degree of confinement, particle size and morphology, and even the dimensions and shape of the charge affect the pressure and temperature behind the detonation front where the products are rapidly expanding and the various equilibria between the products are being quenched. Experimentally determined values for heat of detonation are preferred for engineering calculations when they can be obtained. When these data are not available, a reasonable estimate of the heat of detonation can be obtained by using the ideal product hierarchy rule of thumb. This rule states that in the detonation reaction (or burning for propellants), the following applies:

1. All nitrogen goes to N_2.
2. All the hydrogen burns with available oxygen to form H_2O.
3. Any oxygen left after step (2) burns carbon to CO.
4. Any oxygen left after step (3) burns CO to CO_2.
5. Any excess oxygen forms O_2.
6. Any excess carbon forms C(s).

As an example of estimating ΔH^0_d for an explosive, let us consider the detonation of RDX. The elemental formula of RDX is $C_3H_6N_6O_6$. Its detonation reaction according to the above hierarchy is:

$$C_3H6N_6O_6 \rightarrow 3N_2 + 3H_2O + 3CO$$

From Table 9.1:

$$\Delta H^0_f(H_2O, l) = -68.3174 \text{ (kcal)/(g mole)}$$

$$\Delta H^0_f(CO, g) = -26.4157 \text{ (kcal)/(g mole)}$$

and from Table 9.2:

$$\Delta H^0_f(RDX, s) = +14.7 \text{ (kcal)/(g mole)}$$

Table 9.4   Comparison of estimates of heats of detonation of pure explosive compounds with experimentally measured values; ΔH_d^0 is in (kcal)/(g mole)

Explosive	ΔH_d^0 Experimental	ΔH_d^0 calculated by first hierarchy	Error (%)	ΔH_d^0 calculated by CO_2 assumption	Error (%)
BFTF	355.5	303.0	−14.8	426.7	+20.0
DATB	238.2	239.6	+0.6	311.8	+30.9
HMX	438.4	396.8	−9.5	479.3	+9.3
NC(13.3% N)	304.6	231.4	−24.0	344.8	+13.2
NM	75.03	88.7	+18.2	99.0	+31.9
PETN	471.1	479.6	+1.8	520.8	+10.5
RDX	335.4	298.9	−10.9	360.7	+7.5
TACOT	380.4	405.6	+6.6	529.3	+39.1
TETRYL	327.2	320.7	−2.0	434.1	+36.7
TNT	247.5	248.2	+0.3	320.4	+29.5
			x = 8.9		x = 22.9
			σ = 7.7		σ = 11.5

Remember that $\Delta H_d^0 = \Delta H_f^0$ (products) $- \Delta H_f^0$ (explosive); therefore:

$$\Delta H_d^0 = 3\ \Delta H_f^0(H_2O,\ l) + 3\ \Delta H_f^0(CO,\ g) - \Delta H_f^0(RDX,\ s)$$

$$\Delta H_d^0(RDX) = 3(-68.3174) + 3(-26.4157) - (+14.7)$$

$$= 298.9\ (kcal)/(g\ mole)$$

The experimental value for $\Delta H_d^0(RDX,\ s)$ is -335.4 (kcal)/(g mole). Our estimate is off by -10.9%.

Some workers in this field use a different hierarchy of products, that is, one in which all the hydrogen burns to H_2O and then all of the remaining oxygen burns with carbon to form CO_2. For some calculations, as we will see later, this is a good assumption. But it often yields a very high estimate for calculating heats of detonation. A comparison of the two methods is shown in Table 9.4. Extensive lists of both experimental and calculated values of heats of detonation of numerous explosive materials can be found in Refs. 2–7.

9.6 Heat of Afterburn

The products of detonation of an underoxidized explosive are themselves fuels. These are normally products such as CO and free carbon. When these expand and mix with air, they eventually reach the lower combustion limit for these materials, and if they are at high enough temperature, or there is some other ignition source present, they will burst into flame. This afterburn or secondary fireball can be very energetic. The heat evolved is equal to the difference between the heat of combustion of the original explosive and its heat of detonation.

$$\Delta H_{AB}^0 = \Delta H_c^0 - \Delta H_d^0$$

where ΔH_{AB}^0 is the standard heat of the afterburn reaction, (kcal)/(g mole).

An example of this is in the detonation and subsequent afterburn of TNT. The elemental formula of TNT is $C_7H_5N_3O_6$.

In the detonation the following products are formed:

$$C_7H_5N_3O_6 \rightarrow 1.5\ N_2 + 2.5\ H_2O + 3.5\ CO + 3.5\ C$$

The CO and C are fuels and will burn with the oxygen in the air:

$$3.5\ CO + 3.5\ C + 5.25\ O_2 \rightarrow 7\ CO_2$$

The heat of this reaction is:

$$\Delta H_{AB}^0 = \Delta H_f^0(CO_2,\ g) - 3.5\Delta H_f^0(CO,\ g) = 7(-94.0518)$$

$$- 3.5(-26.4157)$$

$$= -566 \text{ (kcal)/(mole of TNT)}$$

This is more than twice the heat liberated in the detonation itself (the heat of detonation of TNT is -247 kcal/mole).

Calculating this from the difference between the heats of combustion and detonation, the combustion of TNT with oxygen is:

$$C_7H_5N_3O_6 + 5.25\ O_2 \rightarrow 1.5\ N_2 + 2.5\ H_2O + 7\ CO_2$$

$$\Delta H_f^0 = 2.5\ \Delta H_f^0(H_2O) + 7\ \Delta H_f^0(CO_2) - \Delta H_f^0(TNT)$$

$$= 2.5\ (-68.3174) + 7(-94.0518) - (-15.0)$$

$$= -814 \text{ kcal/mole}$$

The heat of detonation of TNT is, as stated before, -247 (kcal)/(gmole).

$$\Delta H_{AB}^0 = \Delta H_c^0 - \Delta H_d^0 = -814 - (-247) = -567 \text{ kcal/g mole TNT}$$

10

Group Additivity

In Chapter 8, along with tables of measured thermophysical data, we saw some fairly simple techniques for estimating these values when experimental results are not available. Among these techniques were Kopp's Rule for the heat capacity of both liquids and solids, and Trouton's ratio for latent heats of fusion and vaporization, along with Kistiakowski's temperature correction for the latter.

There are also several additional techniques for estimating the heat capacity of organic gases as well as the heats of formation of organic gases, liquids, and solids. These methods, like Kopp's Rule, assume that the various thermophysical and thermochemical properties of a compound are the sum of the like properties of each of its individual atoms and the particular bonds between them. The simplest of these methods for estimating heats of formation assigns a particular enthalpy value to each type of bond, such as, C—H, C—C, C=C, C—C, C—N, C—O, N—O, etc. One merely counts up the number of each type of bond, multiplies by the appropriate bond value, and then adds all these together. This method suffers from the fact that it does not take into account the interactive forces of nearby or adjacent atoms, or the geometry of the molecule. As you recall, the energy stored in bonds is kinetic energy of flexing, vibration, and rotation, and neighboring groups of atoms influence and severely limit these bond motions. Therefore, the additive bond energy method is rather inaccurate. The most accurate method would be one in which all the effects of geometry and weak interactive forces were taken into account. This, however, would be very cumbersome for a method of estimation and would require extremely detailed knowledge of the exact structure of the molecule. The best compromise between accuracy and simplicity currently available is the method of group additivity. This method was developed independently by several workers in the field including Benson (Ref. 8) and Shaw (Ref. 3).

10.1 Group Additivity Notation

In the method of group additivity, groups of atoms within the molecule are assigned enthalpy values. By doing this, much of the problem of the effect of interactive neighboring atoms is minimized. Only the interactive effects the groups have upon each other are left unaccounted for. These effects are taken into account somewhat by the use of structural family corrections, but in general they are relatively small.

The method involves a kind of shorthand, or notation, for describing the groups. It is quite simple and easily mastered.

First, a core atom is designated followed by a dash, which in turn is followed by a number of other atoms or subgroups in parentheses. The latter are the atoms attached to the core atom. For instance, C—$(C)_2(H)_2$ means a carbon atom core attached by single bonds to two other carbon atoms and two hydrogen atoms. Likewise, C—$(C)(H)_3$ would be a carbon atom core attached to another carbon atom and three hydrogens. Carbons that are double bonded are designated as C_d and are attached to two other atoms or subgroups, not including the double bond. Carbons that are triple bonded are designated as C_t and are attached to only one other atom or subgroup, not including the triple bond. Aromatic carbons, those within a benzene ring, are designated as C_B and are also considered to be attached to only one other atom or subgroup. The best way to grasp this is by way of a few examples.

The molecule in Figure 10.1 has four core atoms, all carbons. The two end carbons are the cores for the C—$(C)(H)_3$ group. The two interior carbons are the cores for the C—$(C)_2(H_2)$ group.

The molecule in Figure 10.2 has four core carbons also. In this case all four are cores for different groups. Starting from the left side of the molecule we have: C—$(C)(H)_3$, then the second carbon gives us C—$(C)(C_d)(H)_2$, the third carbon gives us C_d—$(C)(H)$, and the fourth carbon C_d—$(H)_2$.

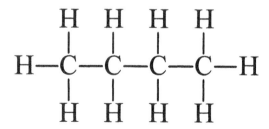

Figure 10.1   N-butane.

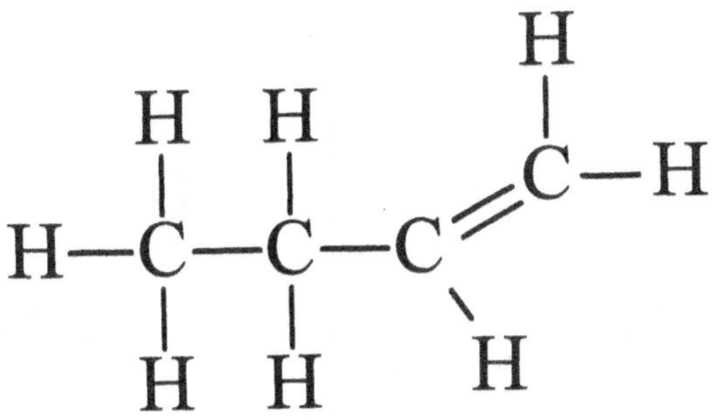

Figure 10.2 N-butene.

Again starting at the left end carbon, in Figure 10.3 we have C—(C)(H)$_3$, C—(C)(C$_t$)(H)$_2$, C$_t$—(C), and C$_t$—(H).

The various groups in the molecule in Figure 10.4 are: C—(C$_B$)(H)$_3$, the methyl group attached to the benzene ring; C$_B$—(C), the aromatic carbon attached to the methyl; three each C$_B$ — (NO$_2$); and two each C$_B$ — (H).

4.2 Data for the Ideal Gas State

In Benson's data (Ref. 8), the aromatic nitro group is broken up and instead of one group, C$_B$—(NO$_2$), there are two groups: C$_B$—(N), and N—(C$_B$)(O)$_2$. Likewise, Benson breaks up the amines such that an aromatic amine as on the TNA or TATB molecule is not C$_B$—(NH$_2$), but instead is C$_B$—(N) and N—(C$_B$)(H)$_2$. Also, the aromatic hydroxyl group instead of C$_B$—(OH), are C$_B$—(O) and O—(C$_B$)(H). Benson has determined values of many groups for estimating the heat, as well as the entropy, of formation of organic compounds in the ideal gas state at standard conditions. Tables 10.1 through 10.5 give selected values of Benson's data for both the heat of formation and entropy of formation for numerous groups. Not included in this table are the correction values for the effects of close neighboring groups, or for ring strain corrections. Both of these effects constitute relatively small corrections compared to the final heat or entropy of

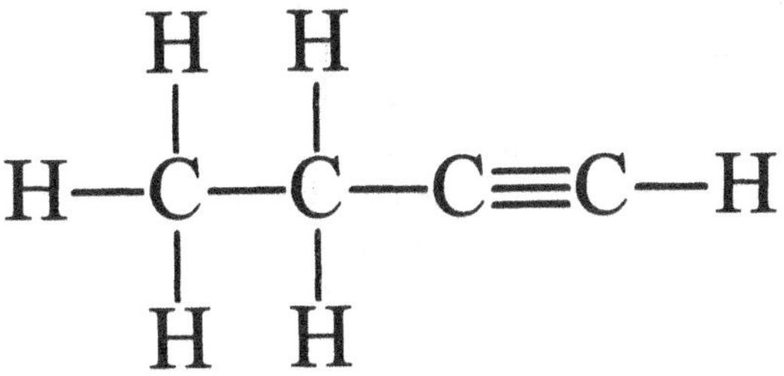

Figure 10.3 N-butyne.

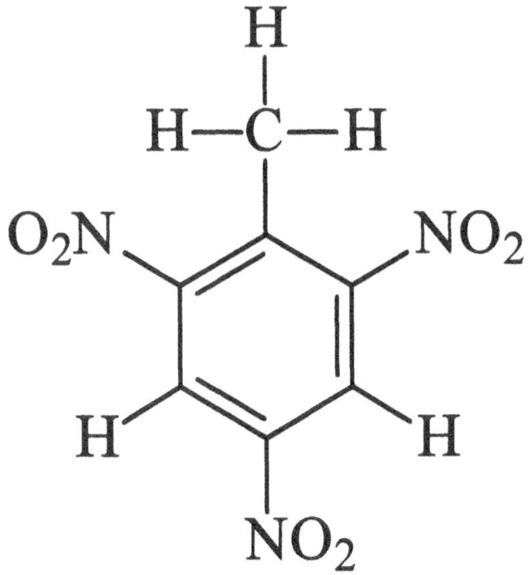

$$H$$
$$|$$
$$H—\overset{|}{C}—H$$

$$O_2N \qquad\qquad NO_2$$

$$H \qquad\qquad H$$

$$NO_2$$

Figure 10.4   TNT.

formation. If more exact estimates are required, these corrections can be found in Ref. 8. Benson's group additivity values for heat capacity of various temperatures are also presented in that same reference.

The problem with the Benson data, as far as most explosives are concerned, is that the data are for the gaseous state. If one had values of the heat of formation of a given compound in both the gas phase and the solid phase, the difference between the two would be the latent heat of sublimation, λ_s. Sublimation is evaporation directly from the solid to the gas phase. The heat of sublimation ideally is the sum (at a given temperature) of the heat of fusion and the heat of vaporization. Heats of sublimation are not generally available for many explosive compounds. When such data are available, very likely the data for the heats of formation would also be available.

By taking Trouton's ratio for both solids and liquids, making some rather broad assumptions about the normal melting and boiling points of typical organic explosives, and estimating the corrections for C_p of the gas and solid phases, a very crude estimate can be derived that the latent heat of sublimation (λ_s) for organic solid explosives at standard state conditions is around 25 (kcal)/(g mole). Comparing this value with what little data could be found in the literature, the average (λ_s) from five explosives (TNT, TATB, DATB, HNS, and TNA) was 32.85 (kcal)/(g mole). Now, armed with an estimate for heat of sublimation, one can use the Benson data and find $\Delta H_f^0(g)$ and then subtract the approximate latent heat of sublimation, 25 (kcal)/(g mole), to obtain $\Delta H_f^0(s)$. As an example of this, let us estimate the standard heat of formation of solid PETN (Fig. 10.5).

Table 10.1 Benson Group Contributions to Ideal-Gas Properties for
Hydrocarbon Groups (copied w/permission from—see reference 8)

Group	$\Delta H^0_{f,298}$ (kcal/g mole)	S^0_{298} (cal/g mole K)
C—(C)(H)$_3$	-10.08	30.41
C—(C)$_2$(H)$_2$	-4.95	9.42
C—(C)$_3$(H)	-9.90	-12.07
C—(C)$_4$	0.5	-35.10
C$_d$—(H)$_2$	6.26	27.61
C$_d$—(C)(H)	8.59	7.97
C$_d$—(C)$_2$	10.34	-12.7
C$_d$—(C)$_d$(H)	7.78	6.38
C$_d$—(C)$_d$(C)	8.88	-14.6
C$_d$—(C$_d$)$_2$	4.6	$\ldots$
C$_d$—(C$_B$)(H)	6.78	6.4
C$_d$—(C$_B$)(C)	8.64	-14.6
C$_d$—(C$_B$)$_2$	8.0	$\ldots$
C$_d$—(C$_1$)(H)	6.78	6.4
C$_d$—(C$_1$)(C)	8.53	$\ldots$
C—(C$_d$)(H)$_3$	-10.08	30.41
C—(C$_d$)$_2$(H)$_2$	-4.29	10.2
C—(C$_d$)$_2$(C)$_2$	1.16	$\ldots$
C—(C$_d$)(C)$_3$	1.68	-34.72
C—(C$_d$)(C)(H)$_2$	-4.76	9.8
C—(C$_d$)(C)$_2$(H)	-1.48	-11.7
C—(C$_d$)$_2$(C)(H)	-1.24	$\ldots$
C—(C$_1$)(H)$_3$	-10.08	30.41
C—(C$_1$)(C)(H)$_2$	-4.73	10.3
C—(C$_1$)(C)$_2$(H)	-1.72	-11.2
C—(C$_B$)(H)$_3$	-10.08	30.41
C—(C$_B$)(C)(H)$_2$	-4.86	9.3
C—(C$_B$)(C)$_2$(H)	-0.98	-12.2
C—(C$_B$)(C)$_3$	2.81	-35.18
C—(C$_B$)$_2$(C)(H)	-1.24	$\ldots$
C—(C$_B$)$_2$(C)$_2$	1.16	$\ldots$
C—(C$_B$)(C$_d$)(H)$_2$	-4.29	10.2
C$_1$—(H)	26.93	24.7
C$_1$—(C)	27.55	6.35
C$_1$—(C$_d$)	29.20	6.43
C$_1$—(C$_B$)	29.20	6.43
C$_B$—(H)	3.30	11.53
C$_B$—(C)	5.51	-7.69
C$_B$—(C$_d$)	5.68	-7.80
C$_B$—(C$_1$)	5.7	-7.80
C$_B$—(C$_B$)	4.96	-8.64

Table 10.2   Benson Group Contributions to Ideal-Gas Properties for Oxygen-Containing Compounds (copied w/ permission—see ref 8)

Group	$\Delta H^0_{f,298}$ (kcal/g mole)	S^0_{298} (cal/g mole K)
CO—(CO)(H)	−26.0	...
CO—(CO)(C)	−29.2	...
CO—(O)(C_d)	−32.5	...
CO—(O)(C_B)	−32.5	...
CO—(O)(C)	−35.1	4.78
CO—(O)(H)	−32.1	34.93
CO—(C_d)(H)	−31.7	...
CO—(C_B)$_2$	−38.1	...
CO—(C_B)(C)	−30.9	...
CO—(C_B)(H)	−34.6	...
CO—(C)$_2$	−31.4	15.01
CO—(C)(H)	−29.1	34.93
CO—(H)$_2$	−26.0	53.67
O—(C_B)(CO)	−32.5	...
O—(CO)$_2$	−50.9	...
O—(CO)(O)	−19.0	...
O—(CO)(C_d)	−46.9	...
O—(CO)(C)	−44.3	8.39
O—(CO)(H)	−58.1	24.52
O—(O)(C)	−4.5	9.4
O—(O)$_2$	−19.0	9.4
O—(O)(H)	−16.27	27.85
O—(C_d)$_2$	−32.8	10.1
O—(C_d)(C)	−31.9	9.7
O—(C_B)$_2$	−21.1	...
O—(C_B)(C)	−22.6	...
O—(C_B)(H)	−37.9	29.1
O—(C)$_2$(C)	−23.7	8.68
O—(C)(H)	−37.9	29.07
C_d—(CO)(O)	9.0	...
C_d—(CO)(C)	9.4	...
C_d—(CO)(H)	8.5	...
C_d—(O)(C_d)	8.9	...
C_d—(O)(C)	10.3	...
C_d—(O)(H)	8.6	...
C_B—(CO)	9.7	...
C_B—(O)	−0.9	−10.2
C—(CO)$_2$(H)$_2$	−7.6	...
C—(CO)(C)$_2$(H)	−1.8	−12.0
C—(CO)(C)(H)$_2$	−5.2	9.6
C—(CO)(C)$_3$	1.6	...
C—(CO)(H)$_3$	−10.1	30.41
C—(O)$_2$(C)$_2$	18.6	...
C—(O)$_2$(C)(H)	−16.3	...
C—(O)$_2$(H)$_2$	−15.1	...
C—(O)(C_B)(H)$_2$	−8.1	9.7
C—(O)(C_B)(C)(H)	−6.08	...
C—(O)(C_d)(H)$_2$	−6.9	...
C—(O)(C)$_3$	−6.60	−33.56
C—(O)(C)$_2$(H)	−7.2	−11.00
C—(O)(C)(H)$_2$	−8.1	9.8
C—(O)(H)$_3$	−10.1	30.41

Table 10.3 Benson Group Contributions to Ideal-Gas Properties for Nitrogen Containing Compounds[a] (copied w/ permission—see Ref 8)

Group	$\Delta H_{f,298}^0$ (kcal/g mole)	S_{298}^0 (cal/g mole K)
C—(N)(H)$_3$	−10.08	30.41
C—(N)(C)(H)$_2$	−6.6	9.8
C—(N)(C)$_2$(H)	−5.2	−11.7
C—(N)(C)$_3$	−3.2	−34.1
N—(C)(H)$_2$	4.8	29.71
N—(C)$_2$(H)	15.4	8.94
N—(C)$_3$	24.4	−13.46
N—(N)(H)$_2$	11.4	29.13
N—(N)(C)(H)	20.9	9.61
N—(N)(C)$_2$	29.2	−13.80
N—(N)(C$_B$)(H)	22.1	. . .
N$_I$—(H)	16.3	12.3
N$_I$—(C)	21.3	. . .
N$_I$—(C$_B$)	16.7	. . .
N$_A$—(H)	25.1	26.8
N$_A$—(C)	32.5	8.0
N—(C$_B$)(H)$_2$	4.8	29.71
N—(C$_B$)(C)(H)	14.9	. . .
N—(C$_B$)(C)$_2$	26.2	. . .
N—(C$_B$)$_2$(H)	16.3	. . .
C$_B$—(N)	−0.5	−9.69
N$_A$—(N)	23.0	. . .
CO—(N)(H)	−29.6	34.93
CO—(N)(C)	−32.8	16.2
N—(CO)(H)$_2$	−14.9	24.69
N—(CO)(C)(H)	−4.4	3.9
N—(CO)(C)$_2$	4.7	. . .
N—(CO)(C$_B$)(H)	0.4	. . .
N—(CO)$_2$(H)	−18.5	. . .
N—(CO)$_2$(C)	−5.9	. . .
N—(CO)$_2$(C$_B$)	−0.5	. . .
C—(CN)(C)(H)$_2$	22.5	40.20
C—(CN)(C)$_2$(H)	25.8	19.80
C—(CN)(C)$_3$	29.0	−2.80
C—(CN)$_2$(C)$_2$	. . .	28.40
C$_d$—(CN)(H)	37.4	36.58
C$_d$—(CN)(C)	39.15	15.91
C$_d$—(CN)$_2$	84.1	. . .
C$_d$—(NO)$_2$(H)	. . .	44.4
C$_B$—(CN)	35.8	20.50
C$_I$—(CN)	63.8	35.40
C—(NO$_2$)(C)(H)$_2$	−15.1	48.4
C—(NO$_2$)(C)$_2$(H)	−15.8	26.9
C—(NO$_2$)(C)$_3$	. . .	3.9
C—(NO$_2$)$_2$(C)(H)	−14.9	. . .
O—(N)(C)	−5.9	41.9
O—(NO$_2$)(C)	−19.4	48.50

[a] N$_I$ represents a double-bonded nitrogen in imines; N$_I$—(C$_b$) represents a pyridine nitrogen; N$_A$ represents a double-bonded nitrogen in azo compounds.

141

Table 10.4 Benson Group Contributions to Ideal-Gas Properties for Halogen Groups (copied w/ permission—see Ref 8)

Group	$\Delta H^0_{f,298}$ (kcal/g mole)	S^0_{298} (cal/g mole K)
C—(F)$_3$(C)	−158.4	42.5
C—(F)$_2$(H)(C)	−109.3	39.1
C—(F)(H)$_2$(C)	−51.5	35.4
C—(F)$_2$(C)$_2$	−97.0	17.8
C—(F)(H)(C)$_2$	−49.0	14.0
C—(F)(C)$_3$	−48.5	. . .
C—(F)$_2$(Cl)(C)	−106.3	40.5
C—(Cl)$_3$(C)	−20.7	50.4
C—(Cl)$_2$(H)(C)	−18.9	43.7
C—(Cl)(H)$_2$(C)	−16.5	37.8
C—(Cl)$_2$(C)$_2$	−22.0	22.4
C—(Cl)(H)(C)$_2$	−14.8	17.6
C—(Cl)(C)$_3$	−12.8	−5.4
C—(Br)$_3$(C)	. . .	55.7
C—(Br)(H)$_2$(C)	−5.4	40.8
C—(Br)(H)(C)$_2$	−3.4	. . .
C—(Br)(C)$_3$	−0.4	−2.0
C—(I)(H)$_2$(C)	8.0	43.0
C—(I)(H)(C)$_2$	10.5	21.3
C—(I)(C)(C$_d$)(H)	13.32	. . .
C—(I)(C$_d$)(H)$_2$	8.19	. . .
C—(I)(C)$_3$	13.0	0
C—(Cl)(Br)(H)(C)	. . .	45.7
N—(F$_2$)(C)	−7.8	. . .
C—(Cl)(C)(O)(H)	−21.6	15
C—(I)$_2$(C)(H)	26.0	54.6
C—(I)(O)(H)$_2$	3.8	40.7
C$_d$—(F)$_2$	−77.5	37.3
C$_d$—(Cl)$_2$	−1.8	42.1
C$_d$—(Br)$_2$	. . .	47.6
C$_d$—(F)(Cl)	. . .	39.8
C$_d$—(F)(Br)	. . .	42.5
C$_d$—(Cl)(Br)	. . .	45.1
C$_d$—(F)(H)	−37.6	32.8
C$_d$—(Cl)(H)	−1.2	35.4
C$_d$—(Br)(H)	11.0	38.3
C$_d$—(I)(H)	24.5	40.5
C$_d$—(C)(Cl)	−2.1	15.0
C$_d$—(C)(I)	23.6	. . .
C$_d$—(C$_d$)(Cl)	−3.56	. . .
C$_d$—(C$_d$)(I)	22.14	. . .
C$_1$—(Cl)	. . .	33.4
C$_1$—(Br)	. . .	36.1
C$_1$—(I)	. . .	37.9
C$_B$—(F)	−42.8	16.1
C$_B$—(Cl)	−3.8	18.9
C$_B$—(Br)	10.7	21.6

Table 10.4 *(Continued)*

Group	$\Delta H^0_{f,298}$ (kcal/g mole)	S^0_{298} (cal/g mole K)
C_B—(I)	24.0	23.7
C—$(C_B)(F)_3$	−162.7	42.8
C—$(C_B)(Br)(H)_2$	−6.9	...
C—$(C_B)(I)(H)_2$	8.4	...
C—$(Cl)_2(CO)(H)$	−17.8	...
C—$(Cl)_3(CO)$	−19.6	...
CO—(Cl)(C)	−30.2	...

Table 10.5 Benson Group Contributions to Ideal-Gas Properties for Organo-sulfur Groups (copied w/ permission—see Ref 8)

Group	$\Delta H^0_{f,298}$ (kcal/g mole)	S^0_{298} (cal/g mole K)
C—$(H)_3(S)$	−10.08	30.41
C—$(C)(H)_2(S)$	−5.65	9.88
C—$(C)_2(H)(S)$	−2.64	−11.32
C—$(C)_2(S)$	−0.55	−34.41
C—$(C_B)(H)_2(S)$	−4.73	...
C—$(C_d)(H)_2(S)$	−6.45	...
C_B—(S)	−1.8	10.20
C_d—(H)(S)	8.56	8.0
C_d—(C)(S)	10.93	−12.41
S—(C)(H)	4.62	32.73
S—$(C_B)(H)$	11.96	12.66
S—$(C)_2$	11.51	13.15
S—$(C)(C_d)$	9.97	...
S—$(C_d)_2$	−4.54	16.48
S—$(C_B)(C)$	19.16	...
S—$(C_B)_2$	25.90	...
S—(S)(C)	7.05	12.37
S—$(S)(C_B)$	14.5	...
S—$(S)_2$	3.04	13.36
C—$(SO)(H)_3$	−10.08	30.41
C—$(C)(SO)(H)_2$	−7.72	...
C—$(C)_3(SO)$	−3.05	...
C—$(C_d)(SO)(H)_2$	−7.35	...
C_B—(SO)	2.3	...
SO—$(C)_2$	−14.41	18.10
SO—$(C_B)_2$	−12.0	...
C—$(SO_2)(H)_3$	−10.08	30.41
C—$(C)(SO_2)(H)_2$	−7.68	...
C—$(C)_2(SO_2)(H)$	−2.62	...
C—$(C)_3(SO_2)$	−0.61	...
C—$(C_d)(SO_2)(H)_2$	−7.14	...
C—$(C_B)(SO_2)(H)_2$	−5.54	...

Table 10.5   *(Continued)*

Group	$\Delta H^0_{f,298}$ (kcal/g mole)	S^0_{298} (cal/g mole K)
C_B—$(SO)_2$	2.3	. . .
C_d—$(H)(SO_2)$	12.53	. . .
C_d—$(C)(SO_2)$	14.47	. . .
SO_2—$(C_d)(C_B)$	−68.58	. . .
SO_2—$(C_d)_2$	−73.58	. . .
SO_2—$(C)_2$	−69.74	20.90
SO_2—$(C)(C_B)$	−72.29	. . .
SO_2—$(C_B)_2$	−68.58	. . .
SO_2—$(SO_2)(C_B)$	−76.25	. . .
CO—$(S)(C)$	−31.56	15.43
S—$(H)(CO)$	−1.41	31.20
C—$(S)(F)_3$	. . .	38.9
CS—$(N)_2$	−31.56	15.43
N—$(CS)(H)_2$	12.78	29.19
S—$(S)(N)$	−4.90	. . .
N—$(S)(C)_2$	29.9	. . .
SO—$(N)_2$	−31.56	. . .
N—$(SO)(C)_2$	16.0	. . .
SO_2—$(N)_2$	−31.56	. . .
N—$(SO)_2(C)_2$	20.4	. . .

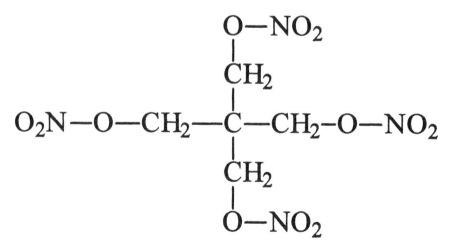

Figure 10.5   PETN.

1. The groups, and their values from Tables 10.1, 10.2, and 10.3, are:

$$1 \text{ each } C—(C)_4 = 1 \ (+0.50)$$
$$4 \text{ each } C—(C)(O)(H)_2 = 4(-8.1)$$
$$4 \text{ each } O—(C)(NO_2) = 4(-19.4)$$
$$\Delta H_f^0 \text{ (PETN, g)} = 0.5 + 4(-8.1) +$$
$$4(-19.4) = -109.5 \text{ (kcal)/(g mole)}$$

2. Subtracting the assumed value of the heat of sublimation:

$$\Delta H_f^0 \text{ (PETN, s)} = \Delta H_f^0 \text{ (PETN, g)} - \lambda_s$$
$$\Delta H_f^0 = -109.5 - 25 = -134.5 \text{ (kcal)/(g mole)}$$

Table 10.6 Group Additivity Values Per Shaw

Group	$\Delta H_{f,298}^0$ Value (kcal) (g mole)		
	Solid	Liquid	Ideal Gal
Aromatics			
$C_B—(H)$	0		
$C_B—(NO_2)$	-3		
$C_B—(NH_2)$	-9		
$C_B—(CH_3)$	-6		
$C_B—(OH)^a$	-45		
Alkanes			
$C—(C)(H)_2(NO_2)$	-22.2	-21.5	-14.4
$C—(C)_2(H)(NO_2)$	-21	-21.2	-13.6
$C—(C)_3(NO_2)$	-17	-18.3	x
$C—(C)(H)(NO_2)_2$	x	-24.0	-9.9
$C—(F)(C)(NO_2)_2$	x	-60.2	-46.9
$C—(C)_2(NO_2)_2$	-21.2	-21.0	-10.2
$C—(C)(NO_2)_3$	-13.6	-13.0	x
$C—(C)(H)_3$	-13.5	-11.6	-10.1
$C—(C)_2(H)_2$	-6.85	-6.1	-5.0
$C_d—(H)(NO_2)$			+7.1
$C_d—(C)(NO_2)$			+4.4
Azo groups[b]			
$N_A—(N_A)(C)$	+56.4		
$N_A—(C)$	+27.0		
$C—(N_A)(H)_3$	-10.1		
$C—(N_A)(C)(H)_2$	-6.0		
$C—(N_A)(C)_2(H)$	-3.4		
$C—(N_A)(C)_3$	-3.0		

[a] Calculated by author from picric acid, trinitrocresol, and styphnic acid.
[b] N_A represents a double-bonded nitro in azo compounds.

The measured value of the standard heat of formation of PETN is -128.7 (kcal)/(g mole). Our error is 4.5%.

Another problem with the Benson data (Table 10.3) is that it does not give a group value for C_B—(NO_2). It does give a value for C_B—(N), but if we use this we need a value for N—$(C_B)(O)_2$, which is also not in the Benson data. By taking the measured data for heat of formation in the gas phase for TNT, TATB, DATB, HNS, and TNA, and all the group values except N—$(C_B)(O)_2$, which are known, we can calculate N—$(C_B)(O)_2$ for each of these explosives. The average value found in this manner is

$$N—(C_B)(O)_2 = 0.5 \text{ (kcal)/(g mole)}$$

10.3 Data for the Solid State

Another, and preferred, method is to compile all the group data for solids directly. Unfortunately very little group additivity data for solids have been derived. Robert Shaw of SRI (Ref. 3) has started to do this, but the data are rather limited. Table 10.6 gives Shaw's data for solids.

An important point to remember is that the heat of formation of an explosive is only one of the terms in the determination of heat of detonation or heat of combustion. It is also one of the smaller terms, as compared to the values of the typical products such as water and carbon dioxide. A large error in the estimation of the ΔH_f^0 of the explosive, say around 25 or 30 (kcal)/(g mole), would only introduce a relatively small error in the estimation of ΔH_d^0. Most explosives have a ΔH_d^0 in the neighborhood of 300 (kcal)/(g mole), and so the 30 (kcal) error in ΔH_f^0 only represents an error of 10% in ΔH_d^0.

CHAPTER

11

Reaction Temperature

The previous chapters presented methods that enable one to calculate the amount of heat liberated by an exothermic chemical reaction carried out at standard conditions. What happens to that heat? If the reaction is conducted so that no heat escapes from the system, it is called an adiabatic reaction. If the system consists only of the materials that are involved in the reaction, then all the heat generated goes into heating the products to some higher temperature.

11.1 Reaction Temperature at Constant Pressure

If the whole process is carried out at constant pressure, then all the heat generated goes into increasing the enthalpy of the products. This internally generated heat is designated as Q, where $Q = n \Delta H_r^0$ (heat generated by the reaction at standard state conditions), and $Q = n \Delta H[\text{products}]$ (heat absorbed by the products of the reaction, at adiabatic conditions).

The change in enthalpy of the products includes any latent heats, if such are involved, such as the vaporization of water (λ_b, H_2O) if that was one of the products, as well as the heating, $n \int C_p \, dT$, of the products. If the products contained nH_2O moles of water as liquid then:

$$Q = n_{H_2O}\lambda_{b,H_2O} + n \int_{T0}^{T_a} C_p \, dT$$

where T_a is the adiabatic flame temperature at constant pressure and T_0 the standard temperature.

For ease of calculation it is often convenient to include the latent heat terms with Q, and call this quantity Q'. This is shown graphically in Figure 11.1.

$$Q' = Q - n_{n_2o}\lambda_{b,H_2O} = n \int_{T0}^{T_a} C_p \, dT$$

and also $Q' = Q - n_{H_2O}\lambda_{b,H_2O} = -n\Delta H_r^0 - n_{H_2O}\lambda_{b,H_2O}$.

Determining the final product temperature, T_a, based upon the above is exemplified by the following example: What is the adiabatic temperature of the product gases from the detonation of PETN? The gases are allowed to expand freely at one atmosphere (a constant-pressure process), but adiabatic conditions are maintained.

1. PETN is $C_5H_8N_4O_{12}$, and using the previously given product hierarchy:

$$C_5H_8N_4O_{12} \rightarrow 2N_2 + 4H_2O + 2\,CO + 3\,CO_2$$

2. For each mole of PETN, there are 4 moles of water in the products, $n_{H_2O} = 4$.

$$Q' = -n \, \Delta H_d^0(\text{PETN}) - n_{H_2O}\lambda_{b,H_2O}$$

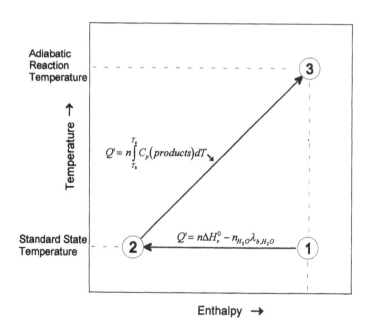

Figure 11.1 Temperature versus enthalpy for an adiabatic reaction.

Table 8.10 gives $\lambda_{b,H_2O} = 9.717$ (kcal)/(g mole).

Table 9.4 gives $\Delta H_d^0(\text{PETN, } H_2O, \text{ liq}) = -471.1$ (kcal)/(g mole). Therefore, $Q' = -(1)(-471.1) - (4)(9.717) = 432.2$ (kcal)/(g mole) PETN.

3. The quadratic weighted average C_p for the detonation products of PETN was given as an example in Chapter 8. It is $C_p = 6.652 + 4.307 \times 10^{-3}T - 0.9758 \times 10^{-6}T^2$, where C_p is in the units of (cal)/(g mole)(K).

$$Q' = n\int_{T_0}^{T_a} C_p\, dT = n\int_{T_0}^{T_a} (a + bT + cT^2)\, dT$$

$$Q' = n\left(a(T_a - T_0) + \frac{b}{2}(T_a^2 - T_0^2) + \frac{c}{3}(T_a^3 - T_0^3)\right)$$

where $n = 11$ moles (from part 1 above); a, b, and c are given above; T_0 is the standard temperature 298 K (25°C); and T_a the adiabatic flame temperature at constant pressure, K.

4. The simplest way to solve this cubic equation in T_a is to set

$$A = n\left(a(T_a - T_0) + \frac{b}{2}(T_a^2 - T_0^2) + \frac{c}{3}(T_a^3 - T_0^3)\right)$$

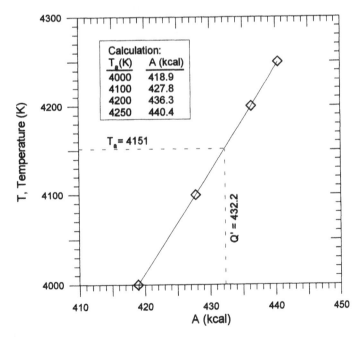

Calculation:	
$T_a(K)$	A (kcal)
4000	418.9
4100	427.8
4200	436.3
4250	440.4

$T_a = 4151$

$Q' = 432.2$

Figure 11.2 Trial and graph method of solving cubic equation in T_a.

Then substitute in several values of T_a, approximately where you think the true answer is and plot T_a versus A. Choose the point on the plot where $A = Q'$ and read the correct value of T_a from the T axis. (This procedure is very easy if one uses a pocket calculator.) In this example $T_a = 4151$ K, as seen from Figure 11.2.

11.2 Reaction Temperature at Constant Volume

If the reaction process were carried out at constant volume, C_v would have had to have been used in lieu of C_p. It is not necessary to correct this early in the calculation. Since $C_p/C_v = \gamma$ (the ratio of specific heats), the final temperature calculated from C_p can be corrected by $T_v = T_a\gamma$, where T_v is the adiabatic

Table 11.1   Values of γ for Various Gases at 1 atm pressure

Gas	Temperature (°C)	γ
Air	−118	1.415
	−78	1.408
	0	1.403
	17	1.403
	100	1.401
	200	1.398
	400	1.393
	1000	1.365
	1400	1.341
	1800	1.316
Ammonia	15	1.310
Argon	15	1.668
Carbon dioxide	15	1.304
Carbon monoxide	15	1.404
Chlorine	15	1.355
Ethane	15	1.22
Ethylene	15	1.255
Helium	−180	1.660
Hydrogen	15	1.410
Hydrogen sulfide	15	1.32
Methane	15	1.31
Neon	19	1.64
Nitric oxide	15	1.400
Nitrogen	15	1.404
Nitrous oxide	15	1.303
Oxygen	15	1.401
Propane	16	1.13
Steam	100	1.324
Sulfur dioxide	15	1.29

American Institute of Physics Handbook, 2nd Edition

flame temperature at constant volume. T_v is also called the isochoric flame temperature. Table 11.1 gives values of γ for various common detonation and combustion product gases. Like C_p, γ can be mole averaged for a mixture of gases by:

$$\gamma = \sum_i n_i \gamma_i$$

In general, γ decreases with increasing temperature and increases with increasing pressure. For the range of temperatures and pressures used in most of these types of calculations (the Interior Ballistics range; $25 < T < 4000°C$, and 1 bar$< P <$1 kbar), the value of γ at 15°C and 1 atm can be considered to be constant throughout the range without introducing a significant error. At detonation conditions typical of a solid high explosive (around 4000°C and 250 kbar), the average γ for the typical detonation product gases is around 3.

We shall see application of this use of γ to convert T_a to T_v in examples in the following chapter.

CHAPTER

12

Closed-Vessel Calculations

The methods of calculation in the previous chapter yield T_v for an adiabatic reaction at constant volume. That method, somewhat modified, also yields the adiabatic temperature for an explosive or propellant or fuel that is burned in a closed vessel. Using that temperature, along with knowing the volume of the vessel, the mass of reactant, and the type of gas (if any) in the vessel along with the reactant, one can calculate the pressure in the vessel at adiabatic conditions. This type of calculation is very important in many situations. It applies not only to a laboratory situation where one needs to predict the pressures that will be developed in a closed-bomb apparatus, but also applies to prediction of pressures from explosions in grain elevators, oil tanker hulls, buildings, mine shafts, etc. (If the system is not a constant volume, say in the case of a gun breech where a projectile is moving while the propellant is burning, this method represents the pressure condition at any one specific instant in time.)

12.1 Effect of Free Volume

In a closed system in which the volume is greater than the initial volume of the reactant, the extra or free volume must be taken into account. If this free volume is filled with air, then the oxygen in the air can react with excess carbon monoxide and free carbon from underoxidized explosives or propellants. A large air-filled free volume causes the explosive to undergo a complete afterburn or secondary fireball reaction. Both the products of this reaction and the additional heat developed by it must be taken into account.

153

The best way to show this is by example. Let us assume that we have 2 kg of TNT (ρ = 1.65 g/cm^3) in a large spherical vessel, 3 m inside diameter. The vessel is filled with air at 25°C and 1-atm pressure, in addition to the TNT.

1. The volume V_0 of the vessel is $(4/3)\pi r^3$, where r is the radius.

$$V_0 = (4/3)\pi r^3 = (4)(\pi)(1.5 \text{ m})^3/3 = 14.137 \text{ m}^3$$

$$= 14.137 \times 10^6 \text{ cm}^3$$

The volume of the TNT is its weight divided by its density:

$$V_{TNT} = (2000 \text{ g}) / (1.65 \text{ g/cm}^3) = 1.21 \times 10^3 \text{ cm}^3$$

Therefore, the volume of air in the vessel is:

$$V_0 - V_{TNT} = 14.137 \times 10^6 - 1.2 \times 10^3 = 14.136 \times 10^6 \text{ cm}^3$$

The volume of 1 g mole of ideal gas at STP (standard temperature and pressure, 25°C, 1 atm) is 24.467×10^3 cm^3. Therefore, the number of moles of air in the vessel is:

$$n_{(air)} = 14.136 \times 10^6 \text{ (cm}^3)/24.467 \times 10^3 \text{ (cm}^3/\text{g mole)}$$

$$= 577.76 \text{ (g moles)}$$

Air is made up of 21% oxygen and 79% nitrogen; so the number of g moles of each of these is:

$$n(O_2) = 0.21 \times 577.76 = 121.33 \text{ g moles}$$

$$n(N_2) = 0.79 \times 577.76 = 456.43 \text{ g moles}$$

2. The elemental formula for TNT is $C_7H_5N_3O_6$. Its molecular weight is 227.1. Since we have 2 kg of TNT, the number of moles is:

$$n_{(TNT)} = 2000 \text{ (g)}/227.1 \text{ (g/mole)} = 8.81 \text{ g moles}$$

The complete reaction, including the secondary fireball, is then:

$$8.81 \text{ } C_7H_5N_3O_6 + Z \text{ } O_2 \rightarrow [(3)(8.81) / (2)] \text{ } N_2$$

$$+ [(5)(8.81) / (2)] \text{ } H_2O + (7)(8.81) \text{ } CO_2$$

$$Z = [(5)(8.81)/(2)(2)] + (7)(8.81) - [(6)(8.81)/(2)]$$

$$= 46.25 \text{ g moles } O_2$$

so

$$8.81 \text{ TNT} + 46.25 \text{ } O_2 \rightarrow 13.22 \text{ } N_2 + 22.03 \text{ } H_2O + 61.67 \text{ } CO_2$$

We used up 46.25 g moles of the original 121.33 g moles of oxygen that were present in the free- volume air. We now have the following g moles of gaseous products in the system:

$$n(N_2) = 13.22 + 456.43 = 469.65 \text{ g moles}$$

$$n(H_2O) = 22.03 \text{ g moles}$$

$$n(CO_2) = 61.67 \text{ g moles}$$

$$n(O_2) = 121.33 - 46.25 = 75.08 \text{ g moles, and therefore,}$$

$$n(\text{total gases}) = 628.43 \text{ g moles}$$

12.2 Heat Produced

Once the materials in the system have been inventoried, that is, the products have been calculated, we know whether or not the secondary fireball reaction has gone to completion. In our example they have. If there were no oxygen in the free volume, then the total heat evolved would be from the heat of detonation. If there is enough oxygen in the free volume to complete the secondary fireball, then the heat evolved is from the heat of combustion (remember $\Delta H_c^0 = \Delta H_d^0 + \Delta H_{AB}^0$). If there were some oxygen in the free volume but not enough to burn all the product gases completely, then the heat evolved would have to be calculated from the heat of reaction for the particular amount of oxygen available.

In our example, the second case is true, the secondary fireball is complete. We can find the heat of combustion of TNT either by calculating it from the heat of formation or by looking up a measured value. Table 9.3 (in this section) gives ΔH_c^0 (H_2O, liq) for TNT $= -821$ (kcal)/(g mole); so $Q = n(\text{TNT})$ $\Delta H_c^0(\text{TNT}) = -(8.81 \text{ g moles})(-821 \text{ kcal/g mole}) = 7233 \text{ kcal}$, and $Q' = Q - n(H_2O) \lambda_b (H_2O) = 7233 - (22.03 \text{ g moles } H_2O)(9.717 \text{ kcal/g mole}) = 7019$ kcal.

12.3 Temperature of the Gases

The easiest to use strategy to find the gas temperature is to first find T_a, the adiabatic flame temperature at constant pressure, and then use γ to correct it to constant-volume conditions.

In order to do this we must first calculate the average heat capacity, C_p, of the mixture of gases.

$$\bar{C}_p = \sum_i n_i C_{pi}$$

$$n(N_2) = 469.65/628.43 = 0.7473$$

$$n(H_2O) = 22.03/628.43 = 0.0351$$

$$n(CO_2) = 61.67/628.43 = 0.0981$$

$$n(O_2) = 75.08/628.43 = 0.1195$$

From these values and the values of a, b, and c from Table 8.2, we find:

$$C_p = 6.429 + 2.504 \times 10^{-3}T - 0.5051 \times 10^{-6}T^2$$

Remembering that $Q' = n\int_{T_0}^{T_a} C_p \, dT$, we integrate and get the cubic equation with T_a as the only unknown.

$$Q' = n\left(a(T_a - T_0) + \frac{b}{2}(T_a^2 - T_0^2) + \frac{c}{3}(T_a^3 - T_0^3)\right)$$

Solving this by the trial and graph method used in the previous chapter, we get:

$$T_a = 1643 \text{ K } (1370°C)$$

Now, in order to use γ for the correction to constant volume, we must calculate γ for this particular mixture of gases. Values of γ for each of the product gases as a function of temperature T_a are found in Table 11.1.

$$\gamma = \sum_i n_i\gamma_i = 1.391$$

and finally:

$$T_v = T_a\gamma = (1643)(1.391) = 2285 \text{ K } (2012°C)$$

12.4 Pressure in the Vessel

We now know the volume of the vessel, the number of moles of gas in the vessel, and the temperature of that gas in the final pressurized state. In order to find the pressure, we put these values into an equation of state (EOS) for the gas. An EOS is a relationship between the volume, pressure, quantity, and temperature of a material in a given state. For gases at low pressure and nominal-to-high temperatures, the ideal gas equation works fine.

$$PV = nRT$$

where P is the pressure of the gas, V the volume of the gas, n the number of moles of gas, T the absolute temperature of the gas, and R the universal ideal gas constant. When P is in atmospheres, V in liters, n in moles, and T in Kelvin, then $R = 0.0821$ (l)(atm)/(g mole)(K). Rocket combustion engineers use the ideal gas equation extensively because in their range of operating conditions ($25 < T < 4000°C$, and $1 < P < 100$ atm) it very closely approximates the behavior of the gases with which they deal. Interior ballisticians, however, dealing with these same gases, use them to higher ranges of pressure, up to 5000 atm. At pressures above 200 atm (3000 psi), the ideal gas equation begins to predict values further and further away from observed experimental data. Therefore, a different EOS must be used. The most common in use in the field of interior ballistics is the Nobel-Able EOS:

$$P(V - \alpha w) = nRT$$

where α is the covolume of the gas, and w the weight of the gas.

Notice that this EOS differs from the ideal gas equation only in the volume term. Here the volume is corrected by an amount essentially equal to the volume of the molecules of gas as if they were compressed together with no space around them. The weight term w is obtained easily by taking the product of the number of moles of the gas times the molecular weight of the gas. The covolume, α, is very close to 1/1000 of the specific volume of the gas (cm^3/g). As with heat capacity and γ, both the molecular weight and density (therefore, also specific volume which is the reciprocal of density) can be used as molar averages for a mixture of gases. Figure 12.1 shows a comparison between the Nobel-Abel EOS and the ideal gas EOS for a mixture of product gases typical of those formed in both explosive and propellant reactions.

An easier version of the Noble-Able EOS to use is where the covolume is expressed on a molar rather than weight basis. In this case the EOS is

$$P(V - 0.025n) = 0.0821nT$$

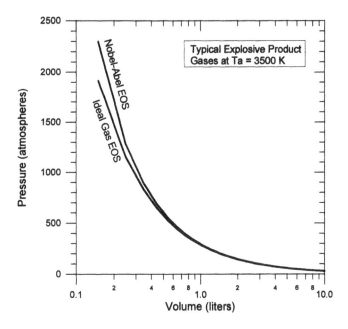

Figure 12.1   Comparison of ideal and Nobel-Abel EOSs for a typical explosive or propellant product gas mixture at a temperature of 3500 K.

where P is in atm, V is in l, n is in g moles, and T is in Kelvin. The gram molar value of α in this form is an empirical average from experimental data of typical propellant product gases in typical proportions and each other.

Applying the molar version of the Noble-Able EOS to the example at hand:

V, the internal volume of the vessel

$$= 14.137 \times 10^6 cm^3 \text{ or } 14.137 \times 10^3 \text{ l}$$
$$n = 628.43 \text{ g moles}$$
$$T = 2285 \text{ K}$$
$$P = \frac{0.0821nT}{V - 0.025n} = \frac{(0.0821)(628.43)(2285)}{14137 - (0.025)(628.43)} = 8.35 \text{ atm}(123 \text{ psi})$$

In this case, the ideal gas equation would have sufficed.

12.5 Summary

We have seen that by applying thermophysics and thermochemistry along with the physical dimensions of a given system that residual gas pressures can be readily calculated. However, it is very important to note that this pressure is not the shock pressure. The residual pressure is the pressure remaining in an enclosed explosive system independent of and after the passage of the shock. It is also the pressure one gets by burning a fuel or propellant in a closed volume, and is the instantaneous pressure at a particular time in an interior ballistic system where the volume is changing.

13

Estimating Detonation Properties

In Section I, Chemistry of Explosives, methods were described that enable one to estimate detonation properties (detonation velocity D and detonation pressure P_{CJ}) from the molecular structure of an explosive. This section gives an alternate method that utilizes the thermochemical properties of an explosive in order to estimate the values of these two output properties. This method was developed by M. J. Kamlet and S. J. Jacobs of the Naval Ordnance Laboratory in White Oak, MD (Ref. 9) and is referred to in this text as the KJ method.

13.1 KJ Assumed Product Hierarchy

This method assumes a different hierarchy of formation of product species from the detonation reaction of a CHNO explosive than the hierarchy used earlier, where CO is assumed to be formed preferentially prior to the formation of CO_2. Here, with the Kamlet-Jacobs method, CO_2 is assumed to be formed as the only oxidation product of carbon. As with the previous hierarchy assumptions, water is still formed first. The generalized reaction for an underoxidized explosive can be written as:

$$C_xH_yN_wO_z \rightarrow (w/2)N_2 + (y/2)H_2O + [(z/2)$$
$$-(y/4)]CO_2 + [x - (z/2) + (y/4)]C$$

For balanced explosive:

$$C_xH_yN_wO_z \rightarrow (w/2)N_2 + (y/2)H_2O + xCO_2$$

And for an overoxidized explosive:

$$C_xH_yN_wO_z \rightarrow (w/2)N_2 + (y/2)H_2O + xCO_2 + [(z/2) - (y/4) - x]O_2$$

13.2 Detonation Velocity

The Kamlet-Jacobs method, using the above product hierarchy, along with the thermochemical property, ΔH_d^0, the heat of detonation, estimates detonation velocity by:

$$D = A\left[NM^{1/2}(-\Delta H_d^0)^{1/2}\right]^{1/2} (1 + B\rho_0)$$

where D is the detonation velocity in mm/μsec; A, a constant, 1.01; N, the number of moles of gas per gram of original explosive; M, the grams of gas per mole of gas; ΔH_d^0, the heat of detonation in cal/g; B, a constant, 1.30; and ρ_0, the density of the original unreacted explosive in g/cm³. *In order to perform this calculation, we must also calculate the molecular weights involved.*

Example 13.1 Using the Kamlet-Jacobs method, estimate the detonation velocity of TNT at a density of 1.64 g/cm³.

1. The elemental formula of TNT is $C_7H_5N_3O_6$. TNT is underoxidized; so the reaction is:

$$C_7H_5N_3O_6 \rightarrow (3/2)N_2 + (5/2)H_2O + (6/2$$

$$- 5/4)CO_2 + (7-6/2+5/4)C$$

or

$$C_7H_5N_3O_6 \rightarrow (1.5)N_2 + (2.5)H_2O + (1.75)CO_2 + (5.25)C$$

2. The number of moles of gas per mole of TNT is $(1.5 + 2.5 + 1.75) = 5.75$ moles gas/mole TNT. The molecular weight of TNT is $(7)(12.01) + (5)(1.008) + (3)(14.00) + (6)(16) = 227.1$ g per mole TNT. Therefore, $N = 5.75/227.1 = 0.02532$ moles of gas per gram of explosive.

3. The grams of gas per mole of gas M is found by adding the products of the number of moles of each gas species times its molecular weight and then dividing by the number of moles of gas:

$$M = [(1.5)(28.016) + (2.5)(18.016)$$

$$+ (1.75)(44.01)] / 5.75 = 28.54$$

4. The heat of detonation for TNT was given in Table 9.4 as -247.5 kcal/g mole. To convert this to cal/g, we must multiply by 1000 and divide by the molecular weight:

$$\Delta H_d^0 = (-247.5)(1000)/(227.1) = -1089.8 \text{ cal/g}$$

5. D is now found by:

$$D = A[NM^{1/2}(-\Delta H_d^0)^{1/2}]^{1/2} (1 + B\rho_0)$$

$$= 1.01[(0.02532)(28.54)^{1/2}(1089.8)^{1/2}]^{1/2}[1+(1.30)(1.64)]$$

$$= 6.68 \text{ mm/}\mu s$$

Data in Ref. 2 give the detonation velocity of TNT at 1.64 g/cm^3 as 6.93 mm/μs. Our estimate is only in error by -3.6%.

13.3 Detonation (CJ) Pressure

Using the same parameters, the CJ pressure is given in the Kamlet-Jacobs method as:

$$P_{CJ} = K\rho_0^2 NM^{1/2}(-\Delta H_d^0)^{1/2}$$

where P_{CJ} is the detonation pressure in kbar, and K is a constant, 15.85. Continuing with the example in which the explosive was TNT, we had as values:

$$\rho_o = 1.64 \text{ g/cm}^3$$

$$N = 0.02532 \text{ moles gas/g HE}$$

$$M = 28.54 \text{ gas/mole gas}$$

$$\Delta H_d^0 = 1089.8 \text{ cal/g HE}$$

$$P_{CJ}(TNT) = (15.58)(1.64)^2(0.02532)(28.54)^{1/2}(1089.8)^{1/2} = 187.1 \text{ kbar}$$

Data in Ref. 2 give P_{CJ} as 210.0 kbar. The error in estimation is only -10.9%.

13.4 Modifications of the KJ Method

Based upon what was previously given, it can be seen that this method of estimation can be modified in several ways. First, one could use the earlier described hierarchy of formation of detonation products. This would yield higher values of N and lower values for M in the case of underoxidized explosives. The difference in the values of both N and M in these two instances (for the two hierarchy assumptions) approach zero as oxygen balance approaches zero, and are identical for explosives with positive oxygen balance. So, one might use the CO-

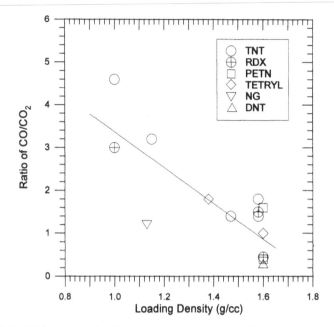

Figure 13.1   Molar ratio of carbon monoxide to carbon dioxide in reaction products versus loading density for six explosives.

first assumption for very low (negative) oxygen balance explosives. From limited experimental data (Ref. 6), it appears that the equilibrium of the gas composition shifts toward CO for explosives loaded at very low density, and toward CO_2 for the same explosive loaded at close to the crystal density. This is supported theoretically by calculations in the same reference. So one might arbitrarily use the CO-favored hierarchy at low densitites and the CO_2-favored hierarchy at high densities. The CO-to-CO_2 ratio could be chosen from the graph shown in Figure 13.1, which is a plot of that ratio versus density. Data for this plot were extracted from Ref. 4.

Another point of modification would be that once D is found, the P_{CJ} could also be estimated by $P_{CJ} = \rho_o D^2/4$ (which was given in Section I, Chemistry of Explosives). This estimate yields values very close to the KJ method.

References

1. Hougen, O. A., Watson, K. M., and Ragatz, R. A., *Chemical Process Principles, Part I,* John Wiley and Sons, New York, 1954.

2. Dobratz, B. M., *LLNL Explosives Handbook,* Lawrence Livermore National Laboratories, Report No. UCRL-52997, March 16, Rev. 1985.

3. *Encyclopedia of Explosives and Related Items,* PATR 2700, Vol. 7, Picatinny Arsenal, Dover, New Jersey, 1975.

4. Cook, M. A., *The Science of High Explosives,* R. E. Krieger Publ., New York, 1971.

5. *Engineering Design Handbook, Properties of Explosives of Military Interest,* AMCP 706-177, U.S. Army Materiel Command, January 1971.

6. *American Institute of Physics Handbook, 2nd Ed.,* McGraw-Hill Publishers, New York, 1963.

7. Meyer, R., *Explosives,* Second Edition, Verlag Chemie, 1981.

8. Reid, R. C., Prausnitz, J. M., and Sherwood, T. K., *The Properties of Gases and Liquids,* 3rd Edition, McGraw-Hill, New York, 1977.

9. Kamlet, M. J., and Jacobs, S. J., Chemistry of Detonations, I, A Simple Method for Calculating Detonation Properties of CHNO Explosives, *J. Chem. Phys.* **48**, p. 23, 1968.

SHOCK WAVES

Introduction

In this section, inert or nonreactive shock waves are discussed. We will learn the behavior of shocks by studying simple, mechanically analogous models and then proceed to develop the basic equations that describe dynamic uniaxial strain (shocks in only one direction). We will see how these equations are supplemented by experimentally derived empirical correlations, which will then allow us to solve them for simple shock wave interactions.

We will study the mechanisms involved in spall and scab formation and review data relating to shock waves.

14

Qualitative Description
of a Shock Wave

We can begin to understand shock phenomena by first considering the compression characteristics of most materials. For the purpose of this course, we will consider solid materials; however, what we describe for solids also applies (in principle) to liquids and gases.

14.1 Stress-Strain

We are all familiar with a typical stress-strain curve as depicted in Figure 14.1. As we know, almost all materials behave linearly. That is, the strain (amount of distortion) produced in a material is directly proportional to the stress placed on the material. This linear behavior holds until a point at which the material, if released, will not return to its exact original shape or dimensions. This point is called the yield point or *elastic* limit. When we strain a material beyond its elastic limit, we cause *plastic* deformation.

In shock behavior, we will only consider compressive stress and strain. Also, to keep the mathematics and models simple, let us only consider uniaxial compressive stress and strain. This means we will be studying these effects along only one axis of a material. We will assume the dimension of the material perpendicular to the strain axis to be infinite. This assumption means that the systems we will study have no edge effects. In Figure 14.2, we follow the same material to a much, much higher level of stress.

From σ_o to σ_1, the material behaves elastically; when we release it, it returns to its original shape and dimensions. From σ_1 to σ_2, the material behaves partly

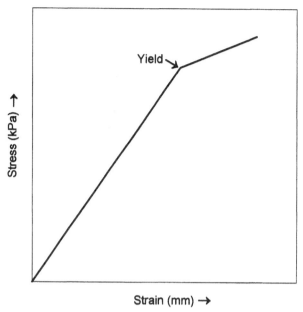

Figure 14.1 A typical stress-strain curve.

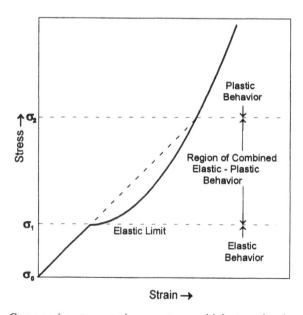

Figure 14.2 Compressive stress-strain curve to very high stress level.

plastic and partly elastic. σ_2 for most materials is broadly in the range of around ten times the elastic limit (σ_1). Above σ_2 the material exhibits plastic behavior similar to a fluid. In the study of strong shock waves, this is the region we can most easily evaluate, the region in which we work most with explosives.

14.2 Sound, Particle, and Shock Velocities

In the elastic region, the sound velocity in the material is constant. The square of the sound velocity, C, is proportional to the ratio of the change in pressure with change in density.

$$C^2 = dP/\rho \qquad (14.1)$$

This means that in the elastic region, pressure and density are linearly related. Beyond the elastic region, the wave velocity increases with pressure or density and P/ρ is not linearly proportional. Wave velocity continues to increase with stress or pressure throughout the region of interest. Therefore, up to the elastic limit, the sound velocity in a material is constant. Beyond the elastic limit, the velocity increases with increasing pressure. Let us look at a major implication of this fact. Consider the pressure wave shown in Figure 14.3.

At point A the pressure is low; therefore, C, the wave velocity, is low. Also the particle velocity or speed to which the material locally has been accelerated

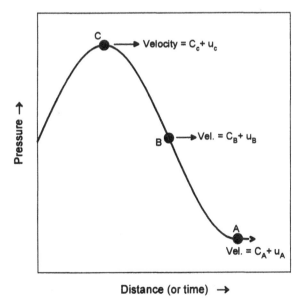

Figure 14.3 A pressure wave at high pressure.

is fairly low. Therefore, the velocity of the pressure wave is low. At point B, the wave velocity is higher than at point A, because as we just pointed out, the wave velocity increases with increasing pressure (remember we are above the elastic limit). The particle velocity is also higher (we shall see more of this later; for now let us just believe that the higher the pressure, the higher the particle velocity). Therefore, the pressure wave at point B is traveling faster than at point A. The same argument holds for point C that has a faster wave velocity than point B. The effect is shown in Figure 14.4, where we see that the wave continues to get steeper and steeper at the front until it approaches a straight vertical line.

When the wave assumes this vertical front, it is called a shock wave. Notice that there is not a smooth transition from matter in front of the wave to matter behind the wave. The material "jumps" from the nonshocked to the shocked state. This is called a *discontinuity*. Let us back up for a moment. We mentioned three different velocities: sound, particle, and pressure wave. We also said that the wave velocity was equal to the sum of the sound and particle velocities.

Sometimes it is difficult to visualize how the wave velocity can be faster than the particle velocity, because the particles are moving. In order to visualize this, let us run a simple experiment. As shown in Figure 14.5, line up eleven popsicle sticks with the space between each stick equal to the width of a stick.

Let us assume that each stick is one-half-inch wide and each space between the sticks is one-half-inch wide. Now start to push the first stick, at constant velocity, toward the others. Continue to push until the last stick is just contacted as shown in Figure 14.6.

Let us say that the time it took to push from the beginning until the last stick was touched was 10 seconds. The first stick moved 5 inches in that time; therefore, its velocity was 5 inches ÷ 10 seconds = 0.5 inch/second. As it contacted each additional stick, that stick was pushed also at this same velocity. If the sticks represent particles in a material, we would say the particle velocity was 0.5 inch/second. The signal that signified the beginning of motion of each stick moved from the first stick to the last stick, a distance of 10 inches in the same

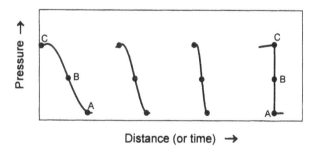

Figure 14.4 Shocking-up of a pressure wave.

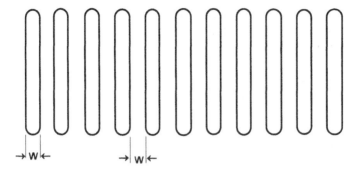

Figure 14.5 Eleven popsicle sticks.

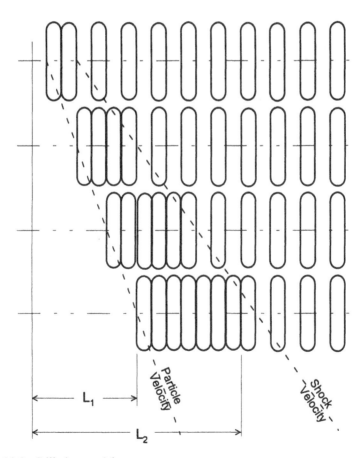

Figure 14.6 Still eleven sticks.

time. That signal is our analogous pressure wave front. It moved at a velocity of 1 inch/second, twice the velocity of the particles. From this we can better grasp the phenomenon of pressure wave motion.

14.3 Attenuation Behind Shock Waves

A smooth front pressure disturbance will "shock-up" because the wave speed increases with increasing pressure. Let us look at another wave, a square-pulse shock wave that has been formed and is traveling through a material (Figure 14.7).

The front of this wave is already a shock; so let us examine the back of the wave. The material front of point A is at high pressure. It is also traveling at a particle velocity u, and is at a high density ρ. The velocity of the back of the wave is, as we saw earlier, the sum of particle velocity and wave speed. Since the back is moving into a higher-density region than the front and is also encountering a faster material or particle speed than the front, it is traveling faster than the front. It therefore tends to catch up rapidly with the front. See Figure 14.8.

Consider point C; it is at ambient or zero pressure. Its velocity is less than either the shock front or point A; it therefore lags farther and farther behind the rest of the wave. So as the shock wave travels along, the back side, or *rarefaction* wave, smears out the back and eventually catches up to the shock front (Figure 14.9).

Still traveling faster that the front, the upper parts of the rarefaction wave now start "whittling" down the pressure in the front, until eventually the pres-

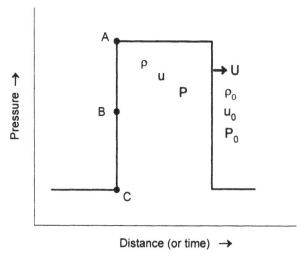

Figure 14.7 A square-wave shock.

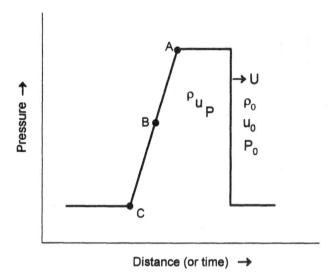

Figure 14.8   The back catching up with the front.

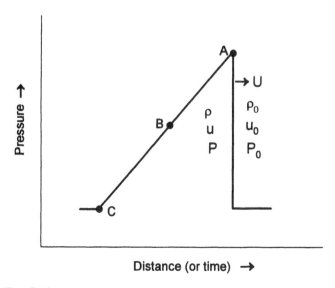

Figure 14.9   Rarefaction wave.

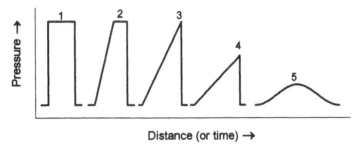

Figure 14.10   Attenuation of a square-wave shock.

sure is reduced to the region of elastic behavior, and the shock has decayed to a sound wave. The whole process is seen in Figure 14.10.

Summarizing what we have seen so far:

1. Shock waves occur when a material is stressed far beyond its elastic limit by a pressure disturbance.
2. Because pressure-wave velocity increases with pressure above the elastic limit, a smooth pressure disturbance ''shocks-up.''
3. Because the rarefaction wave moving into the shocked region travels faster than the shock front, the shock is attenuated from behind.

15

The Bead Model

In order to examine the relationships of the various shock parameters, let us look at another analogous model, the "Bead Model." This model was given in class notes on a course "Introduction to Stress Wave Phenomena" by Dennis B. Hayes at Sandia National Laboratories, internally published as SLA-773-0801, August 1973. Hayes stated in those notes that he got the model from Duval and Bond, at that time unpublished.

15.1 Arrangement of the Model

In this model we consider the case where we have a taut wire or string, on which we have strung a number of beads (Figure 15.1).

The beads are rather special in that they have no friction when moving along the string and also in that though their mass is equal to m per bead, their diameter is zero. (The diameter does not have to be zero to make this model work; it just simplifies the mathematics.)

15.2 Wave and Particle Velocity

As shown in Figure 15.2, a large solid wall moving at a constant velocity V_w impacts the beads. The wall impacts the first bead at time $t = 0$. To find the

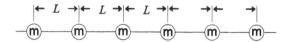

Figure 15.1   Beads on a string.

velocity of the first bead after impact, we apply the conditions that both momentum and energy (in this case, kinetic energy) must be conserved.

If we call the mass of the wall, M; the mass of the bead, m; the initial velocity of the wall, V_w; the velocity of the wall after impact, V_w* (this is virtually the same as V_w since the mass of the wall is huge compared to the mass of the bead); and the final velocity of the bead, V_b, then from conservation of momentum, we conclude

$$MV_w - MV_w^* = mV_b \tag{15.1}$$

The momentum lost by the wall is equal to the momentum gained by the bead. And from conservation of kinetic energy, we see

$$\frac{1}{2} MV_w^2 - \frac{1}{2} MV_w^{*2} = \frac{1}{2} mV_b^2 \tag{15.2}$$

The kinetic energy loss by the wall is equal to the kinetic energy gained by the bead. From Eq. (15.1)

$$V_w^* = V_w - \frac{m}{M}V_b \tag{15.3}$$

And by combining Eqs. (15.2) and (15.3),

$$V_b\left(1 + \frac{m}{M}\right) = 2V_w \tag{15.4}$$

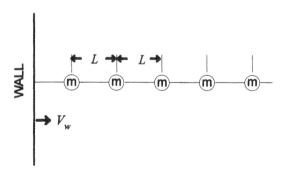

Figure 15.2   Wall moving at V_w impacts beads.

For all practical purposes, the mass of the wall is so much greater than the mass of the bead that we can say $m/M = 0$, and therefore,

$$V_b = 2V_w \tag{15.5}$$

The first bead slides up the wire at velocity $2V_w$ and impacts the second bead. The distance between the beads is L; so the time it took to travel from its initial position to bead number 2 is

$$t = \frac{L}{2V_w} \tag{15.6}$$

At this point, when the first bead strikes the second bead, it transfers all its momentum to bead 2 and stops. The second bead now travels up the wire at velocity $2V_w$. After another time period of $L/2V_w$, the second bead hits the next or third bead. The second bead is stopped, and the third bead proceeds at $2V_w$. Meanwhile, the wall caught up with the first bead and hit it again. This occurred at time L/V_w, and the whole process is repeated. From this we can draw a graph showing the time-velocity history of bead number 1 (Figure 15.3).

Notice that for half the time, the bead is moving at velocity $2V_w$, and for the other half of the time, it is at rest; velocity equals zero. Therefore, its average velocity is equal to V_w, the same as the wall's. This is the particle velocity in this model.

The disturbance, or shock velocity, moved at $2V_w$. So we see in this system, the shock velocity equals twice the particle velocity. In most real materials, as we will see later, the shock velocity is:

$$U = C_0 + su \tag{15.7}$$

where U is the shock velocity; C_0, the bulk sound speed for the particular material; s, an empirical constant for that material; and u, the particle velocity.

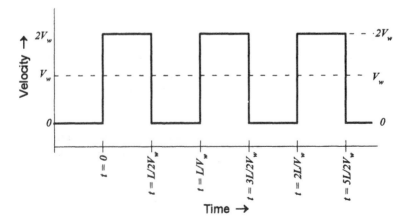

Figure 15.3 Time-velocity history of Bead No. 1.

15.3 Energy Partition

Now that we have seen the velocities, let us look at the energies involved in this bead-model system. The kinetic energy, based on the mean or average velocity of the particles, T, is

$$T = \frac{1}{2} mV_w^2 \tag{15.8}$$

The total kinetic energy, H, is

$$H = \frac{1}{2}\left[\frac{1}{2} m(2V_w)^2 + \frac{1}{2} m(0)^2 \right] = mV_w^2 \tag{15.9}$$

This is the average of the kinetic energies of the fastest and slowest state of each particle. The difference between T and H is the internal energy, E; therefore,

$$\text{internal energy, } E = \frac{1}{2} mV_w^2 \tag{15.10}$$

$$\text{kinetic energy, } T = \frac{1}{2} mV_w^2 \tag{15.11}$$

In both the case of the beads on a wire and the real case behind a shock front, the energy is partitioned equally between specific internal energy and specific kinetic energy. The term *specific* merely means on a per-unit-mass basis.

15.4 Density Changes

Now look at the density of this system. In front of the disturbance, there is one bead of mass m each distance L along the wire; so we can say that the density of the undisturbed material, ρ_0, is

$$\rho_0 = m/L \tag{15.12}$$

The density has increased behind the disturbance. If the disturbance moves through n beads at a velocity of $2V_w$, the original n beads had occupied nL length of wire. But in this same time, the wall has been moving at V_w velocity and therefore has traveled a distance of $nL/2$. The same number of beads now occupies a length of $nL-nL/2$, or $nL/2$ length; so the density behind the shock is

$$\rho = nm/nL/2 = 2m/L \tag{15.13}$$

The density behind the shock is twice the original density, or the compression ratio equals 2. For most solid real materials, even at very high pressures, densities cannot be doubled.

The preceding gave us a fairly realistic look at how the various shock parameters, U, u, ρ, and E, are related to and affect each other. Now we will get closer to the real world and look at the equations that describe the actual conditions we can calculate and test for in real materials.

16

Rankine-Hugoniot Jump Equations

In the previous figures of shock waves, and in the popsicle stick and bead models, we looked at the phenomenon from a fixed position and the phenomenon traveled past us. We refer to this view as ''laboratory'' coordinates. Another example of laboratory coordinates would be if we observed a train passing in front of us from left to right. Observing and representing phenomena in this manner is also referred to as ''Eulerian'' coordinates, named after the great mathematician Euler. We could view the system from another perspective, for instance, as a rider on the train we just mentioned. From his position, the train would be standing still, and we, along with the scenery, would be speeding past him from front to back. This system of coordinates is called ''Lagrangian'' and has some mathematical advantages over Eulerian coordinates in some calculational systems. Lagrangian coordinates are usually fixed on a particular particle in shock systems. We could also use a system of ''shock'' coordinates, where we view the system as if we were sitting on the crest of the shock wave. Let us use this ''shock'' coordinate point of view to derive the basic ''jump'' equations that describe shock phenomena.

Remember the description of a shock, where the front was referred to as discontinuous, then the original states of particle velocity u_0, density ρ_0, internal energy E_0, and pressure P_0, suddenly change across the shock front. They do not change gradually along some gradient or path but discontinuously jump from unshocked to shocked values (Figure 16.1). We have five variables to deal with; so we will need five relationships or related equations to solve for all those variables.

The first three relationships can be derived from the fact that we must con-

SHOCKED MATERIAL	UNSHOCKED MATERIAL
u_1	u_0
ρ_1	ρ_0
P_1	P_0
E_1	E_0

Figure 16.1   Shock parameters in front of and behind a shock-wave front.

serve mass, momentum, and energy across the shock front. These balances do not depend upon a process path but merely upon the initial and final states of the material in question. We call these three conservation or balance relationships the "Rankine-Hugoniot jump equations."

Orienting ourselves in shock coordinates, we would see material rushing toward us at shock velocity minus initial particle velocity. If we turned around, we would see that same stream of material at higher density, pressure, and energy rushing away from us at the shock velocity minus the imparted particle velocity (Figure 16.2). The mass shown entering the shock front is the same mass shown leaving the shock front.

16.1 Mass Balance

The mass balance implies that we are neither creating nor destroying mass. What goes in comes out, or mass entering equals mass leaving. Mass (m) is equal to density (ρ) times volume (v)

$$m = \rho v$$

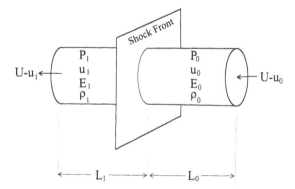

Figure 16.2   Control volume or mass passing through a shock front.

Volume (v) is equal to area (A) times length (L)

$$v = AL$$

(A is the cross-sectional area of the control volume, and remains constant throughout the process. We are looking at uniaxial strain only.)

Length in this system can be found as the distance a particle travels relative to the shock front, which is the velocity relative to the front times the time it took to travel that length.

$$L = t \times \text{velocity, relative}$$

Based on the above, we then have for mass entering, $m = \rho_0 V_0$ and for mass leaving, $m = \rho_1 V_1$. Since $v = AL$,

$$m = \rho_0 A L_0 \quad \text{and} \quad m = \rho_1 A L_1$$

and since $L = t(U - u)$,

$$m = \rho_0 A t(U - u_0) \quad \text{and} \quad m = \rho_1 A t(U - u_1)$$

Mass in equals mass out; so $\rho_1 A t(U - u_1) = \rho_0 A t(U - u_0)$, or $\rho_1(U - u_1) = \rho_0(U - u_0)$. This is the mass equation; it can also be written as

$$\frac{\rho_1}{\rho_0} = \frac{U - u_0}{U - u_1}$$

In most cases we encounter $u_0 = 0$; the material is standing still before it is shocked. So where $u_0 = 0$,

$$\frac{\rho_1}{\rho_0} = \frac{U}{U - u_1}$$

Another point to consider is that density can also be represented by the reciprocal of specific volume, v.

$$v = \frac{1}{\rho}$$

It is sometimes more convenient to use specific volume; so

$$\frac{\rho_1}{\rho_0} = \frac{U - u_0}{U - u_1} = \frac{v_0}{v_1}$$

16.2 Momentum Balance

The momentum balance implies that the rate of change of momentum, for the control mass to go from the state before the shock to the state after the shock, must be equal to the force applied to it. The force applied is simply the pressure difference across the shock front times the area over which it is applied, our control volume cross-sectional area

$$F = (P_1 - P_0)A \qquad (16.1)$$

The rate of change of momentum is

rate $= (mu_1 - mu_0)/t$

While deriving the mass-balance equation, we saw that $m = \rho At(U\text{-}u)$; so

$$\text{rate} = [\rho_1 Atu_1(U - u_1) - \rho_0 Atu_0(U - u_0)]/t \qquad (16.2)$$

Cancelling t out of Eq. (16.2) and equating this to Eq. (16.1), we get

$$(P_1 - P_0)A = \rho_1 Au_1(U - u_1) - \rho_0 Au_0(U - u_0)$$

and cancelling out A leaves

$$P_1 - P_0 = \rho_1 u_1(U - u_1) - \rho_0 u_0(U - u_0)$$

and from the mass equation we get

$$\rho_1(U - u_1) = \rho_0(U - u_0)$$

Combining these equations yields

$$P_1 - P_0 = \rho_0(u_1 - u_0)(U - u_0)$$

This is the momentum equation. For the common case where $u_0 = 0$, we have

$$P_1 - P_0 = \rho_0 u_1 U$$

16.3 Energy Balance

The implication of the energy balance is that the rate of energy increase of the control mass is equal to the rate of work being done on it. The rate of work done on the control mass would be the change in the pressure-volume product divided by the time required for the process. The volume divided by time is the same as area times velocity; so the rate at which work is done on the control mass is

$$w/t = P_1 Au_1 - P_0 Au_0$$

The rate of energy increase of the control mass is the sum of the rate of change of internal energy plus the rate of change of kinetic energy.

The internal energy, E, is the mass times the specific internal energy, e

$$E = me = \rho ALe$$

Therefore, the rate of change of internal energy is

$$E/t = (\rho_1 AL_1 e_1 - \rho_0 AL_0 e_0)/t$$

The rate of change of kinetic energy is

$$\frac{KE}{t} = \left(\frac{1}{2} \rho_1 AL_1 u_1^2 - \frac{1}{2} \rho_0 AL_0 u_0^2 \right)/t$$

Repeating the above, the rate of work done is equal to the rate of change of energy; therefore,

$$P_1 Au_1 - P_0 Au_o = (\rho_1 AL_1 e_1 - \rho_0 AL_0 e_0)/t$$

$$+ \left(\frac{1}{2} \rho_1 AL_1 u_1^2 - \frac{1}{2} \rho_0 AL_0 u_0^2 \right)/t \quad (16.3)$$

As we saw in the mass balance

$$L = t(U - u)$$

Substituting this in Eq. (16.3) and cancelling the As

$$P_1 u_1 - P_0 u_0 = \rho_1 (U - u_1) \left(e_1 + \frac{1}{2} u_1^2 \right) - \rho_0 (U - u_0) \left(e_0 + \frac{1}{2} u_0^2 \right)$$

Remembering from the mass equation that $\rho_0(U - u_0) = \rho_1(U - u_1)$, this could be written as

$$e_1 - e_0 = \frac{P_1 u_1 - P_0 u_0}{\rho_0(U - u_0)} - \frac{1}{2}(u_1^2 - u_0^2)$$

This is the energy equation. Again, for that common case where $u_0 = 0$, and remembering the mass equation

$$\frac{\rho_1}{\rho_0} = \frac{U}{U - u_1} = \frac{v_0}{v_1}$$

and momentum equation

$$P_1 - P_0 = \rho_0 u_1 U$$

we get

$$e_1 - e_0 = \frac{1}{2}(P_1 + P_0)(v_0 - v_1)$$

Summarizing the above, we have the following equations, where $u_0 \ne 0$, $\rho_o \ne 0$, $P_o \ne 0$, and $e_0 \ne 0$:

mass equation: $\dfrac{\rho_1}{\rho_0} = \dfrac{U - u_0}{U - u_1} = \dfrac{v_0}{v_1}$

momentum equation: $P_1 - P_0 = \rho_0(u_1 - u_0)(U - u_0)$

energy equation: $e_1 - e_0 = \dfrac{P_1 u_1 - P_0 u_0}{\rho_0(U - u_0)} - \dfrac{1}{2}(u_1^2 - u_0^2)$

And in the case where $u_0 = 0$

mass equation: $\dfrac{\rho_1}{\rho_0} = \dfrac{U}{U - u_1} = \dfrac{v_0}{v_1}$

momentum equation: $P_1 - P_0 = \rho_0 u_1 U$

energy equation: $e_1 - e_0 = \dfrac{1}{2} (P_1 + P_0)(v_0 - v_1)$

Now we have three equations for the five variables. In order to solve shock problems, we need two more relationships. We will see in the next chapter that these are derived from experimental data of material properties and from specification of boundary conditions.

17

The Hugoniot Planes, *U-u*, *P-v*, *P-u*

In the previous chapter we had developed three equations to deal with the five variables in shock-wave problems. In this chapter we will examine an empirical relationship called the Hugoniot, and will see how to select the proper boundary conditions such that we will be able to deal with all the shock variables. We will investigate how shock waves behave as they pass solid-solid interfaces, the meeting of two shocks, and shocks produced by the impact of two bodies.

17.1 The Hugoniot

We found in the previous session that five basic parameters were involved to describe fully, and therefore calculate, a shock wave: P (pressure), U (shock velocity), u (particle velocity), ρ (density), and e (specific internal energy). We had derived the three equations based upon mass, momentum, and energy conservation:

mass equation: $\dfrac{\rho_1}{\rho_0} = \dfrac{U - u_0}{U - u_1} = \dfrac{v_0}{v_1}$

momentum equation: $P_1 - P_0 = \rho_0(u_1 - u_0)(U - u_0)$

energy equation: $e_1 - e_0 = \dfrac{P_1 u_1 - P_0 u_0}{\rho_0(U - u_0)} - \dfrac{1}{2}(u_1^2 - u_0^2)$

The subscripts 0 and 1 refer to the states just in front of and just behind the shock front, respectively.

What we need at this point is another relationship relating these same parameters. One such relationship is called an equation of state (EOS). The EOS gives all of the equilibrium states in which a material can exist and is written in terms of specific internal energy, pressure, and specific volume. We do not have a general EOS that can be derived for all materials. There is, of course, the ideal gas equation, $PV = nRT$, where RT is related to the specific internal energy, but we are not dealing with ideal gases here. Our main interest is in solids. But if there were such an EOS,

$$e = f(P,v)$$

then it could be combined with the energy jump equation and the energy, e, term could be eliminated, giving us the relationship

$$P = f(v)$$

This is the Hugoniot equation. In the mass and momentum equations we have two equations in four variables. If we could determine some relationship involving any two of these that was also specifically descriptive of the states of a material, then we would have an alternative to the EOS-derived Hugoniot. Out of the four variables we could determine a Hugoniot equation relating any two. Our choices would be among the following:

$$P\text{-}U,\ P\text{-}u,\ P\text{-}v,\ U\text{-}u,\ U\text{-}v,\ \text{and}\ u\text{-}v$$

In the past, many experiments were conducted to determine such a relationship, and it was found that the shock velocity was linearly related to the particle velocity, for most materials.

$$U = C_0 + su$$

So a Hugoniot equation could be determined from experimental data. Out of the possible six variable pair ''planes'' mentioned above, three are found to be especially useful: the U-u plane, the P-v plane, and the P-u plane.

17.2 The U-u Plane

The first value of the U-u plane is that the U-u relationship is linear. Therefore, if we determine U and u experimentally for a particular material, at a number of shock states, we could plot the data on this plane, strike a straight line through the data points and easily find the U-u Hugoniot equation. This is shown in Figure 17.1, the data and plot for the Hugoniot of type 6061 aluminum.

The equation for this linear relationship is again $U = C_0 + su$. The constant C_0 is called the bulk sound speed, but this is misleading. C_0 has no real physical meaning other than the fact that it is the y-axis intercept on a straight line drawn through the data points. The units in this expression are mm/μs (or km/s) for the terms U, C_0, and u, and the term s is dimensionless. Table 17.1 lists U-u

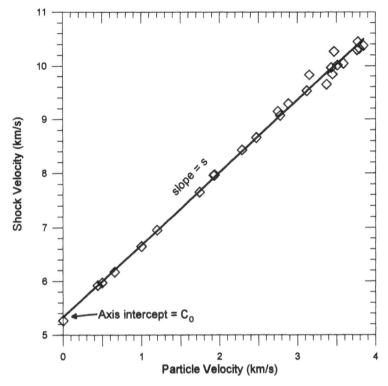

Figure 17.1   *U-u* Hugoniot data for 6061 aluminum, ρ_0 = 2.703 g/cm^2 (Ref.2).

Hugoniot values for a number of inert materials, and Table 17.2 lists *U-u* Hugoniots for a number of unreacted explosives.

In the tables you will notice that the Hugoniot is given as $C_0 + su + qu^2$. This might lead you to think that the *U-u* Hugoniot is not really linear. Those values were derived from least-squares fits to the data. What probably really is happening is that the data are composed of two or more straight-line segments with a transition region between them. The reason for the shift in slope of the *U-u* Hugoniot is most likely that a phase change or a shift in crystal lattice has occurred at that point. This is the case, however, in relatively few materials. Figure 17.2(a) and (b) (data from Ref. 2) show extreme cases of this behavior. Unfortunately, there is no simple correlation of the constants C_0 or *s* with any other material properties. Therefore, you may not be able to find the Hugoniot for a material in any references.

What you can do, however, is use the *U-u* Hugoniot values of another material that is similar to the one of interest. Try to match the materials by chemical and physical type, that is, same chemical family and similar crystal structure and habit. Also try to match by mechanical state, that is, a pressed powder, a casting, a foam, etc. When the *U-u* Hugoniot is used in calculations, it is usually included

Table 17.1 *U-u* Hugoniot Values for Inert Materials

Material	ρ_0 (g/cm^3)	c_0 (km/sec)	s	q (sec/km)	Comments
Elements					
Antimony	6.700	1.983	1.652		
Barium	3.705	0.700	1.600		Above $P = 115$ and $u_s = 2.54$
Beryllium	1.851	7.998	1.124		
Bismuth	9.836	1.826	1.473		
Cadmium	8.639	2.434	1.684		
Calcium	1.547	3.602	0.948	1.20	
Cesium	1.826	1.048	1.043	0.051	
Chromium	7.117	5.173	1.473	1.19	
Cobalt	8.820	4.752	1.315		
Copper	8.930	3.940	1.489		
Germanium	5.328	1.750	1.750		Above $P = 300$ and $u_s = 4.20$
Gold	19.240	3.056	1.572		
Hafnium	12.885	2.954	1.121		Below $P = 400$ and $u_s = 3.86$
Hafnium	12.885	2.453	1.353		Above transition
Indium	7.279	2.419	1.536		
Indium	22.484	3.916	1.457		
Iron	7.850	3.574	1.920	-0.068	Above $u_s = 5.0$
Lead	11.350	2.051	1.460		
Lithium	0.530	4.645	1.133		
Magnesium	1.740	4.492	1.263		
Mercury	13.540	1.490	2.047		
Molybdenum	10.206	5.124	1.233		
Nickel	8.874	4.438	1.207		
Niobium	8.586	4.438	1.207		
Palladium	11.991	3.948	1.588		
Platinum	21.419	3.598	1.544		
Potassium	0.860	1.974	1.179		
Rhenium	21.021	4.184	1.367		
Rhodium	12.428	4.807	1.376		
Rubidium	1.530	1.134	1.272		
Silver	10.490	3.229	1.595		
Sodium	0.968	2.629	1.223		
Strontium	2.628	1.700	1.230		Above $P = 150$ and $u_s = 3.63$
Sulfur	2.020	3.223	0.959		
Tantalum	16.654	3.414	1.201		
Thallium	11.840	1.862	1.523		
Thorium	11.680	2.133	1.263		
Tin	7.287	2.608	1.486		
Titanium	4.528	5.220	0.767		Below $P = 175$ and $u_s = 5.74$
Titanium	4.528	4.877	1.049		Above transition
Tungsten	19.224	4.029	1.237		
Uranium	18.950	2.487	2.200		
Vanadium	6.100	5.077	1.201		
Zinc	7.138	3.005	1.581		
Zirconium	6.505	3.757	1.018		Above $P = 260$ and $u_s = 4.63$
Zirconium	6.505	3.296	1.271		Above transition

(Continued)

Table 17.1   *(Continued)*

Material	ρ_0 (g/cm³)	c_0 (km/sec)	s	q (sec/km)	Comments
Alloys					
Brass	8.450	3.726	1.434		
2024 Aluminum	2.785	5.328	1.338		
921-T Aluminum	2.833	5.041	1.420		
Lithium-magnesium alloy	1.403	4.247	1.284		
Magnesium alloy AZ-31B	1.775	4.516	1.256		
304 Stainless steel	7.896	4.569	1.490		
U-3 wt % Mo	18.450	2.565	2.200		
Synthetics					
Adiprene	0.927	2.332	1.536		
Epoxy resin	1.186	2.730	1.493		Below $P = 240$ and $u_s = 7.0$
Epoxy resin	1.186	3.234	1.255		Above transition
Lucite	1.181	2.260	1.816		
Neoprene	1.439	2.785	1.419		
Nylon	1.140	2.570	1.849	−0.081	
Paraffin	0.918	2.908	1.560		
Phenoxy	1.178	2.266	1.698		
Plexiglass	1.186	2.598	1.516		
Polyethylene	0.915	2.901	1.481		
Polyrubber	1.010	0.852	1.865		
Polystyrene	1.044	2.746	1.319		
Polyurethane	1.265	2.486	1.577		Below $P = 220$ and $u_s = 6.5$
Silastic (RTV-521)	1.372	0.218	2.694	−0.208	
Teflon	2.153	1.841	1.707		
Compounds					
Periclase (MgO)	3.585	6.597	1.369		Above $P = 200$ and $u_s = 7.45$
Quartz	2.204	0.794	1.695		Stishovite above $P = 400$
Sodium chloride	2.165	3.528	1.343		Transition ignored
Water	0.998	1.647	1.921	0.096	
Gases					
	—	0.899	0.939		Approximate for all gases

Reference 3.

in an expression with other terms, such as density, that usually have more weight in the calculation. By judicious matching you can usually be confident that your ''guessed'' Hugoniot will yield calculational values within 10 to 15% of reality.

17.3   The *P-v* Plane

If we combine the U-u Hugoniot equation with the momentum and mass equations, and let $P_0 = 0$ and $u_0 = 0$, we can eliminate the particle and shock velocity terms and get an expression $P = f(v)$

Table 17.2 *U-u* Hugoniot Values for Unreacted Explosives

Explosive	Density (g/cm^3)	c_0 (km/s)	s	Range limitations (U, Shock Velocity)
AN	0.86	0.84	1.42	
	1.73	2.20	1.96	
Baratol	2.611	2.40	1.66	2.4–3.66
		1.5	2.16	3.66–4.0
		2.79	1.25	
Comp B	1.70	2.95	1.58	
	1.710	1.20	2.81	4.40–5.04
Comp B (cast)	1.700	2.49	1.99	3.57–5.02
Comp B-3	1.70	3.03	1.73	
	1.70	2.88	1.60	4.24–7.01
	1.72	2.71	1.86	3.42–4.45
	1.723	1.23	2.81	4.42–5.07
Comp B-3 (cast)	1.680	2.710	1.860	3.387–4.469
Cyclotol (75/25)	1.729	2.02	2.36	4.67–5.22
DATB	1.780	2.449	1.892	3.159–4.492
H-6 (cast)	1.760	2.832	1.695	2.832–4.535
	1.76	2.654	1.984	<3.7
HBX-1 (cast)	1.750	2.936	1.651	
HBX-3 (cast)	1.850	3.134	1.605	
HMX	1.903	2.74	2.6	
	1.891	2.901	2.058	
HNS	1.38	0.61	2.77	1.44–1.995
	1.57	1.00	3.21	1.00–3.18
HNS-II	1.47	1.10	3.48	
	1.58	1.98	1.93	
LX-04-1	1.860	2.36	2.43	2.61–3.24
LX-09-0	1.839	2.43	2.90	
LX-10-1		1.178	2.779	
LX-17-0	1.90	2.33	2.32	
NC	1.59	2.24	1.66	
NM	1.13	2.00	1.38	2.83–4.40
NQ[a]		3.544	1.459	
(C)		3.048	1.725	
Octol	1.80	3.01	1.72	
(cast)	1.803	2.21	2.51	3.24–4.97
PBX-9011-06	1.790	2.225	2.644	4.1–6.1
PBX-9404-03	1.721	1.89	1.57	2.4–3.7
	1.84	2.45	2.48	2.45–6.05
PBX-9404	1.84	2.310	2.767	<3.2
PBX-9407	1.60	1.328	1.993	2.11–3.18
PBX-9501-01	1.844	2.683	1.906	2.9–4.4
PBX-9604	1.491	0.987	2.509	
Pentolite 50/50	1.67	2.83	1.91	
	1.676	2.885	3.20	4.52–5.25

(Continued)

Table 17.2   *(Continued)*

Explosive	Density (g/cm³)	c_0 (km/s)	s	Range limitations (U, Shock Velocity)
PETN	0.82	0.47	1.73	
	1.59	1.33	2.18	1.40–2.14
		0.64	4.19	1.86–2.65
	1.60	1.32	2.58	1.89–2.56
	1.72	2.326	2.342	2.83–3.18
		1.83	3.45	2.52–3.87
	1.75	2.53	1.88	
	1.77	2.42	1.91	
		2.811	1.73	<4.195
Polystyrene	1.05	2.40	1.637	3.87–6.493
RDX	1.0	0.4	2.00	
	1.64	1.93	0.666	2.00–2.16
		0.70	4.11	2.14–2.63
	1.799	2.78	1.9	
	1.80	2.87	1.61	4.21–5.45
TATB	1.847	2.340	2.316	3.125–5.629
	1.876	1.46	3.68	1.5–3.23
		2.037	2.497	3.23–5.9
	1.937	2.90	1.68	<3.404
Tetryl	0.86	0.35	1.75	
	1.70	2.4763	1.416	3.08–4.17
	1.73	2.17	1.91	
TNT (pressed)	0.98	0.366	1.813	1.05–3.26
	1.643	2.372	2.16	2.78<
(cast)	1.614	2.390	2.050	3.034–5.414
	1.62	2.274	2.652	<3.7
		2.987	1.363	3.7<
	1.63	2.57	1.88	
(liquid)(82°C)	1.472	2.14	1.57	3.49–4.65
Tritonal (cast)	1.73	2.313	2.769	<3.8
XTX-8003	1.53	1.49	3.30	2.38–4.06

[a]   Reference 4.
[b]   Large grain.
[c]   Commercial grain.

$$P = C_0^2(v_0 - v)[v_0 - s(v_0 - v)]^{-2}$$

This is the Hugoniot in the *P-v* plane. Plotted, it is as appears in Figure 17.3.

Remember that we stated earlier that the Hugoniot is not the path along which a material is stressed, but is the locus of all the possible equilibrium states in which a particular material can exist. We even had to find it experimentally by conducting a separate experiment for each state value (refer to the *U-u* Hugoniot data in Figure 17.1 for aluminum).

The isentrope, the path function that describes a continuity and not a jump, is different from the Hugoniot. Remember that a relief wave is a continuous

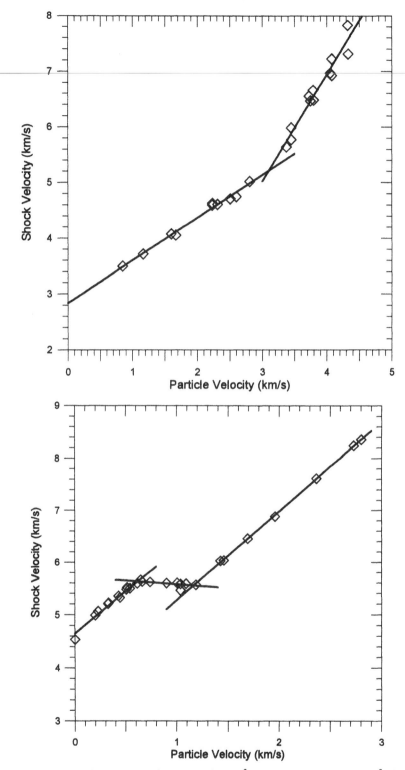

Figure 17.2 (a) Quartz, ceramic, $\rho_0 = 1.9$ g/cm^3, average $\rho_0 = 1.877$ g/cm^3. (b) Iron-40.0 wt % cobalt, average $\rho_0 = 8.091$ g/cm^3.

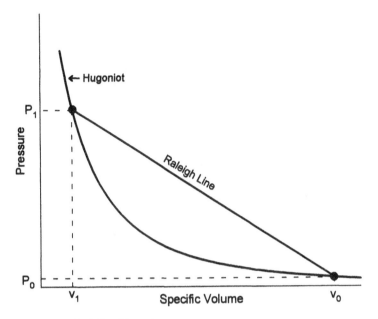

Figure 17.3   A typical *P-v* Hugoniot.

process and its path would be along the "unloading" isentrope. However, for engineering purposes, the values along the isentrope are so close to those along the Hugoniot that the Hugoniot can be used to approximate the isentrope closely.

Since the Hugoniot represents the locus of all possible states behind the shock front, then a line joining the initial and final states on the *P-v* Hugoniot represents the jump condition. This line is called the Raleigh line and is shown in Figure 17.3. If we eliminate the particle velocity term *u* by manipulating the mass- and momentum-jump equations, and let $u_0 = 0$, we get

$$P_1 - P_0 = \frac{U^2}{v_0} - \frac{U^2}{v_0^2} v_1$$

This is the equation of the Raleigh line, and we see that the slope of this line is $-U^2/v_0^2$, or $-\rho_0^2 U^2$. If we know the initial and final *P-v* states of a shock, then we can calculate the shock velocity by taking the slope of the Raleigh line: $U = -(\text{slope})^{1/2}/\rho_0$.

Conversely, if we knew the initial *P-v* state and the shock velocity, we could calculate the final *P-v* state. We have now fixed the fifth variable by specifying it as a boundary condition in our calculation.

Now we will examine the lower-pressure region of the Hugoniot on the *P-v* plane (Figure 17.4). The straight segment at the lower end of the Hugoniot (P_0, v_0 to P_1, v_1) is the familiar linear elastic stress-strain relationship and P_1 is the elastic limit. Since the slope of this region is constant, all pressure waves going from

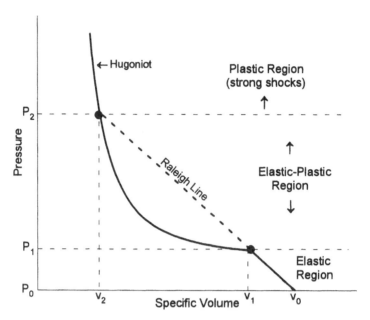

Figure 17.4 Lower part of the P-V Hugoniot.

state P_0,v_0 to any other state where P is less than P_1 will have the same velocity. This velocity is the characteristic velocity of a uniaxial elastic wave, and is called the longitudinal sound speed, C_L. Do not confuse this with the constant C_0 from the U-u Hugoniot. C_L for most materials is higher than the value of the constant C_0, usually by anywhere from 15 to 20%. In some materials it is lower than C_0.

Any pressure wave going from state P_0,v_0 to any pressure higher than P_2 will have a Raleigh-line slope greater than the slope of the elastic Hugoniot, and hence will have a shock velocity higher than C_L. In the region above P_2, called the region of strong shocks, no pressure waves can be faster at the shock front than the shock velocity. Shocks in this region have the sharp, discontinuous front.

Any pressure wave going from state P_0,v_0 to any pressure between P_1 and P_2 will have a Raleigh line slope less than that of the elastic Hugoniot, and hence will have a shock velocity below C_L. This region between the elastic limit and P_2 is called the elastic-plastic region, and now we see why. Any shock going from P_0,v_0 to any pressure $P_2>P>P>_1$ will have two distinct wave velocities. It will have an elastic wave traveling at C_L and trailing farther and farther behind, a plastic shock wave whose velocity is given by the Raleigh-line slope to that pressure. Therefore, in this region the front of the wave will "smear out" by the faster outrunning "elastic precursor." Of course, the rarefaction wave catching up with the back of the shock is smearing it out there, and we can now grasp the mechanism by which a shock decays to a sound or acoustic wave once the pressure has dropped into the elastic-plastic region.

The U-u Hugoniot shown previously is only valid in the region of strong shocks. Between the elastic limit and P_2, some transition form of the Hugoniot is needed, but relatively little data exist for this area. For our purposes, however, we are interested in the strong shock region and are only viewing the lower-pressure region qualitatively. Table 17.3 gives longitudinal sound speeds for a number of inert materials. Table 17.4 gives longitudinal sound speeds for a number of unreacted explosives.

Return your attention again to the region of strong shocks, but now to observe how energy changes in the shock process can be visualized on the P-v plane.

Table 17.3 Longitudinal Sound Speed for Some Inert Materials

Materials	C_L (km/s)
Aluminum, rolled	6.420
Beryllium	12.890
Brass, yellow 70 Cu, 30 Zn	4.700
Constantan	5.177
Copper, rolled	5.010
Duralunium 17S	6.320
Gold, hard-drawn	3.240
Iron, cast	4.994
Iron, electrolytie	5.950
Armco	5.960
Lead, rolled	1.960
Magnesium, drawn, annealed	5.770
Monel metal	5.350
Nickel	6.040
Nickel silver	4.760
Platinum	3.260
Silver	3.650
Steel K9	5.941
347 Stainless steel	5.790
Tin, rolled	3.320
Titanium	6.070
Tungsten, drawn	5.410
Tungsten, carbide	6.655
Zinc, rolled	4.210
Fused silica	5.968
Pyrex glass	5.640
Heavy silicate fluid	3.980
Light borate crown	5.100
Lucite	2.680
Nylon 6-6	2.620
Polyethylene	1.950
Polystyrene	2.350

Reference 5.

Table 17.4 Longitudinal Sound Speed for Some Unreacted Explosives

Material (preparation)	Density (g/cm³)	C_L (km/s)
AP (bulk, 500 m)	1.20	0.57
	1.55	1.79
	1.90	2.18
Baratol	2.61	2.90
(cast)	2.611	2.95
Comp B-3	1.70	3.00
(cast)	1.726	3.12
Cyclotol 75/25	1.752	3.12
DATB (pressed)	1.78	2.99
H-6	1.75	2.46
HNAB	1.577	0.853
LX-15-0	1.58	1.749
LX-17-0	1.899	2.815
Octol (cast)	1.80	3.14
PBX-9010-02	1.78	2.72
PBX-9011-06	1.790	2.89
PBX-9404-03	1.840	2.90
PBX-9407	1.78	3.04
	1.608	1.922
PBX-9501	1.82	2.97
PBX-9502	1.88	2.74
TATB	1.868	1.907
	1.87	2.00[a]
	1.87	2.55[b]
(isotropic purified)	1.876	1.98
Tetryl (pressed)	1.68	2.27
TNT (cast)	1.63	2.68
(creamed, cast)	1.624	2.48
(pressed)	1.61	2.48
(pressed)	1.632	2.58

Reference 4.
[a] Parallel to pressing direction.
[b] Perpendicular to pressing direction.

Figure 17.5 shows a P-v Hugoniot in the strong shock region for a given material. The Raleigh line is also drawn in for a shock going from some initial state P_0, v_0 to a state behind the shock of P_1, v_1.

By eliminating the U term from the mass and momentum equations (let $u_0 = 0$), we can derive

$$\frac{1}{2} u_1^2 = \frac{1}{2} (P_1 - P_0)(v_0 - v_1)$$

The particle-velocity term in the above equation is the kinetic energy per unit mass of the material at the state behind the shock (KE = $\frac{1}{2}mu^2$), also called the

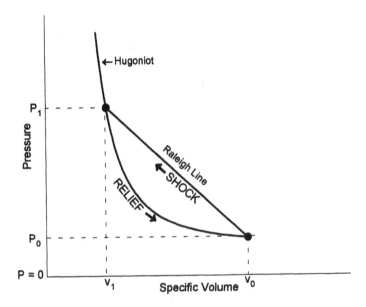

Figure 17.5   A shock shown on the *P-v* Hugoniot.

specific kinetic energy. The term $\frac{1}{2}(P_1 - P_0)(v_0 - v_1)$ is the area on the *P-v* plane of the triangle bounded by the Raleigh line, $P = P_0$ and $v = v_1$. Therefore, the increase in specific kinetic energy equals the area of the triangle in Figure 17.5.

We can also slightly modify the form of energy equation thus,

$$e_1 - e_0 = \frac{1}{2}(P_1 + P_0)(v_0 - v_1) \text{ energy equation (when } u_0 = 0)$$

$$= \frac{1}{2}(P_1 + 2P_0 - P_0)(v_0 - v_1)$$

$$= \frac{1}{2}(P_1 - P_0)(v_0 - v_1) + P_0(v_0 - v_1)$$

which, in Figure 17.5, is the area of triangle plus the area of a rectangle bounded by $P = P_0$, $P = 0$, $v = v_0$, and $v = v_1$. Therefore, the change of specific internal energy is equal to the total area under the Raleigh line.

Now remember that we said that a relief wave unloads along the path of the isentrope, which we earlier said could be closely approximated by the Hugoniot. If we took this material, which we have just shocked to P_1, v_1, and allow a relief wave to bring it back to P_0, v_0, then, by the same arguments, the change in specific internal energy for the relief wave is the total area under the Hugoniot segment from P_1 to P_0. In the process of shocking and then relieving the material, we increased the final specific internal energy by the amount equal to the difference between the area under the Raleigh line and the area under the Hugoniot segment. This difference is the area between these two lines.

What does that mean? Changes in internal energy are changes in thermal state; therefore, the material must have increased in temperature; it got hot. If you have ever been present at an explosive event and picked up a metal fragment thrown from the explosive charge, you must have noticed that the fragment was hot. If we shocked the material to a high enough pressure, the area between the Hugoniot and the Raleigh line, for some materials, gets sufficiently large that the final internal energy change is great enough to melt or vaporize it. Shock pressures, relative to one atmosphere, needed to cause incipient melting (to bring the material to the melting point), complete melting, and vaporization for several materials are listed in Table 17.5.

We see now that the value of the P-v plane is that it allows:

1. Calculation of the shock velocity if the initial and final P-v states are specified;
2. Calculation of the final P-v state if the initial state and shock velocity are specified;
3. Calculation of the final-state specific kinetic and internal energies if either the final P-v state or the shock velocity is specified;
4. Calculation of the relief-wave energy changes (as in 3); and
5. Visualization of the above processes and the mechanisms of two-speed behavior in the elastic-plastic region and the process of shock front decay at lower pressures.

Table 17.5 Shock Heating Effects

Material	Melting Temperature (°C)	Vaporization Temperature (°C)	Pressure to Cause Incipient Melting (Mbar)	Pressure to Cause Complete Melting (Mbar)	Pressure to Cause Vaporization (Mbar)
Aluminum	600	2057	0.6	0.9	—
Cadmium	321	767	0.4	0.46	0.8
Copper	1083	2336	1.4	>1.8	—
Gold	1063	2600	1.5	1.6	—
Iron	1535	3000	—	2.0	—
Lead	327	1620	0.3	0.35	1.0
Magnesium	651	1107	—	—	—
Nickel	1455	2900	>1.5	—	—
Titanium	1800	>3000	>1.0	—	—

Reference 6.

17.4 The *P-u* Plane

Starting again with the momentum and *U-u* Hugoniot equations, this time we will manipulate the equations to eliminate *U*, leaving *P-u*

$$P_0 = 0, \; u_0 = 0$$

$$P_1 = \rho_0 u_1 U$$

$$U = C_0 + su$$

$$P_1 = \rho_0 u_1 (C_0 + s u_1)$$

Figure 17.6 shows this *P-u* Hugoniot plotted on the *P-u* plane.

Now let us bring an additional factor into play. Until now we have looked at calculations on the *P-v* Hugoniot, always allowing u_0, the initial material particle velocity, to be zero. This, in effect, meant looking at the shock, relative to the material, in Lagrangian terms. Now, allow the material to be in motion before shock arrival, that is, $u_0 \neq 0$, by changing u_1 (Lagrangian) to $(u_1 - u_0)$, the Eulerian transform for the particle velocity.

$$P_1 = \rho_0 C_0 u_1 + \rho_0 s u^2 1 \text{ (Lagrangian) becomes} \tag{17.1}$$

$$P_1 = \rho_0 C_0 (u_1 - u_0) + \rho_0 s (u_1 - u_0)^2 \text{ (Eulerian)}$$

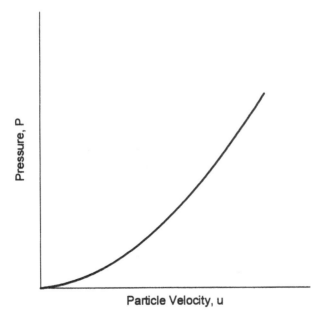

Figure 17.6   *P-u* Hugoniot with $u_s = 0$.

Now we can plot a series of P-u Hugoniots, each one with a different u_0, as in Figure 17.7.

We have here a series of parallel Hugoniots, or looking at it in a slightly different way, a Hugoniot that can "slide" back and forth along the u axis, depending upon the initial particle velocity.

Looking at a single shock on any of these parallel Hugoniots, jumping from state P_0,u_0 to P_1,u_1, is, as in the P-v plane, a straight line connecting these two states. The slope of this line (see Figure 17.7 for an example) is $\Delta P/\Delta u$.

$$(\Delta P/\Delta u) = \text{slope of jump} = (P_1 - P_0)/(u_1 - u_0) = \rho_0(U - u_0)$$

Now we see that the slope of the jump condition on the P-u plane is also a function of shock velocity. The interesting part to note is that U in this equation is the shock velocity in laboratory or Eulerian coordinates. The quantity $(U - u_0)$ is the shock velocity in Lagrangian coordinates, or relative to the material. So we see that for the jump condition on the P-u plane

$$\text{(slope of jump)}/\rho_0 + u_0 = U_{\text{LAB}} \text{ (laboratory coordinates)}$$
$$\text{(slope of jump)}/\rho_0 = U_{\text{MAT}} \text{ (material coordinates)}$$

The next implication we must note is that, since we are dealing with velocity on one axis, we must remember that velocity is a vector quantity; it has direction. What we described and by inference have done in Figures 17.6 and 17.7 and in Eq. (17.1) was plot the Hugoniot for a material where the shock travels from left to right (a right-going shock). When the shock travels from right to left (a left-going shock), then in the convention we established above, the particle velocity would be $-(u_1 - u_0)$. For that case, the Hugoniot for a right-going shock in Eq. (17.1) would not be correct. In order to find the Hugoniot for a

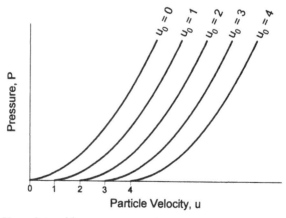

Figure 17.7 P-u Hugoniots with u_0 as a parameter.

left-going shock, let us go back again to the P-u Hugoniot and change ($u_1 - u_0$) to $-(u_1 - u_0)$, or $-(u_0 - u_1)$.

$$P_R = \rho_0 C_0 (u_1 - u_0) + \rho_0 s(u_1 - u_0)^2$$

right-going becomes

$$P_L = \rho_0 C_0 (u_0 - u_1) + \rho_0 s(u_0 - u_1)^2$$

left-going! This is the left-going shock Hugoniot on the P-u plane. It is the left-to-right mirror image of the right-going Hugoniot, and it too "slides" back and forth along the u axis depending upon the value of u_0. The slope of the line connecting two states, before and behind a shock, on the left-going Hugoniot is $-\rho_0(U - u_0)$, where the minus sign tells us that the shock is moving toward the left.

Why do we want these particular properties: the Lagrangian slope and the vector qualities along the u axis? The answer is that these allow us to solve interactions of shocks. In essence, when we deal with shock interactions, we will be using two different P-u Hugoniots along with the three jump relationships. Then we will have five equations in the five variables and be able to solve shock problems specifying only the initial conditions.

18

Interactions of Shock Waves

In this chapter we will examine three basic types of shock interactions. These are

1. Impact of two different materials;
2. Behavior of shock at a material interface where
   a. material A has a lower impedance than material B;
   b. material A has a higher impedance than material B; and
3. Interaction of two colliding shocks.

In dealing with interactions, we are suddenly faced with some potentially cumbersome bookkeeping problems. We have the shock and what it will interact with before, during, and after the interaction. The materials involved will have been moving all this time; therefore, their positions, as well as those of the shock (or shocks), will be changing in time and space. To deal with these problems, we will introduce two more graphical planes: the P-x plane, a kind of snapshot of the shock at one or more discrete times, and the x-t plane, a bookkeeping device showing the relative positions of shocks and material surfaces in time. We start with Case 1, the impact of one material upon another at high velocity.

18.1 Impact of Two Slabs

In this example, we have two slabs of solid materials, A and B. The slabs are very thick, so we only see one edge of each. Slab A is flying toward slab B at a velocity of $u_0 = u_{0A}$, as shown in Figure 18.1.

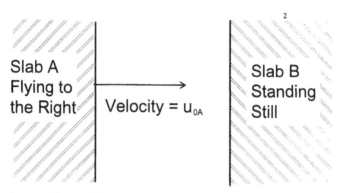

Figure 18.1 Impact of two slabs.

When impact occurs, a pressure or shock pulse is formed. Slab A continues to press upon slab B, sustaining the pressure. The shock moves into B toward the right, and also into A toward the left. As long as the slabs are in contact, the pressure, as well as the particle velocity on both sides of the interface, must remain the same and equal. This is shown in the x-t diagram in Figure 18.2.

In the x-t plane in Figure 18.2, we see that at zero time the front of slab A is at position x_1 and the rear or back of slab B at x_2. B is standing still, and, as time progresses, its position is unchanged and is therefore indicated by a vertical line. A is moving toward the right, and, as time progresses, the position of A continues toward the right and is represented by the sloped straight-line segment from (x_1, t_0) to (x_2, t_1). The slope of this line is $\Delta t/\Delta x$, the reciprocal of its velocity.

Note that on the x-t plane, the slopes of lines are the reciprocal of velocity; therefore, the higher the slope, the lower the velocity, and vice versa. Also, the x-t diagram shows the phenomena in laboratory coordinates; so shock velocities indicated by these slopes are laboratory, or Eulerian velocity.

At time t_1, A has reached and impacts B. A shock is created by the impact and travels into B toward the right and into A toward the left. Conditions at the interface must preserve the restraints of conservation. Therefore, the pressure as well as particle velocity must be the same in both materials. Here is the key! If pressure and particle velocity are the same in both materials, then we can equate the Hugoniots for each and solve for both the final pressure and particle velocity.

The shock in B is right-going, and B has an initial particle velocity of zero; so we can find its P-u Hugoniot by

$$P_R = \rho_0 C_0(u_1 - u_0) + \rho_0 s(u_1 - u_0)^2$$

$$\rho_0 = \rho_{0B}, \, C_0 = C_{0B}, \, s = s_B, \, u_0 = 0 \tag{18.1}$$

$$P_1 = \rho_{0B}C_{0B}u_1 + \rho_{0B}s_B u_1^2$$

The shock in A is left-going, and A has an initial particle velocity of u_{0A}; so we can find its P-u Hugoniot by

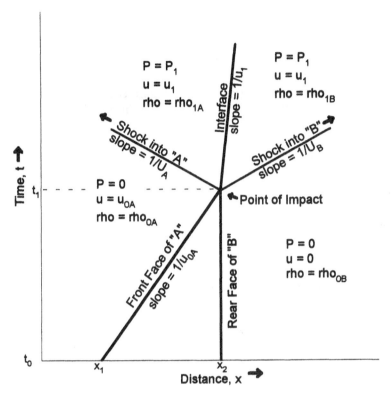

Figure 18.2   *x-t* diagram of impact of two slabs.

$$P_L = \rho_0 C_0 (u_0 - u_1) + \rho_0 s (u_0 - u_1)^2$$
$$\rho_0 = \rho_{0A},\ C_0 = C_{0A},\ s = s_A,\ u_0 = u_{0A} \qquad\qquad (18.2)$$
$$P_1 = \rho_{0A} C_{0A} (u_{0A} - u_1) + \rho_{0A} s_A (u_{0A} - u_1)^2$$

As we stated above, pressure and particle velocity in both materials must be the same at the interface; so we can set Eq. (18.1) equal to Eq. (18.2), and solve for u_1. We could then use that value of u_1, and find P_1 with either Eq. (18.1) or (18.2).

Also at this point we can solve for the shock velocity in both materials.

$$P_0 = 0,\ u_{0A} = u_{0A},\ u_0 = 0$$

$$\frac{P_1 - P_0}{u_1 - u_0} = \rho_0 (U - u_0)$$

$$U_{LAB} = \left(\frac{P_1 - P_0}{u_1 - u_0} \right) \frac{1}{\rho_0} + u_0$$

$$U_{\text{LAB,B}} = \frac{P_1}{u_1 \rho_{0B}}$$

$$U_{\text{LAB,A}} = \frac{P_1}{\rho_{0A}(u_1 - u_{0A})} + u_{0A}$$

What we have done is shown graphically in Figure 18.3.

We constructed the P-u Hugoniot for the right-going shock in material B from the initial conditions $P_0 = 0$, $u_0 = 0$. Then we constructed the P-u Hugoniot for the left-going shock in material A from initial conditions $P_0 = 0$, $u_0 = u_{0A}$. The intersection of the two curves provides the solution P_1, u_1. The slopes of the lines connecting the jump conditions for each material give us their respective shock velocities.

Example 18.1 Suppose a slab of 2024 aluminum alloy flying through the air at 1.8 km/s (5900 ft/s) strikes a slab of 304 stainless steel. What particle velocity would be generated in the two materials at the impact interface? What shock pressure would be generated? How fast would the shock be traveling into each material?

Solution From Table 17.1 we have the following:

2024 aluminum: $\rho_0 = 2.785$ g/cm³, $C_0 = 5.328$ km/s, $s = 1.338$

304 stainless steel: $\rho_0 = 7.896$ g/cm³, $C_0 = 4.569$ km/s, $s = 1.490$

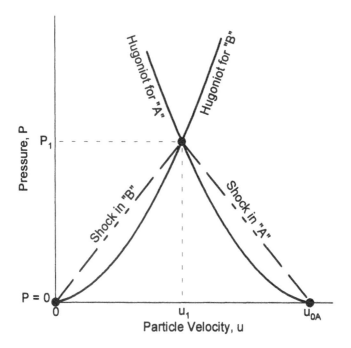

Figure 18.3 *P-u* Hugoniot solution of impact problem.

The impact will form a right-going shock in the steel whose P-u Hugoniot is

$$P = \rho_0 \, C_0 \, u + \rho_0 s u^2$$
$$= (7.896)(4.569) \, u + (7.896)(1.490)u^2$$
$$= 36.077 \, u + 11.765 \, u^2$$

The impact will also form a left-going shock in the aluminum, whose P-u Hugoniot is

$$P = \rho_0 \, C_0 \, (u_0\text{-}u) + \rho_0 s(u_0\text{-}u)^2$$
$$= (2.785)(5.328)(1.8\text{-}u) + (2.785)(1.328)(1.8\text{-}u)^2$$
$$= 38.782 - 28.253 \, u + 3.726 \, u^2$$

Since the pressure and particle velocity are the same at the interface for both materials, we can equate the two Hugoniot equations and obtain

$$u^2 + 8.002 \, u - 4.824 = 0$$

and solving this quadratic equation gives us

$$u = 0.563 \text{ km/s}$$

Using this value in either of the above P-u Hugoniots to find P yields

$$P = 24 \text{ GPa}$$

The shock velocity in the steel target would be found by using the above particle velocity in the U-u Hugoniot for the steel, yielding

$$U_{steel} = 5.41 \text{ km/s}$$

The shock velocity running back into the aluminum (relative to the oncoming material) is found from this same particle velocity and its U-u Hugoniot

$$U_{aluminum} = 6.08 \text{ km/s (left-going)}$$

It is interesting that the graphical solution of this problem could be made much simpler by using a clever device demonstrated by Dr. Orville Jones, of Sandia Laboratories, in a short course on shock waves that he taught in the mid-1960s (Ref. 4). Suppose you had a plot of the right-going, $u_0 = 0$, P-u Hugoniot of B on a piece of graph paper. You also have a plot of the right-going, $u_0 = 0$, P-u Hugoniot of A printed on a sheet of clear plastic. If you turned over the sheet of plastic, you would have the mirror image of A's Hugoniot (the left-going one). If you now placed this over the graph of B's Hugoniot, lined up the two u axes, you could slide A back and forth until you matched A's u_0 with the proper value, u_{0A}, on B's u axis. You have now duplicated Figure 18.3. The intersection is the solution. Dr. Jones plotted the Hugoniots of several common materials on the same graph, replicated these on an acetate sheet, and had in effect a shock wave interaction slide rule. Jones's original plot is shown in Figure 18.4. This graphical technique is an easy alternative to solving the quadratic equations.

18.2 Shock at a Material Interface Case a, $Z_A < Z_B$

In this example we have two slabs of different materials, A and B. The slabs are in contact and are at rest. A right-going shock is passing through A and approaching the interface. When the shock reaches the interface, some change will occur, but at this point we must stop and consider an additional factor.

Recall the momentum equation where we have set P_0 and u_0 equal to zero,

$$P = \rho_0 u U$$

This contains an interesting factor, the product $\rho_0 U$. This product is called the *shock impedance* and is designated by the letter Z.

$$Z = \rho_0 U$$

Of course we know that ρ_0 is constant; U is not. The shock velocity increases with pressure. However, the product $\rho_0 U$ or Z, although increasing with pressure, increases rather slowly, and we can consider it to be "somewhat constant" within reasonable ranges of interest. It is constant enough to let us differentiate between a low-impedance material and a high-impedance material. When a shock passes from a low to high impedance across a material interface, the shock pressure will be increased; the converse also holds.

In Case a, we will observe this effect, as the impedance of A is very much lower than that of B. First we will look at the P-x diagram, a few diagrams in time of the shock pressure and the interface, Figure 18.5. The x-t diagram is

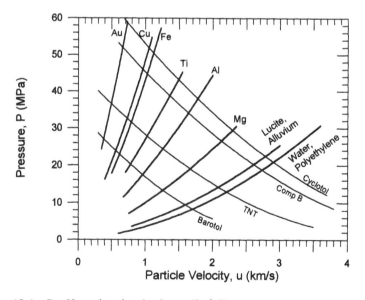

Figure 18.4 P-u Hugoniot plots by Jones (Ref. 7).

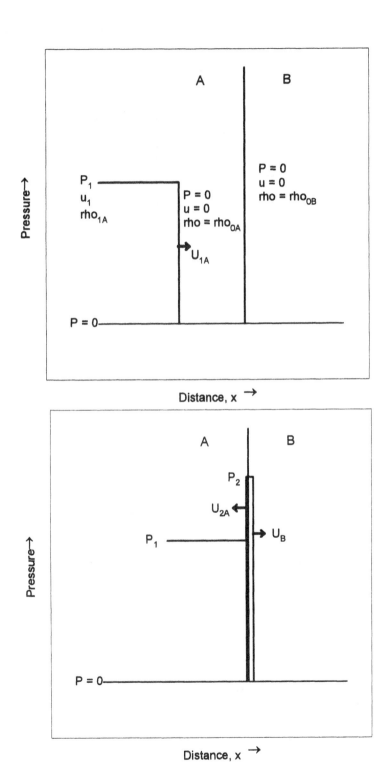

Figure 18.5 (a) *P-x* diagram, before interaction. (b) *P-x* diagram, at interaction. (c) *P-x* diagram, after interaction.

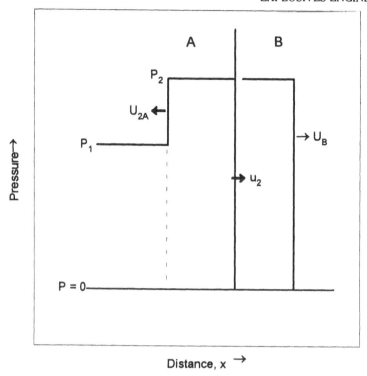

Figure 18.5 *(continued)*

shown in Figure 18.6. For the purpose of this discussion, and also in the interest of easier bookkeeping, the various states have been listed in each area of the x-t diagram. This type of labeling will make it easier to keep track of things. The right-going shock through material A is raising or jumping material in area A_0 to the state in area A_1. When the shock reaches the interface, a new shock is formed at pressure P_2 and is right-going into B, raising the state of area B_0 to the state of area B_2. This shock also is traveling back into A, left-going, raising the state of material in A_1 to that of A_2. Pressures and particle velocities across the interface are equal.

To solve this problem, find the values of P_2, u_2, U_{AZ}, and U_{BZ}, first we will construct the right-going Hugoniot for material A around $u_0 = 0$. This is for the permissible shock states in material A before the interface is reached, as shown in Figure 18.7. Since we know the shock in A is at P_1, u_1 (this was specified as a boundary condition), we know that the left-going wave Hugoniot must pass through this point. So we can merely reflect the Hugoniot around this point to obtain the left-going Hugoniot. We know the reflected Hugoniots must be sym-

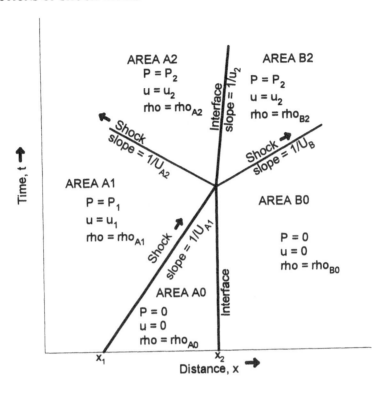

Figure 18.6   *x-t* diagram for shock across a material interface.

metrical; therefore, it must be centered on the *u* axis at a velocity twice that at the reflection point. This would make the reflected Hugoniot for *A*

$$P = \rho_o C_{0A}(2u_1 - u) + \rho_o S_A(2u_1 - u)^2$$

Either way, we now have the Hugoniot for the left-going wave from the point p_1, u_1. In Figure 18.8, we construct over this the right-going ($u_0 = 0$) Hugoniot for *B*. As in the previous example, the solution is the interception of the two curves, in this case at P_2, u_2. The velocity of the shock into *B* is found from the slope of the line connecting 0,0 to P_2, u_2, and the shock velocity of the wave back into *A* from the slope of the line connecting P_1, u_1 to P_2, u_2.

Indeed we find that the shock pressure of a wave going from a low to a high impedance does increase in pressure.

Example 18.2   Let us assume that we have a slab of 921-T aluminum (material *A*) in contact with a slab of copper (material *B*). A long-pulse shock wave traveling through the aluminum encounters the interface. The initial shock pressure in the aluminum was 25 GPa. What pressure does this change to when the shock interacts at the interface?

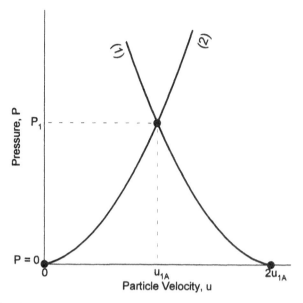

Figure 18.7   Constructing left-going Hugoniot from P_1, u_1. (1) Left-going Hugoniot for A, centered on the u axis at $u_0 = 2u_{1A}$ and passing through P_1, u_{1A}; (2) Right-going Hugoniot for A, centered on the u axis at $u_0 = 0$.

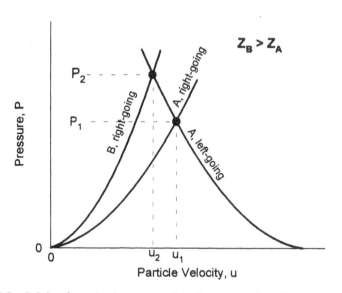

Figure 18.8   Solving for a shock across an interface where $Z_B > Z_A$.

Solution First lets find the U-u Hugoniot values for both materials from Table 17.1.

921-T Aluminum: $\rho_o = 2.833$ g/cm³, $C_0 = 5.041$ km/s, $s = 1.420$

Copper: $\rho_o = 8.930$ g/cm³, $C_0 = 3.940$ km/s, $s = 1.489$

Now let us find the particle velocity in the oncoming shock in the aluminum before it encounters the interface (this will establish the point we have to reflect around to find the left-going wave Hugoniot after the interaction).

$$P = \rho_0 C_0\, u + \rho_0 s u^2$$

$$(25) = (2.833)(5.041)u + (2.833)(1.420)u^2$$

Solving this quadratic, we get

$$u = 1.285 \text{ km/s}$$

Now we can write the values for the left-going wave Hugoniot in the aluminum remembering that u_0 will equal two times the particle velocity in the oncoming wave.

$$P = \rho_o C_0(u_0 - u) + \rho_o s(u_0 - u)^2$$

$$= (2.833)(5.041)(2 \times 1.285 - u) + (2.833)(1.420)(2 \times 1.285 - u)^2$$

The right-going wave Hugoniot for the shock in the copper after the interaction is

$$P = \rho_o C_0\, u + \rho_o s u^2$$

$$= (8.93)(3.94)\, u + (8.93)(1.489)u^2$$

Since the shock pressure and particle velocity in both materials must be the *same at the interface* at the time of the interaction, we can equate these two Hugoniot equations and solve the resulting quadratic for u, and in so doing we find

$$u = 0.814 \text{ km/s}$$

Using this value in either Hugoniot, we find that the shock pressure at this interaction is

$$P = 37.5 \text{ GPa}$$

18.3 Shock at a Material Interface Case b, $Z_A > Z_B$

The P-x snapshots for this example are shown in Figure 18.9. The x-t diagram is virtually identical to that in Case a. The only expected differences will be in the P-x snapshots and the P-u Hugoniots. The left-going and right-going Hugoniots are found exactly the same way as in Case a. The difference is that since $Z_A > Z_B$ the B Hugoniot will be lower. This is shown in Figure 18.10.

Again, the solution is found at the intersection of the left- and right-going Hugoniots. This time, however, the wave going back into material A is reducing its pressure from P_1 down to P_2. This is a rarefaction wave, not a shock going left into A. A shock is moving to the right into B. We will see more about rarefactions in the next chapter.

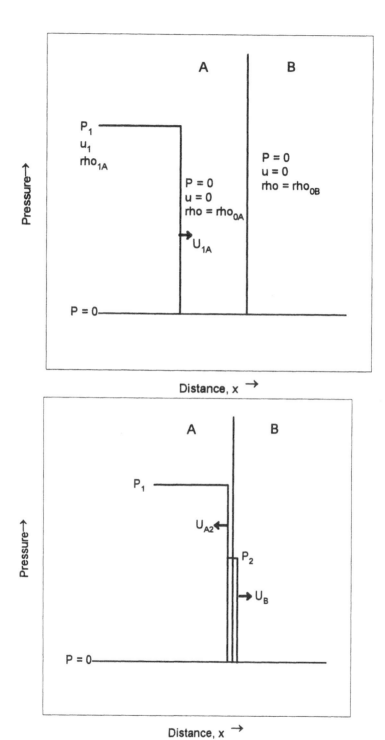

Figure 18.9 (a) *P-X* diagrams of shock at a material interface where $Z_A > Z_B$, before interaction. (b) *P-X* diagrams of shock at a material interface where $Z_A > Z_B$, at interaction. (c) *P-X* diagrams of shock at a material interface where $Z_A > Z_B$, after interaction.

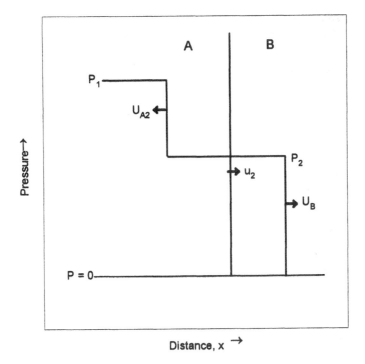

Figure 18.9 *(continued)*

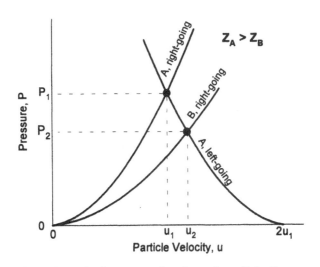

Figure 18.10   Solving for a shock across an interface where $Z_A > Z_B$.

Example 18.3 Let us look at the opposite of the previous example. We will have a shock of 25 GPa traveling right through copper and interacting with 921-T aluminum at the interface.

Solution Again, first we find the particle velocity of the oncoming wave before the interaction, so that we know the point to reflect it around. The right-going wave P-u Hugoniot for copper is

$$P = \rho_o C_0\, u + \rho_o s u^2$$

$$25 = (8.93)(3.94)\, u + (8.93)(1.489)u^2$$

Solving for u, we get

$$u = 0.582 \text{ km/s}$$

Now, again remembering that u_0 will be twice this particle velocity, we can write the left-going wave P-u Hugoniot for the copper

$$P = \rho_o C_0(u_0 - u) + \rho_0 s(u_0 - u)^2$$
$$= (8.93)(3.94)(2 \times 0.582 - u) + (8.93)(1.489)(2 \times 0.582 - u)^2$$

The P-u Hugoniot for the resulting right-going shock in the aluminum is

$$P = \rho_o C_0\, u + \rho_o s u^2$$

$$= (2.833)(5.041)u + (2.833)(1.42)u^2$$

Equating these, solving for u and the P at the interface yields

$$u = 0.809 \text{ and } P = 14.2 \text{ GPa}$$

18.4 Collision of Two Shock Waves

In this example, two shock waves of unequal amplitude approach each other head-on. When they meet they produce a much higher pressure shock that is reflected back in each direction. You will see, as we solve this problem, that the final shock pressure produced is greater than the sum of the pressures of the initial two shocks. The P-x snapshots are shown in Figure 18.11, and the x-t diagram in Figure 18.12.

This interaction produces two shocks, each going back in the opposite direction. To solve this we will need left- and right-going Hugoniots. The position of these Hugoniots is not at first obvious. Let us start first with the left-going Hugoniot, raising area 1 states (on the x-t plane) to the states at area 3. We know that this Hugoniot is coming from state P_1, u_1, and that u_1, is positive. Plot that point first (Figure 18.13).

We know that this state was arrived at by a right-going shock into $u_0 = 0$ material; so the left-going resulting Hugoniot must be rotated around this point and will intercept the $P_0 = 0$ or u axis at $2u_1$. This is also plotted in Figure 18.13.

Now let us consider the Hugoniot of the right-going shock that raises area 2

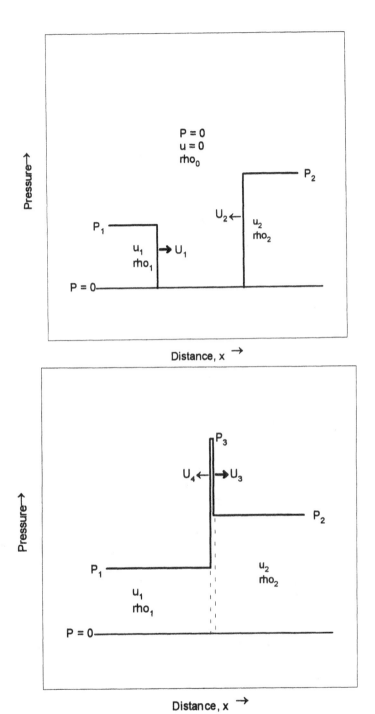

Figure 18.11 (a) *P-X* diagrams of two colliding shocks, before interaction. (b) *P-X* diagrams of two colliding shocks, at interaction. (c) *P-X* diagrams of two colliding shocks, after interaction.

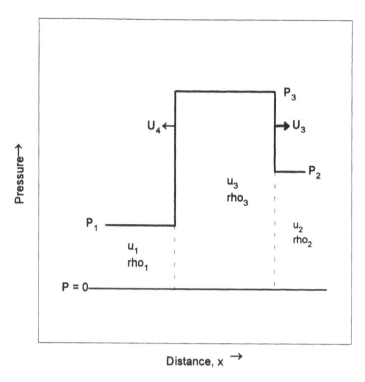

Figure 18.11   *(continued)*

(Figure 18.12) from P_2,u_2,ρ_2 to P_3,u_3,ρ_3. The material in area 2 is moving toward the left; it was shocked by a left-going wave; therefore, u_2 is negative. We will have to expand our P-u plane to include the negative particle velocities. Let us plot the state P_2,u_2 on the plane in Figure 18.14.

We know that the right-going Hugoniot we are looking for must pass through P_2,u_2, and that it is a reflection of the left-going shock Hugoniot that raised P_0,u_0 to state P_2,u_2. That Hugoniot had to have been centered at the origin, and so our reflected right-going Hugoniot passing through P_2,u_2, must have a u-axis intercept at $-2u_2$. This is plotted in Figure 18.15.

Our solution, again, is the intercept of the right- and left-going Hugoniots. Note that if the Hugoniots were straight lines, then it would be easy to show that $P_3 = P_2 + P_1$. But the Hugoniots are not straight lines; they curve upward, and therefore, $P_3 > (P_1 + P_2)$. Also note that the final particle velocity, u_3, is negative. This should not be surprising since the left-going shock coming into this collision was the stronger (higher pressure) of the two.

Example 18.4   For this case, let us assume that in a slab of brass there is a shock traveling toward the right and its pressure is 12 GPa. In the same slab there is also a shock traveling toward the left on a head-on collision course with the other shock. This

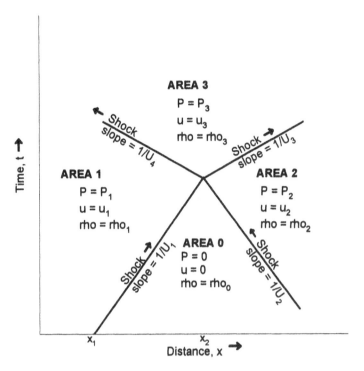

Figure 18.12 *X-T* diagram for collision of two head-on shock waves.

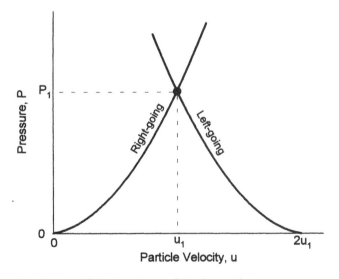

Figure 18.13 Right and left Hugoniots reflected around P_1, u_1.

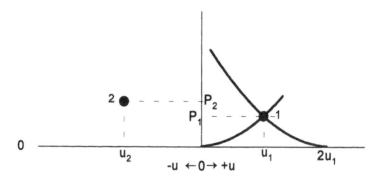

Figure 18.14   *P-u* plane including negative velocity region.

shock has a pressure of 18 GPa. When the two shocks collide, what will the resultant particle velocity and pressure be?

Solution From Table 17.1 the *U-u* Hugoniot values for brass are

$$\rho_0 = 8.450 \text{ g/cm}^3, \ C_0 = 3.726 \text{ km/s, and } s = 1.434$$

1. After the interaction of the two shocks, there will be a left-going wave whose *P-u* Hugoniot is the reflection around the *P-u* state of the original right-going shock. To find this we first must find the particle velocity of that original shock whose pressure was given as 12 GPa.

$$P_1 = \rho_0 C_0 \, u_1 + \rho_0 s u_1^2$$

$$12 = (8.45)(3.726)u_1 + (8.45)(1.434)u_1^2$$

$$u_1 = 0.337 \text{ km/s}$$

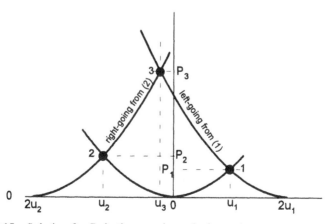

Figure 18.15   Solution for P_3 in the negative velocity region.

The resultant left-going wave Hugoniot is

$$P = \rho_0 C_0(2u_1 - u) + \rho_0 s(2u_1 - u)^2$$

$$= (8.45)(3.726)(2 \times 0.337 - u) + (8.45)(1.434)(2 \times 0.337 - u)^2$$

2. After the interaction there will be a right-going wave whose P-u Hugoniot is the reflection of the original left-going shock whose pressure was 18 GPa. Since this original wave was left-going into still material, its Hugoniot was

$$P = \rho_0 C_0(u_0 - u_2) + \rho_0 s(u_0 - u_2)^2$$

where $u_0 = 0$, and so

$$18 = (8.45)(3.726)(-u_2) + (8.45)(1.434)(-u_2)^2$$

$$u_2 = -0.482 \text{ km/s}$$

The right-going reflection around this point is

$$P = \rho_0 C_0(u - 2u_2) + \rho_0 s(u - 2u_2)^2$$

$$= (8.45)(3.726)(u + 2 \times 0.482) + (8.45)(1.434)(u + 2 \times 0.482)^2$$

The solution for the particle velocity after the interaction is obtained from equating the two resultant Hugoniots, which yields

$$u_3 = -0.145 \text{ km/s}$$

Then P at the interaction is found by using this particle velocity in either of the two final P-u Hugoniots,

$$P = 33.9 \text{ GPa}$$

Note that P is indeed larger than the sum of the two original shock waves, as we had previously predicted qualitatively.

18.5 Summary of Shock Waves and Interactions

Having previously developed the three jump equations, we now have added a fourth relationship, the Hugoniot. This leaves only one variable out of the original five (U, u, ρ, P, and e) to be specified by a boundary condition.

We examined the Hugoniot in three different variable pair planes and found that each of these planes enables us to visualize certain aspects of shock behavior. These are the U-u, P-v, P-u planes.

In the U-u plane we discovered an empirical commonality among most solids, namely, the linear relationship between shock and particle velocities. We also were able to speculate that nonlinearities on this plane represented phase shifts in the structure of the materials.

In the P-v plane, we found we could visualize the mechanisms that cause shocks to decay in the elastic-plastic region and could graph the longitudinal sound speed. We could also utilize this plane, if we specified one boundary condition, to calculate the shock pressure or the shock velocity across a given jump condition. We were able to calculate, through use of areas on this plane, the thermodynamic changes involved in both shock and rarefaction.

On the P-u plane, we found that by solving two Hugoniots simultaneously, we could calculate the values of all of the state variables without having to fix one as a boundary condition. We solved interaction problems representing one material striking another, a shock crossing a material interface, and the head-on collision of two shock waves.

In the next chapter we will examine the properties and behavior of rarefaction, or relief, waves. We will solve problems, on the P-u plane, involving rarefactions and interactions of rarefactions. We will see how these interactions lead to material failure as in the cases of spall and scabbing (multiple spall).

19

Rarefaction Waves

In the previous sections, we dealt in detail with the properties at the shock front, the jump process that takes material in front of the shock to the state behind the shock. We showed that this is indeed a discontinuous process, and that pressure disturbances cannot outrun the shock (in the strong shock region). We stated, but did not demonstrate, that the rarefaction wave (also called relief, or unloading wave; they are all synonymous) is continuous, that it follows a path function, not a jump condition. Let us look into this statement now.

19.1 Development of a Rarefaction Wave

When we shocked a material, we increased its internal energy, as expressed mathematically in the energy equation, $(e_1 - e_0) = 0.5(P_1 + P_0)(v_0 - v_1)$. If we allow a shocked material to unload by moving a rarefaction wave through it, relieving the stress and returning it to the ambient pressure state, we assume this happens so fast that no heat (energy) is lost or transferred to its surroundings. This (recall Section II on thermochemistry) is called an adiabatic process ($dQ = 0$, no loss of heat energy). From thermodynamics, we also know that

$$dE = T \, dS - P \, dv \qquad (19.1)$$

that the energy change is equal to the product of the change of entropy times the absolute temperature, minus the work done, $P \, dv$. We just stated that the process was adiabatic, $dq = 0$, and also from thermodynamics, that the heat change, $dQ = T \, dS$.

When dQ is zero, then $T\,dS$ must be zero, and since we know we did not carry out this process at absolute zero, dS must be zero. This means that during the process, the entropy must have remained constant—aha!—an isentropic process. So the path of the changes in our state variables P,v must be along an isentrope. This means that Eq. 19.1 becomes

$$dE = -P\,dv \qquad \text{or} \qquad E = f(P,v)$$

Look familiar? In order to solve for all these parameters, we need an EOS in order to eliminate E and leave us an expression, $P = f(v)$. Since we do not have the EOS, we again resort to the Hugoniot. So the rarefaction unloads isentropically, and we assume that the isentrope is the same as the values along the Hugoniot. Let us take a look at this process on the P-v plane. To start with, Figure 19.1, a P-x snapshot, shows a square-wave pressure or shock pulse.

We now will treat the rarefaction as if it were a shock; that is, we will apply the jump equations such that we will let the high-pressure material "jump" down to a lower pressure state. We also are going to allow this to happen in two steps. The first step, or rarefaction wavelet, relieves the material from state P_1,v_1 (the shock pressure) to P_2,v_2 (half way down to ambient).

The second wavelet drops the pressure from P_2 to P_3 (ambient or P_0). These two steps are shown on the P-v plane in Figure 19.2. We allow wavelet 1 to jump from P_1 to P_2, likewise wavelet 2 from P_2 to P_3. Look at the Raleigh lines for each of these jumps. Remembering that the slope of the Raleigh line is $-(U^2/v_0^2)$, we see that wavelet 1 has a higher slope than at the shock front; it is traveling faster than the shock! The slope of wavelet 2 Raleigh line is lower than

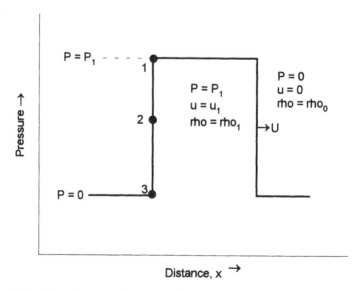

Figure 19.1   A P-x diagram of a square shock.

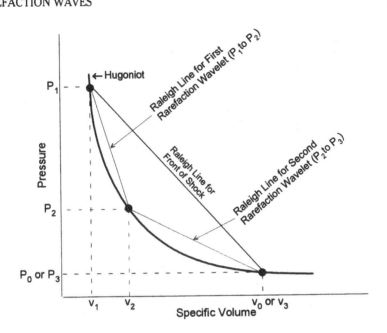

Figure 19.2 *P-V* diagram showing two rarefaction wavelets jumping down the Hugoniots.

that of the shock front; it is traveling slower! Let us take a look at the next few *P-x* snapshots (Figure 19.3) a short time later than that in Figure 19.1.

Now let us repeat the same process, but this time we arbitrarily will take the shock pressure down to ambient in not two steps but ten. Figure 19.4 shows the *x-P* snapshots in time sequence and Figure 19.5 shows these same wavelets on the *P-v* plane. Ten steps appear to follow the Hugoniot closely. If we let the steps get smaller and smaller, approaching zero, then indeed we can unload right

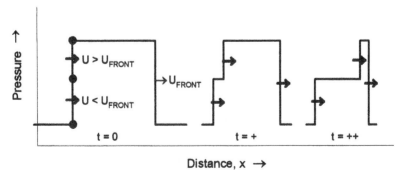

Figure 19.3 *P-x* shapshots (later in time) of the progress of the shock and the two rarefaction wavelets.

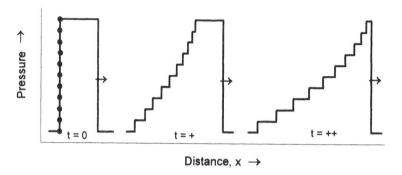

Figure 19.4 *x-t* diagrams, 10 wavelets relieving shock.

along the Hugoniot. This, in essence, is where computer codes start to come in the picture. You recall how cumbersome it was, handling just two shocks and two materials for a shock-front interaction problem. Imagine the bookkeeping to follow ten wavelets on each wave, with all variables (U,u,P,v) different on each one.

How should we treat rarefactions then on graphical or algebraic simple interaction problems? Referring back to Figure 19.5, we see that the highest-pressure

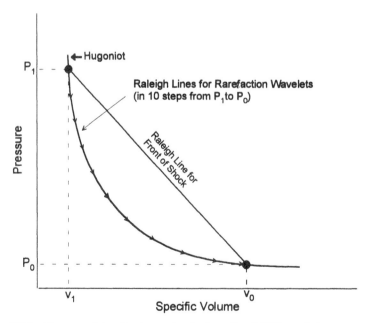

Figure 19.5 Ten wavelets relieving a shock along the *P-V* Hugoniot.

rarefaction wavelet is the fastest and the lowest rarefaction pressure wavelet the slowest. On an x-t plane diagram, we could then show only the leading and trailing edges of the rarefaction waves (see Fig. 19.6). In order to do this, we would have to know the velocities. A problem crops up here as to what is the leading rarefaction velocity? If we treat the wave as a series of wavelets, then it is obvious that the step size will affect the calculated velocity. What we would have to do is to take the slope on the Hugoniot itself (the limit as step size → zero) and calculate the wave velocity from that. The slope of the curve is the value of dP/dv at the particular pressure being relieved. In order to find dP/dv, we need the equation of the P-v Hugoniot. We derived that in a previous chapter (Section 17.3),

$$P = C_0^2(v_0 - v)[v_0 - s(v_0 - v)]^{-2}$$

Differentiating this P-v Hugoniot equation and rearranging the terms yields

$$dP/dv = C_0^2[v_0 + s(v_0 - v)]^{-2}[v_0 - s(v_0 - v)]^{-3} = -U^2/v_0^2$$

This is cumbersome. If, instead, we look at the jump condition for a tiny wavelet jumping down from the shock pressure to the next lower differential pressure on the P-u Hugoniot, the algebra becomes simpler.

Since the slope of the line joining two states on the P-u Hugoniot is $\rho_0 U$ (Lagrangian, relative to the particles, or material), equating this to dP/du at the pressure of interest will give us the rarefaction-wave velocity relative to the material into which it is moving. The equation on the Hugoniot of the P-u plane is

$$P = \rho_0 C_0(u - u_0) + \rho_0 s(u - u_0)^2, \quad \text{right-going wave}$$

$$dP/du = \rho_0 C_0 + 2\rho_0 s(u - u_0)$$

Equating this to $\rho_0 U$ (Lagrangian) yields

$$U_{\text{rarefaction Lagrangian right-going}} = C_0 + 2s(u - u_0) \qquad (19.2)$$

For the left-going rarefaction we similarly find

$$U_{\text{rarefaction Lagrangian left-going}} = -C_0 - 2s(u_0 - u) \qquad (19.3)$$

Remember that the u_0 in Eqs. (19.2) and (19.3) is not the u of the material into which this rarefaction is moving but merely a constant equal to the value of the u-axis intercept for the Hugoniot. The u calculated for the front of the rarefaction is Lagrangian; it is the velocity of the rarefaction relative to the material into which the rarefaction is moving. Therefore, when plotting this velocity on the x-t plane, it must be corrected for the additional particle velocity of the material ahead, into which it is moving. The x-t plane, again, shows slopes of Eulerian velocity reciprocals.

The longitudinal sound speed is appropriate for the tail of the rarefaction. If you do not have a value for this speed, C_L can be estimated by using a value for a similar material in a similar state of aggregation. The values used for the leading and trailing edges of the rarefactions are approximations, since the real unloading isentrope is not known.

On the *x-t* diagram we show the leading- and trailing-edge velocities, and sometimes throw in a few values in between to show this is a rarefaction. This *x-t* representation is called the *rarefaction fan*. Figure 19.6 shows this as a example based on the square pulse we previously saw in Figure 19.4

19.2 Interactions Involving Rarefactions

When rarefaction waves are formed or interact, we find some rather surprising results. In order to see these effects, let us examine four basic interactions that involve rarefaction waves.

1. Interaction of a shock with a free surface; we will see how this generates a rarefaction wave.
2. The impact of a finite-thickness flyer on a thick target; we will see how the flyer thickness and the relative shock impedances of the flyer and target affect duration and shape of the target shock that is produced.
3. Collision of two rarefactions; we will see how this generates tension in a material and can lead to material failure in the form of spall.
4. Interaction of a nonsquare shock pulse with a free surface; we will see how this can lead to multiple spall or scabbing.

As in the shock-front interactions we studied earlier, we will make use of $P - x$ snapshots, *x-t* plane bookkeeping, and will solve the interaction on the $P = u$ plane.

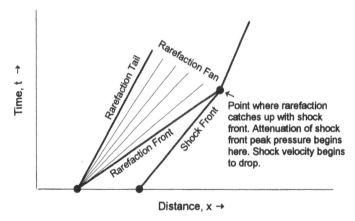

Figure 19.6 The rarefaction fan on an *x-t* plane. (If the entire rarefaction fan does not play a significant role in the problem at hand, then just plot the leading edge of the rarefaction.)

19.2.1 Interaction of a Shock with a Free Surface

If we compressed a piece of sponge between our hands and then suddenly removed our hands, the unrestrained sponge would immediately expand. This process "relieved" the stress of compression in the sponge. The surfaces of the sponge, being suddenly unrestrained, jump outward, or actually accelerate away from the compressed area. This is exactly what a rarefaction wave is. The wave is the progression of particles being accelerated away from a compressed (or pressurized) or shocked zone. We can see then that, if the particles are being accelerated away from a shock and the rarefaction is moving into the shock, a rarefaction travels in the *opposite* direction to the acceleration of the particles. This is exactly opposite to a shock wave, where the particles are accelerated in the direction of the shock. Figure 19.7 shows the P-x snapshots at different times during the interaction and the results of a shock arriving at a free surface.

When the shock reaches the surface, the material at the surface is highly compressed and wants to jump away, relieving the compression, or shock pressure, back to zero. In so doing the particles behind the resulting rarefaction have been accelerated to twice the particle velocity in the shocked region. The x-t diagram shows this a bit more clearly (Figure 19.8). The P-u plane, on which the appropriate Hugoniots are plotted, shows this process (Figure 19.9). The right-going shock is the jump condition from $P = 0$ to P_1. Since the rarefaction wave is left-going and is jumping material from P_1 down to 0, it must be along the left-going Hugoniot that the right one intersects at P_1, u_1. Since this process is occurring in the same material, the Hugoniots are exact mirror images of each other, and by reason of symmetry, the left-going Hugoniot must intersect the u axis at $2u_1$.

We see the leading edge rarefaction velocity is dP/du at P_1, or the tangent to the Hugoniot at this point. The trailing edge is shown as the tangent at $P = 0$, $u = 2u_1$. Note that the material behind the rarefaction has a particle velocity of $+2u_1$, while the velocity of the rarefaction wave is negative.

Example 19.1 A constant-pressure shock wave is traveling to the right in a thick slab of polysytrene. The shock pressure is 6.5 GPa.

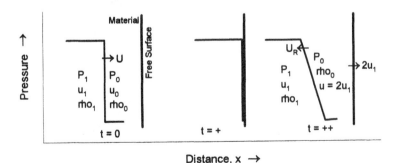

Figure 19.7 P-x diagrams of a shock at a free surface.

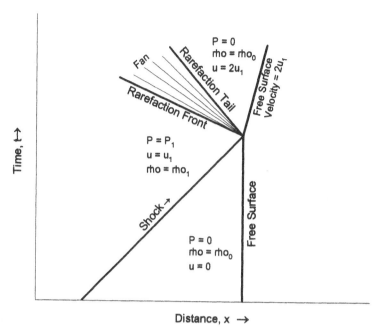

Figure 19.8 *x-t* diagram of shock at free surface.

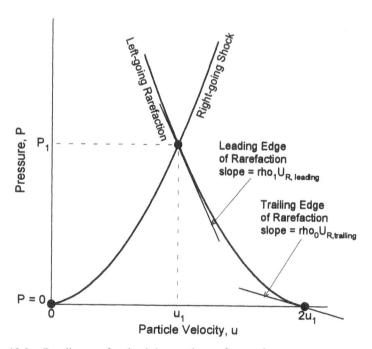

Figure 19.9 *P-u* diagram for shock interaction at free surface.

1. What is the particle velocity in the shock?
2. What is the shock velocity?
3. The shock encounters a free surface. What is the particle velocity after this interaction?
4. What is the Lagrangian velocity of the leading edge of the rarefaction (relative to the material into which it is traveling)?
5. What is its Eulerian velocity?

Solution The U-u Hugoniot values for polysytrene are found from Table 17.1.

$$\rho_0 = 1.044 \text{ g/cm}^3, \ C_0 = 2.746 \text{ km/s, and } s = 1.319.$$

1. The shock wave prior to the interaction with the free surface was right-going and into material with zero particle velocity; therefore, its P-u Hugoniot is

$$P_1 = \rho_0 C_0 u_1 + \rho_0 \ su_1^2$$
$$6.5 = (1.044)(2.746))u_1 + (1.044)(1.319)u_1^2$$
$$u_1 = 1.37 \text{ km/s}$$

2. The shock velocity is found from this particle velocity, and the U-u Hugoniot is

$$U_1 = C_0 + su$$
$$= 2.746 + (1.319)(1.37)$$
$$= 4.55 \text{ km/s}$$

3. The interaction causes a left-going wave to go back into the material. The P-u Hugoniot of this wave is the reflection of the incoming Hugoniot around P_1, u_1, and we saw that the particle velocity after the reflection must be 2 times u_1 or $u_2 = (2)(1.37) = 2.74$ km/s.
4. The Lagrangian velocity of the leading edge of the rarefaction is the slope of the P-u Hugoniot at P_1, u_1 divided by ρ_0. Therefore, we have to find the derivative of the left-going wave Hugoniot and evaluate it at P_1, u_1.

$$P = \rho_0 C_0 (2u_1 - u) + \rho_0 s(2u_1 - u)^2$$
$$dP/du = 2\rho_0 su - (\rho_0 C_0 + 4\rho_0 su_1)$$

evaluated at $u = u_1$

$$dP/du = -\rho_0 C_0 - 2\rho_0 su_1$$
$$= -(1.044)(2.746) - (2)(1.044)(1.319)(1.37)$$
$$= -6.64 \text{ km/s}$$

5. The Eulerian rarefaction is the Lagrangian plus the particle velocity of the material into which the rarefaction is moving and is therefore $R_{\text{Eulerian}} = -6.64 + 1.37 = -5.27$ km/s.

19.2.2 Impact of Finite-Thickness Flyer

Here, as in the problem involving a shock at an interface of two materials, we will get different behavior when the relative values of the shock impedances of target and flyer are reversed. So we must break this example into three cases:

1. $Z_{flyer} < Z_{target}$;
2. $Z_{flyer} = Z_{target}$; and
3. $Z_{flyer} > Z_{target}$.

19.2.2.1 Case 1, $Z_{flyer} < Z_{target}$

First, consider the *x-t* diagram shown in Figure 19.10. At point (1) the flyer has produced a shock, by impact, going to the right in the target and to the left back into the flyer. This interaction is shown on the *P-u* plane in Figure 19.11. We see that at impact, the interaction produces a right-going shock in the target and a left-going shock in the flyer. The state behind both shocks is P_1, u_1.

Following the left-going shock in the flyer, the next interaction to occur is the shock with the free surface at the rear of the flyer, [point (2) in the *x-t* diagram, Figure 19.10]. This interaction is shown on the *P-u* plane in Figure 19.12.

When the left-going shock interacts with the rear face or free surface of the flyer, it forms a right-going rarefaction wave. This rarefaction is along the right-

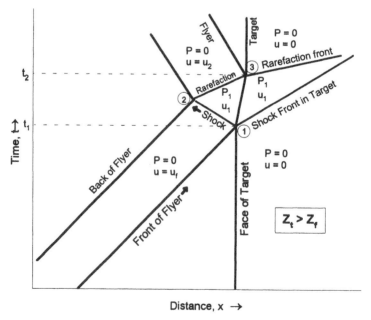

Figure 19.10   Finite-thickness flyer impacting a thick target, $Z_t > Z_f$.

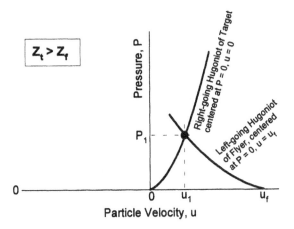

Figure 19.11   Interaction at point (1) (refer to Figure 19.10).

going Hugoniot of the flyer, through P_1, u_1, since it relieves material from P_1, u_1
down to P_0. The intersection of this Hugoniot with the u axis, $P_0 = 0$, is at $u =$
u_2 (negative). Because of symmetry we can easily see that $u_2 = u_1 - u_f$. Follow
this rarefaction toward the right to where it meets and interacts with the material
interface, the flyer pushed up against the target. This interaction is shown on the
P-u plane in Figure 19.13, and is point (3) on the x-t diagram (Figure 19.10).

Here we see something new. At this interaction, the right-going rarefaction
with the flyer-target interface (at P_1, u_1), we must produce a left-going wave back
into the flyer. It is going from $P_2 = 0$, $u = u_2$ to some other value. We must
produce a right-going wave in the target. It is going from P_1, u_1 to some other
value where the intersection is. The only point where these two Hugoniots can
cross, or intersect, is at a negative pressure (P_3). A negative pressure is a tension.

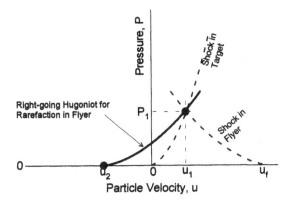

Figure 19.12   Interaction at (2) (refer to Figure 19.10).

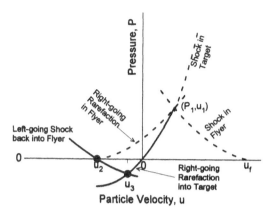

Figure 19.13   Interaction at (3) (refer to Figure 19.10).

Tension is caused where the materials in a given plane or cross section have particle velocities of opposite relative direction on either side. The material at that plane is being pulled apart. If the material has some finite tensile strength, it can still hold together, but instead of being under a compressive stress, it is under a tensile stress.

We do not have good data for the values along the Hugoniot in the tensile region. Therefore, we assume that the Hugoniot is a straight line with a slope of the $\rho_0 C_L$ or $-\rho_0 C_L$.

This is not a bad assumption, since the material under tension will fail somewhere in the region just beyond the end of the tensile elastic limit. Therefore, the elastic or longitudinal wave velocity is quite appropriate.

In our case, the interface at the interaction at point (3) (Figure 19.10) has no tensile strength because we have just two materials pressed together; so the interface separates. The flyer has bounced off the target at this point (at velocity u_3) and is no longer of interest. The face of the target, which was at P_1, u_1, now is a free surface, and a right-going rarefaction wave proceeds into the shocked region. This rarefaction drops the state values from P_1, u_1 to $P = 0$, where we see on that Hugoniot that $u = 0$.

We now see the entire process of forming a square-wave pulse in a target by collision of a flyer. The pulse starts at t_1 (Figure 19.10) and ends at t_2 (Figure 19.10) The constant-pressure portion of the shock pulse initially then has a duration of $t_2 - t_1$, and then gets narrower in time due to attenuation.

Example 19.2   Let us suppose that a polyethylene flyer, 5 mm thick, traveling at 2.5 km/s impacts a thick slab of PBX9404-03 explosive. What pressure shock wave will be driven into the explosive, and what is its initial time duration?

Solution First let us find the U-u Hugoniot values for both of these materials from Tables 17.1 and 17.2.

Polyethylene $\rho_0 = 0.915$ g/cm^3, $C_0 = 2.901$ km/s, $S = 1.481$

PBX-9404-03 $\rho_0 = 1.84$ g/cm^3, $C_0 = 2.450$ km/s, $s = 2.480$

To find P_1 at the impact interface, we will need to equate the left-going wave Hugoniot in the flyer to the right-going wave Hugoniot in the target.

Flyer: $P = \rho_0 C_0(u_0 - u_1) + \rho_0 s(u_0 - u_1)^2$

$\qquad = (0.915)(2.901)(2.5\text{-}u) + (0.915)(1.481)(2.5 - u)^2$

Target: $P = \rho_0 C_0 u + \rho_0 s u_2$

$\qquad = (1.84)(2.45)u + (1.84)(2.48)u_2$

and $u_1 = 0.898$ km/s.

P at the interaction and on into the explosive is then found from this particle velocity and either of the two interacting Hugoniots as

$P = 7.73$ GPa

The time that this pressure is applied at the face of the explosive is equal to the time that it takes for the shock in the flyer to get to the rear surface of the flyer and the resulting rarefaction to return through the flyer to the flyer-explosive interface. So let us first find U of the left-going shock in the flyer. This shock is the jump from $P = 0$, $u = 2.5$ to $P = 7.73$, $u = 0.898$, and remembering that the shock velocity is the slope of the jump on the P-u Hugoniot divided by the initial density, we have

$$U_{\text{in flyer}} = \left(\frac{P_1 P_0}{u_1 - u_0} \right) \rho_0$$

$$= \left(\frac{7.73}{0.898 - 2.5} \right) 0.915$$

$$= -5.27 \text{ km/s (mm/}\mu\text{s)}$$

The flyer is 5.0 mm thick; so the shock took $(5)/(5.27) = 0.95$ μs to reach the back of the flyer.

Now we need the leading edge rarefaction-wave velocity back through the flyer. Referring to Figure 19.12, we can see by inspection that the slope of the reflected flyer Hugoniot at P_1, u_1 is the negative of the slope of the left-going flyer Hugoniot, and we have that. So the slope of the Hugoniot at P_1, u_1 is the negative of

$P = \rho_0 C_0(u_0 - u_1) + \rho_0(su_0 - u_1)^2$

$dP/du = 2\rho_0 su - \rho_0 C_0 - 2\rho_0 su_0$

evaluated at $u_1 = 0.898$ and $u_0 = 2.5$

$dP/du = (2)(0.915)(1.481)(0.898) - (0.915)(2.901)$
$\qquad\qquad\qquad\qquad\qquad - (2)(0.915)(1.481)(2.5) = -6.996$

$(dP/du)/\rho_0 = -7.65$ km/s

So the rarefaction velocity is the negative of that value, or

$R = 7.65$ km/s

It takes $(5)/7.65 = 0.65$ μs to traverse the flyer. The total time from impact until a relief wave reaches the interface is 0.95 (shock) $+ 0.65$ (rarefaction).

$$\Delta t = 1.60 \; \mu s$$

19.2.2.2 Case 2, $Z_{flyer} = Z_{target}$

The x-t diagram for this case is shown in Figure 19.14. In this case, when we follow the P-u diagrams, the interactions are exactly the same as in the previous case, except that since both materials are the same, we are working exact mirror images and find that the velocities u_2 and u_3 of the previous case will now both equal zero. Also we will find that P_3, which was a tension in the previous case is now zero. So the flyer and the target remain in contact, but neither under pressure nor tension. They stay together merely because both have a final particle velocity of $u = 0$. The pulse width is, again, $t_2 - t_1$, but the rarefaction fan started at the back of the flyer, and therefore the earliest back of the pulse in the target already has some smearing out and is therefore not square on the target's back side.

Since these interactions on the P-u plane are so similar to the previous case, we will not show them again, but instead leave it to the reader to go through the

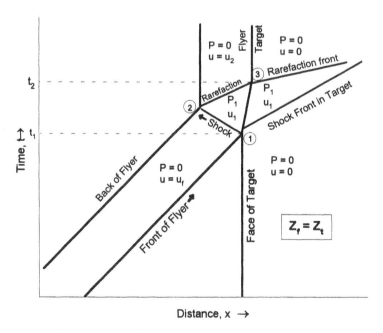

Figure 19.14   Finite–thickness flyer impacting target where $Z_f = Z_t$.

mechanics and prove to him/herself that the above description is correct. A far more interesting case is our third one, where Z_{flyer} is greater than Z_{target}.

19.2.2.3 Case 3, $z_{flyer} > z_{target}$

The x-t diagram for this case is shown in Figure 19.15, where we will see that the target and flyer will remain together by virtue of both having the same (zero) final particle velocity, but a square wave is not obtained. Instead we obtain a pulse with a square front and a stepped back.

The first interaction we will consider is the one labeled (1) on the x-t diagram, where the flyer first impacts the target. This is shown on the P-u plane in Figure 19.16. The flyer impacts the target, a right-going shock is formed in target and a left-going shock in the flyer. The state behind both these shocks is P_1, u_1.

Following the shock in the flyer to interaction (2) on the x-t diagram, the shock encounters a free surface at the back of the flyer. This will drop the

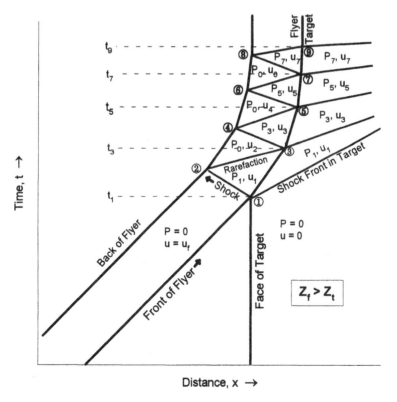

Figure 19.15 Flyer impact where $Z_f > Z_t$.

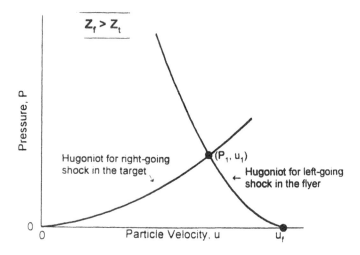

Figure 19.16   Interaction at (1) (refer to Figure 19.15).

pressure P_1 to zero, and a right-going rarefaction will be formed. This is shown on the P-u plane in Figure 19.17.

The rarefaction formed at the rear of the flyer drops P_1 to zero, then proceeds to the right where it encounters the interface between the flyer and target. This is shown as point (3) on the x-t diagram (Figure 19.15) and on the P-u plane in Figure 19.18.

The interaction (3) must produce a left-going shock wave in the flyer, coming from $P = 0$, u_2, and a right-going wave in the target coming from P_1, u_1. The

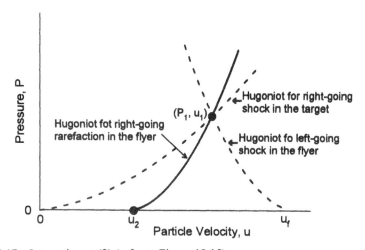

Figure 19.17   Interaction at (2) (refer to Figure 19.15).

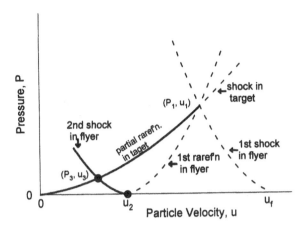

Figure 19.18 Interaction at (3) (refer to Figure 19.15).

Hugoniots for these waves cross at P_3, u_3. So a shock from 0 to P_3 is produced going back into the flyer, and a partial rarefaction dropping P_1 to P_3 goes right, into the target. The P-x snapshot a few instants after this interaction is shown in Figure 19.19. Now following that second shock back into the flyer, it encounters the rear free surface at interaction (4). The P-u plane solution for this interaction is shown in Figure 19.20.

When the second shock encounters the free surface at the back of the flyer (4), it produces a right-going rarefaction that dropped P_3 to zero and reduced

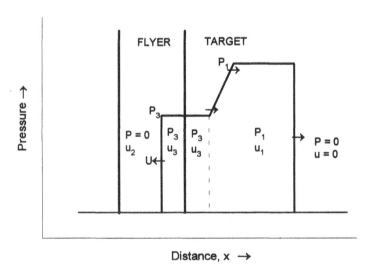

Figure 19.19 P-x diagram just after interaction at point (3) (refer to Figure 19.15).

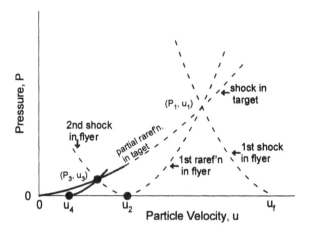

Figure 19.20 Interaction at second shock in flyer with free surface at (4) (refer to Figure 19.15).

the particle velocity to u_4. This rarefaction, the second in the flyer, proceeds toward the interface, point (5), and interacts with it forming a left-going shock wave back into the flyer, and a right-going rarefaction wave into the target. This interaction is shown on the P-u plane in Figure 19.21.

The interaction of the rarefaction coming from $P = 0$, u_4 and the shock at

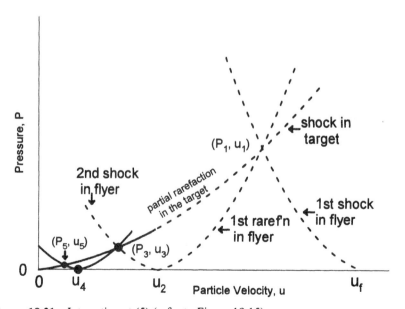

Figure 19.21 Interaction at (5) (refer to Figure 19.15).

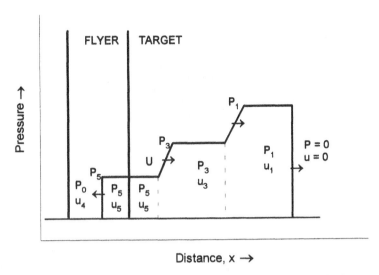

Figure 19.22   *P-x* diagram just after interaction at (5) (refer to Figure 19.15).

the interface, P_3, u_3 form a left-going shock back into the flyer, and another partial rarefaction right-going into the target. The state behind these is P_5, u_5. The *P-x* snapshot taken just after this interaction is shown in Figure 19.22.

By following this process in Figure 19.21, you can see that the final state of the flyer-target interface will converge at $P = 0$, $u = 0$, and that the flyer will remain in contact with the target. You will also surmise that both the number and size of the steps produced on the back of the shock wave in the target will depend upon both the flyer thickness and the relative values of the shock imped-ance, Z, of the flyer and target.

19.2.3 Head-On Collision of Two Rarefactions

In a previous example, section 19.2.2.1, we found a condition where a material can be brought into tension. In this example, we will delve a bit further into that phenomenon. We will consider two rarefaction waves approaching each other. This condition, where a material is under a shock pressure and is being relieved simultaneously from both sides, can be created in several ways.

An example might be a square-wave shock traveling through a material that then encounters a free surface. The interaction at the surface produces a rar-efaction wave that travels back into the shock. It is on a collision course with the rarefaction wave coming up from the back of the shock. If we observed this shock in Lagrangian or material coordinates relative to the material at the center of the shock, then the shock would be standing still and the two rarefactions would be encroaching upon it from both sides. In this case, we would have the situation shown in the *P-x* diagram in Figure 19.23.

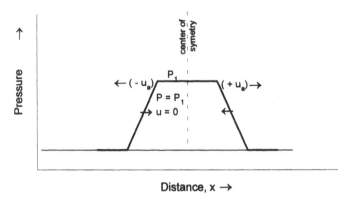

Figure 19.23 *P-x* diagram of two rarefactions on a collision course.

Both waves are relieving the shock pressure from P_1 down to zero, and remembering that the rarefaction accelerates the particles in the opposite direction to the wave velocity vector, we have materials moving away from the shocked region in both directions at velocity magnitude u_a. The interaction is shown in the *x-t* plane in Figure 19.24, and in the *P-u* plane in Figure 19.25.

In the *P-u* plane we see that a tension is formed at the center of symmetry. If the tension exceeds the dynamic tensile strength (spall strength) of the material, then the material will fail and part at this plane. This is called *spalling*. The two rarefactions are dropping the shock pressure before they interact from P_1 to zero. The Hugoniots of these two waves are the positive pressure Hugoniots.

When the rarefactions interact they must produce a right-going shock wave

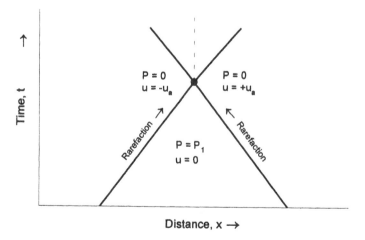

Figure 19.24 *x-t* diagram of collision of two rarefactions.

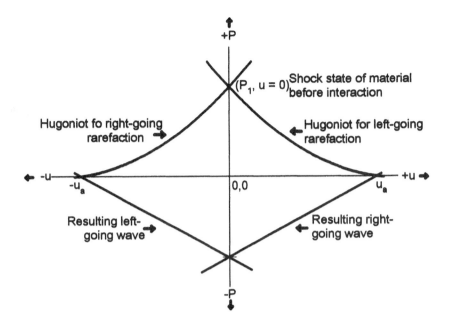

Figure 19.25   *P-u* interaction of two rarefactions.

coming from $P = 0$, u_a and a left-going shock wave from $P = 0$, $-u_a$. These Hugoniots, as was explained earlier, are straight lines with slope $\rho_0 C_L$ or $-\rho_0 C_L$. As always, their intersection is the state of the material after the interaction, which is $P = P_2$ (a tension) and $u = 0$.

If the tension, P_2, is not sufficient to spall the material, a tension wave travels both to left and right. If the tension is sufficient to spall the material, then the material parts, dropping the pressure back to $P = 0$, and forms two free surfaces that fly away in each direction at the particle velocity magnitude u_a.

We do not have much data on the spall strength of many materials. Even when we do have data, the local conditions and history of that one particular specimen come heavily into play. Table 19.1, however, lists approximate values that can be expected for the spall behavior of several materials of common interest. The values for spall strength shown in Table 19.1 are considerably higher than the equivalent static tensile strengths of these same materials, ranging from a factor of 2.5 to as high as 10.

Example 19.3   A thick slab of stainless steel has been impacted by a flyer plate. The impact formed a square-wave shock pulse in the steel that traveled to the opposite surface (which was free). The interaction of the shock pulse at the free surface produced a rarefaction wave that is traveling back into the square-wave shock pulse. The shock pressure is 7.5 GPa. When the rarefactions meet they will form a tension. Will the steel spall at this point?

Table 19.1 Spall Strength Estimates for Several Materials

Material	Condition	Spall Strength	
		(GPa Tensile)	(psi)
Al (pure)	Annealed	−1.3	189,000
Al (1100)	Annealed	−1.3	189,000
Al (2024)	−0	−1.3	189,000
Al (2024)	−T4	−1.6	232,000
Al (2024)	−T6	−2.0	290,000
Cu (pure)	Annealed	−2.4	348,000
Be/Cu	Full hard	−3.7	537,000
Brass (60/40)	Annealed	−2.1	305,000
Steel	Annealed	−1.6	323,000
Steel	(unknown)	−3.8, −2.1, −2.3	336,000
Steel (4340)	Annealed	−3.0	435,000
Ag (pure)	?	−2.1	305,000
Pb (pure)	?	−0.9	131,000

Reference 6.

Solution We know that for this problem we will need values for both the U-u Hugoniot as well as the longitudinal sound speed for stainless steel. We can get these from Tables 17.1 and 17.3. From Table 17.1 we have for 304 stainless steel

$$\rho_0 = 7.896 \text{ g/cm}^3, \; C_0 = 4.569 \text{ km/s, and } s = 1.49$$

From Table 17.3, we do not have the same stainless steel, but we are looking at sound speed and that certainly does not change appreciably from one type of stainless to another, so using the value given for 347 stainless

$$C_L = 5.79 \text{ km/s}$$

Now let us find (referring to Figure 19.25) the particle velocity intercepts on the u axis of the P-u Hugoniots. The oncoming right-going rarefaction was dropping the shock state from $P = 7.5$, $u = 0$ to $P = 0$, $u = u_a$; so its p-u Hugoniot was

$$P = \rho_0 C_0(u\text{-}u_0) + \rho_0 s(u\text{-}u_0)^2$$

where $u = 0$ and $u_0 = u_a$; therefore

$$7.5 = (7.896)(4.569)(0\text{-}u_a) + (7.896)(1.49)(0\text{-}u_a)^2$$

$$u_a = -0.195 \text{ km/s}$$

By arguments of symmetry we know that the oncoming left-going rarefaction is then dropping the shock state from $P = 7.5$, $u = 0$ to $P = 0$, $u = +0.195$ km/s.

The interaction of these two rarefactions will produce the right- and left-going shock waves reflected around the two shock states at $P = 0$, $u = -0.195$ and $P = 0$, $u = +0.195$. Again referring to Figure 19.25, we see that these shock waves will be tensile and will have slopes $-\rho_0 C_L$ and $+\rho_0 C_L$, respectively.

Again by argument of symmetry, we know that these latter two Hugoniots will meet at some point $P = P_2$ and $u = 0$.

So we can obtain the value of P_2 from either one.

$$\Delta P/\Delta u = \rho_0 C_L$$
$$\Delta P = \rho_0 C_L \, \Delta u$$
$$= (7.896)(5.79)(-0.195)$$
$$= -8.9 \text{ GPa (tensile)}$$

From Table 19.1 we see that the strongest of the steels has a spall strength of only 3 GPa; therefore the slab of stainless steel in this example should spall.

19.2.4 Interaction of Nonsquare Shock Wave at a Free Surface

We realize now that the square-wave shock pulse does not remain square. As soon as the rear square face of the shock is formed, it immediately begins to tip forward due to the effects we have seen from the rarefaction wave. Shock waves can also be formed that start with a fully sloped back, as when the shock is induced in a material from detonation of an adjacent explosive. Figure 19.26 shows a sawtoothed wave in the P-x diagram. This is an idealized form of a partially attenuated shock pulse identical to the shock pulse received from an adjacent detonation.

To handle this pulse graphically on the P-u plane, we will break the rarefaction at the back of the pulse into a series of small rarefaction wavelets, as shown in Figure 19.27. On the x-t plane, the series of wavelet interactions is shown in

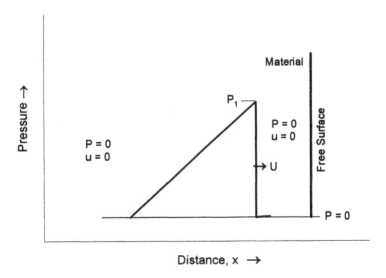

Figure 19.26 P-x diagram of a sawtooth shock pulse.

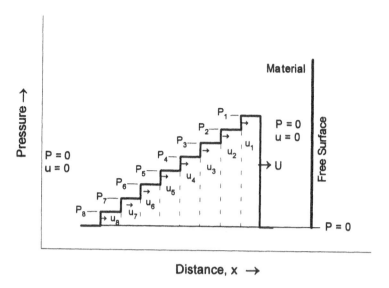

Figure 19.27   Coarse wavelet model of sawtooth wave.

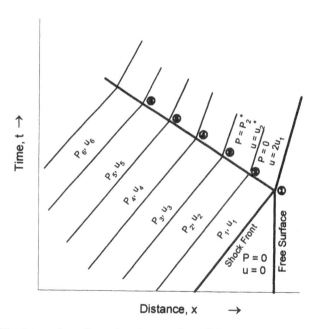

Figure 19.28   Interaction of wavelets (sawtooth model) at a free surface.

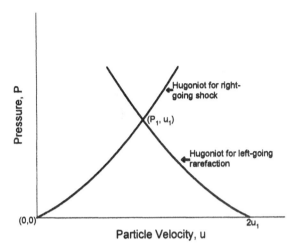

Figure 19.29 Interaction at (1) (reference Figure 19.28).

Figure 19.28. The first interaction we consider is the shock front at P_1, u_1 interacting with the free surface (Figure 19.29). This interaction should produce a left-going rarefaction dropping P_1 to zero. The rarefaction coming from (1) will interact with the first wavelet (P_2, u_2) forming a right-going wave (from $P = 0$, $2u_1$) and a left-going wave (from P_2, u_2). We see in Figure 19.30 that the state achieved is a tension, P_2^*, and particle velocity u_2^*. The P-x diagrams just after interactions (1) and (2) are very interesting; they are shown in Figure 19.31. Now, following the rarefaction moving from (2) toward a collision with wavelet

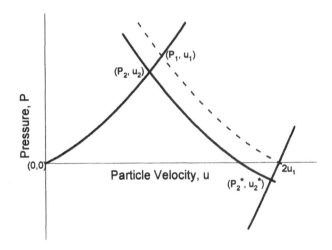

Figure 19.30 Interaction at (2) (reference Figure 19.28).

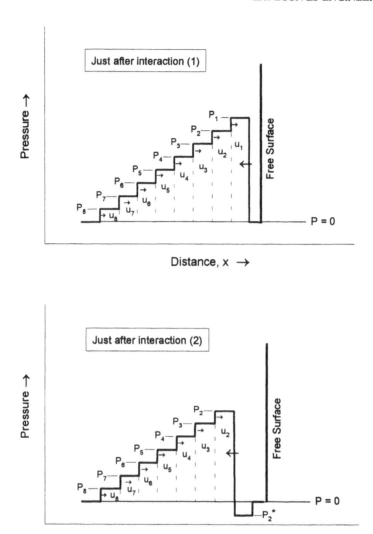

Figure 19.31 (a) and (b) *P-x* diagrams after (1) and (2) interactions.

at P_3, u_3, they interact at (3). A right-going wave must be formed coming from P_2^*, u_2^*, and a left-going wave is formed coming from P_3, u_3. These interact at P_3^*, u_3^*, and we see we have produced an even greater tension wave at this point, as shown in Figure 19.32. This same process repeats back to P_4, u_4, P_5, u_5, and so on. The series of *P-x* shapshots in time that would be produced look like those in Figure 19.33. As the rarefaction wave continues to work toward the left, interacting with the sloped rarefaction at the back of the shock, the tension continues to increase. If spall occurs at some intermediate point, the material parts, exposing a new free surface that still has a sawtoothed shock wave

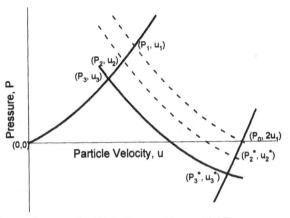

Figure 19.32 Interaction at point (3) (reference Figure 19.27).

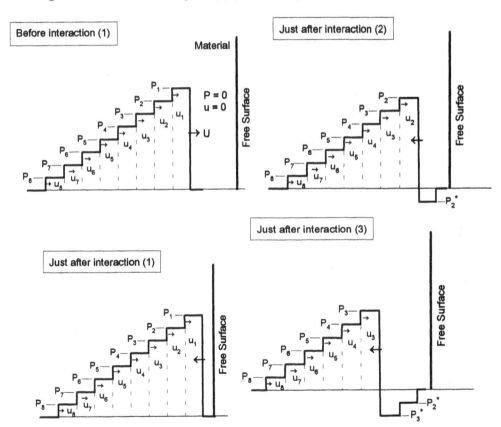

Figure 19.33 (a) *P-x* diagram of sawtoothed wave just before interaction with free surface; (b) *P-x* diagram of sawtoothed wave just after interaction with free surface; (c) *P-x* diagram of sawtoothed wave just after interaction (2); (d) *P-x* diagram of sawtoothed wave just after interaction (3); (e) *P-x* diagram of sawtoothed wave just after interaction (4); (f) *P-x* diagram of sawtoothed wave (with the step size taken to the limit) after interaction with a free surface.

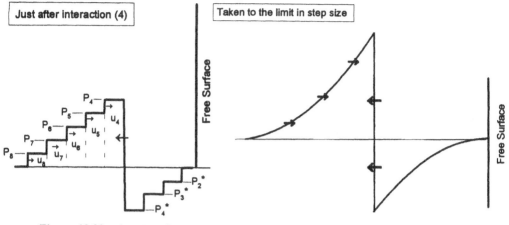

Figure 19.33   *(continued)*

approaching it like that in the last diagram in Figure 19.33. Then the whole process starts over. If the initial shock pressure were sufficiently high, then a second spall could occur when the process worked back from the free surface of the first spall, and maybe even a third and a fourth. Such multiple spalling is called *scabbing*.

19.3 Summary of Rarefactions

We examined the form and behavior of rarefaction waves and determined they indeed unload along a continuum that is approximated by the Hugoniot. We found methods to approximate the leading and trailing edge velocities of a rarefaction wave, and how they are represented on the *x-t* plane. We studied the basic interactions that involve rarefactions, seeing how they are formed and how they cause tensile waves that can lead to material failure.

In the next section we will examine detonation waves, a special form of a shock where chemical energy is being added to the front, and the solid explosive is changed to a gas.

References

1. Hayes, D. B., *Introduction to Stress Wave Phenomena*, SNL/NM Classnotes, 1973.

2. Marsh, S. P., *LASL Hugoniot Data*, 1980.

3. *Selected Hugoniots*, LASL Report LA-4167-MS.

4. Dobratz, B.M., *LLNL Handbook of Explosives*, UCRL-529997, 1981.

5. *American Institute of Physics Handbook*, 2nd Ed. 1963.

6. Gover, J. E., *Shock Wave Physics*, SNLA Classnotes, 1980.

7. Jones, O. E., *Shock Waves*, SNLA Classnotes, 1966.

IV

DETONATION

A detonation is a shock wave with a rapid exothermic chemical reaction occurring just behind the shock front. If we observe the detonation process, we find some general phenomena that seem to apply to all explosives. The first thing we notice is that in a given explosive sample, the wave speed forward is constant; the shock velocity does not speed up or slow down after the material has been initiated. Another observation is that if we detonate cylinders along the axial direction, the detonation-wave speed increases with increasing diameter until at some maximum diameter it seems no longer to increase no matter how large we have made the cylinder. We also would notice that this diameter, above which wave speed no longer increases, is different for each explosive. Based on these observations, we can divide detonation phenomena into two broad categories: ideal detonation, where the cross section of explosive is large enough to have no diameter effect; and nonideal detonation, where the dimensions of the charge affect the detonations' characteristics. For many of the most common military and DOE explosives, especially those closer to their maximum density, the diameter that divides these two categories is quite small, on the order of from a millimeter to a few tens of millimeters. For many commercial explosives, this diameter may be in the range of several centimeters, and for many blasting agents, a meter or more.

20

Detonations, General Observations

In this chapter we will only consider the ideal detonation case. We shall start by examining a simple model of detonation. We will then go on to see methods for estimating steady-state detonation parameters and from these to estimating the Hugoniots of detonation product gases. We will then look at interactions of detonation waves with other materials with which the explosive is in contact.

20.1 Simple Theory of Steady Ideal Detonation

For the purpose of this text, we will confine our analyses to the ideal detonation case, and look qualitatively at the nonideal area as is appropriate. Even though the ideal case is simpler to analyze, it is not in itself a simple phenomenon. In order fully to describe the simplest of detonations mathematically, we would have to quantify the chemical-reaction thermodynamics and kinetics; we would have to treat not only the shock hydrodynamics, but also fluid dynamics governing the expanding gas flow behind the detonation. We can, however, model the ideal detonation in such gross terms that the mathematics become tractable and we can solve first-order engineering problems with the same level of algebraic effort that we found sufficient with nonreactive shocks. We call this model the simple theory or ZND model, after Zeldovich, Von Neumann, and Deering, who all developed it independently in the early 1940s.

This simple theory makes a few assumptions that agree with the gross observations we noted above. These assumptions are:

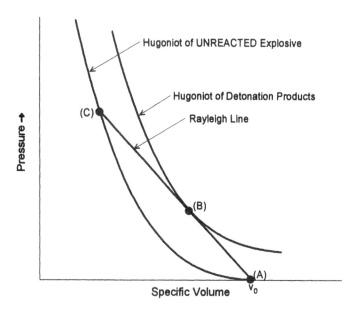

Figure 20.1   *P-v* plane representation of detonation.

1. The flow is one dimensional, which is the same as the uniaxial assumption we made when dealing with nonreactive shock waves.
2. The front of the detonation is a jump discontinuity and therefore can be handled in the same manner as the one we used with nonreactive shock waves.
3. The reaction-product gases leaving the detonation front are in chemical and thermodynamic equilibrium and the chemical reaction is completed.
4. The chemical reaction-zone length is zero.
5. The detonation rate or velocity is constant; this is a steady-state process; the products leaving the detonation remain at the same state independent of time.
6. The gaseous reaction products, after leaving the detonation front, may be time dependent and are affected by the surrounding system or boundary conditions.

With these constraints, the detonation is seen as a shock wave moving through an explosive. The shock front compresses and heats the explosive, which initiates chemical reaction. The exothermic reaction is completed instantly. The energy liberated by the reaction feeds the shock front and drives it forward. At the same time the gaseous products behind this shock wave are expanding, a rarefaction moves forward into the shock. The shock front, chemical reaction, and the leading edge of the rarefaction are all in equilibrium; so they are all moving at the same speed, which we call the *detonation velocity*, *D*. Therefore, the front of the shock does not change shape (pressure remains constant) with time and the detonation velocity does not change with time.

Let us look at the detonation jump condition on the P-v plane, Figure 20.1. As you can see in this figure, we are dealing with two materials in a detonation jump condition, the unreacted explosive and the completely reacted gaseous detonation products. Not only are we jumping from one physical state to another, but also to a new chemical state. In this figure, we see the initial state at point A, the unreacted explosive; we see also the state at point C the jump condition to the fully shocked but as yet unreacted explosive; and on another Hugoniot, the state B of the reaction products.

You will notice that the state of the reaction products is at the point where the Rayleigh line is tangent to the products' Hugoniot. This point is called the Chapman-Jouguet (CJ) point. Chapman and Jouguet hypothesized this as the steady-state detonation condition, hence the name (they did this in the late 1800s, working on gas-phase detonation problems). If the jump condition were such that the Rayleigh line lay below the Hugoniot for the reaction products, then the jump would not involve these products, since we know the gaseous products are formed in a detonation, and we specified earlier that the reaction is traveling at D, the detonation velocity; then the products' Hugoniot must be intersected somewhere along the Rayleigh line. If the Rayleigh line intersected the products' Hugoniot at a slope greater than that at the tangent, then two states would be possible for the products, one at each of the two points where the Hugoniot was intersected (Figure 20.2).

At state F in Figure 20.2, the rarefaction wave velocity, $(dP/dv)^{1/2}$, is greater

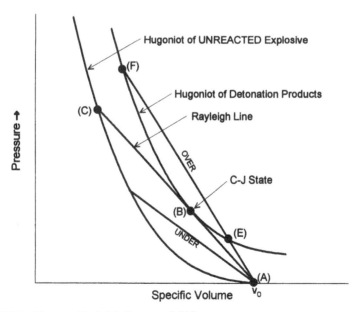

Figure 20.2   Various Rayleigh line possibilities.

than the slope of the jump Rayleigh line and the reaction zone and rarefaction would be overtaking the shock front, thus violating our statement above that these are all at the same velocity. So this state is not possible.

At point E (Figure 20.2), the slope of the Hugoniot, and hence the rarefaction wave velocity, is lower than that of the Rayleigh line; therefore, the rarefaction would be slower than the shock front, making the reaction zone continuously spread out in time. We know that this cannot be possible according to our constraints. The only place on the Hugoniot of the products where the slope of the Hugoniot equals the slope of the Rayleigh line, the reaction zone, rarefaction front, and shock front are all at the same velocity, is at the tangent point, the CJ state.

So the CJ point is the state of the products behind the detonation front. What about point C on Figure 20.1, the Rayleigh line intersection with the unreacted explosive?

Picture this as the shock state that brings on reaction, but the reaction zone is so short, and the reaction so fast that the energy involved in this pressure spike is negligible compared to the energy in the fully reacted products. This point, by the way, is referred to as the Von Neumann spike, and is seen more clearly if we view the detonation wave in the P-x plane, Figure 20.3.

For the purpose of the simple model, the Von Neumann spike is ignored and the reaction zone thickness is assumed to be zero. The gas expansion or rar-

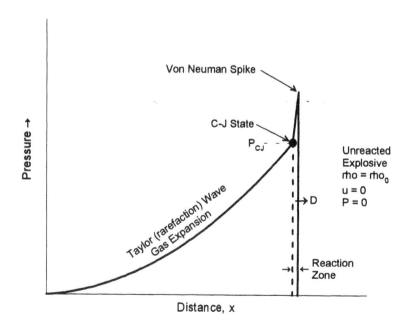

Figure 20.3 P-x diagram of a detonation wave.

efaction wave behind the CJ point is not a fixed characteristic of the explosive. This wave is named after Taylor, who developed EOSs to describe this wave. If the explosive has heavy rear and side confinement, the gases cannot expand as freely as unconfined gases; thus the Taylor wave is higher and longer (Figure 20.4). When the explosive is very thick (along the detonation axis), the Taylor wave is higher. When the explosive is very thin and there is little rear or side confinement, the Taylor wave is lower. The actual shape of the Taylor wave is governed then by a combination of the isentrope for expansion of the detonation gases, the charge size, and the degree of confinement.

20.2 Estimating Detonation Parameters

Before we can do any design calculations or analyses of explosives, we now know that we will require some special data. Specifically we will need to be able to either find or estimate the parameter values at the CJ state (P_{CJ}, D, u_{CJ}, ρ_{CJ}) and we will need to know the Hugoniot equations of the detonation product gases on the P-u plane.

An abundance of CJ-state data is available in the cited literature (Refs. 1–6) but seldom at the densities of interest to us. Also, much of the "data" are actually calculated or estimated, and we do not know how accurate their methods are. Therefore, we should arm ourselves with a tool kit of estimating techniques for which we understand and know the accuracy and limitations.

First we will start with some real data. Table 20.1 gives experimentally

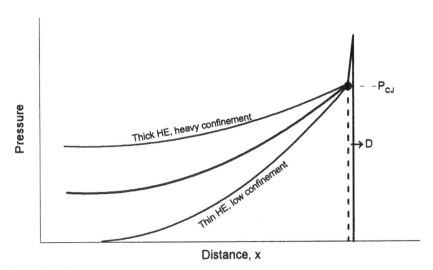

Figure 20.4   The Taylor wave.

Table 20.1 Experimental Data of Detonation Parameters at the CJ State

Explosive	ρ_0 (g/cm^3)	Detonation Velocity (km/s)	P_{CJ} (GPa)	ρ_{CJ} (g/cm^3)
CHNO solids				
ANFO (5.8% FO)	0.82	4.55	5.5	1.213
ANFO (5.8% FO)	0.84	4.74	6.14	1.245
AN/ADNT (2/1 molar)	1.64	7.892	26.1	2.203
AN/ADNT/EDD (3/1/1 molar)	1.607	7.664	24.2	2.161
AN/ADNT/NQ (1.38/1/1.83 molar)	1.654	8.16	25.5	2.152
AN/ADNT/RDX (1.38/1/1.5 molar)	1.717	8.455	31.7	2.315
AN/ADNT/RDX (5/1/1 molar)	1.699	7.712	24	2.228
AN/ADNT/TATB (2/1/1.3 molar)	1.765	7.845	28.3	2.387
AN/TNT (50/50)	1.53	5.795	12.6	2.027
AN/TNT (50/50)	1.58	5.975	14.67	2.135
BH-1	1.673	8.26	28.7	2.235
$C_{20}H_{18}N_{20}O_{40}$	1.47	7.39	21.5	2.008
$C_{3.3}H_{6.2}N_{5.6}$	1.748	8.436	31.6	2.343
Comp-B	1.67	7.868	27.2	2.266
Comp-B	1.671	7.69	25.65	2.257
Comp B	1.674	7.89	26.7	2.251
Comp-B	1.692	7.84	26.75	2.278
Comp-B	1.7	7.85	28.3	2.329
Comp-B	1.703	7.75	27.2	2.320
Comp-B	1.712	8.022	29.3	2.332
Comp-B	1.729	7.98	29.77	2.370
Comp B	1.73	7.95	26.3	2.278
Comp-B	1.73	7.886	27.5	2.324
Comp-B	1.733	8	30	2.376
Comp-B, Grade A	1.717	7.985	29.04	2.337
Comp-B (64/36)	1.715	8.02	29.2	2.332
Cyclotol (75/25)	1.743	8.252	31.3	2.367
Cyclotol (75/25)	1.757	8.3	32.33	2.397
Cyclotol (75/25)	1.76	8.3	31.6	2.380
Cyclotol (77/23)	1.752	8.274	31.58	2.378
Cyclotol (77/23)	1.755	8.29	31.3	2.370
DATB	1.79	7.585	25.7	2.385
EA	1.592	7.34	23	2.175
EAR	1.607	7.51	25	2.219
EDD	1.563	7.45	21	2.062
HMX	1.89	9.11	39	2.515
HMX/EDNP (71/29)	1.66	7.77	27	2.272
HMX/inert (94/6)	1.835	8.778	37.5	2.497
HMX/inert (95/5)	1.776	8.76	33	2.343
HMX/inert (95/5)	1.783	8.73	33.5	2.366
HMX/PB (86/14)	1.66	8.29	27.5	2.187
HMX/polyurethane (95/5)	1.787	8.76	36	2.423
HMX/TNT/inert (68/30/2)	1.776	8.213	31.15	2.400
HNB	1.973	9.335	40	2.571
LX-04-01	1.858	8.46	35.13	2.525

Table 20.1 *(Continued)*

Explosive	ρ_0 (g/cm^3)	Detonation Velocity (km/s)	P_{CJ} (GPa)	ρ_{CJ} (g/cm^3)
LX-04-01	1.867	8.48	34.5	2.513
LX-07	1.85	8.59	37.73	2.557
LX-09	1.861	8.82	36.63	2.491
LX-10	1.841	8.81	37.2	2.489
Nitroguanidine	0.195	2.7	0.63	0.350
Octol (75/25)	1.8	8.55	30.65	2.347
Octol (77.6/22.4)	1.821	8.494	34.18	2.461
PBX-9404	1.84	8.8	37	2.485
PBX-9404	1.84	8.72	34.7	2.447
PBX-9404	1.844	8.81	37.2	2.492
PBX-9404	1.845	8.835	33.4	2.402
PBX-9404	1.846	8.82	37.5	2.498
PBX-9404	1.846	8.776	35.6	2.463
PBX-9502	1.895	7.706	28.9	2.550
Pentolite (50/50)	1.644	7.52	25.2	2.255
Pentolite (50/50)	1.644	7.52	25.63	2.270
Pentolite (50/50)	1.66	7.448	24.1	2.248
PETN/(superfine)/suspended in air	2.03E-03	1.410	1.92E-03	3.88E-03
PETN (regular)/air	2.13E-03	1.450	2.36E-03	4.50E-03
PETN (superfine)/air	2.33E-03	1.510	2.30E-03	4.11E-03
PETN (superfine)/air	2.80E-03	1.92	4.67E-03	5.12E-03
PETN	0.2	1.2	0.06	0.253
PETN	0.24	0.93	0.051	0.318
PETN	0.25	2.83	0.7	0.384
PETN	0.287	2.95	1.1	0.513
PETN	0.48	3.6	2.4	0.782
PETN	0.885	5.08	6.95	1.272
PETN	0.93	5.26	7.33	1.300
PETN	0.95	5.33	8.5	1.387
PETN	0.99	5.48	8.7	1.400
PETN	1.23	6.368	13.87	1.704
PETN	1.38	6.91	17.3	1.871
PETN	1.45	7.18	20.17	1.986
PETN	1.53	7.49	22.5	2.074
PETN	1.597	7.737	26.37	2.205
PETN	1.703	8.082	30.75	2.354
PETN	1.762	8.27	33.7	2.446
PETN	1.77	8.27	33.5	2.447
RDX	0.56	4.05	3.16	0.854
RDX	0.7	4.65	4.72	1.017
RDX	0.95	5.8	9.46	1.349
RDX	1.07	6.26	11.6	1.479
RDX	1.1	6.115	11.27	1.515
RDX	1.1	6.18	12	1.540
RDX	1.13	6.62	13.25	1.543
RDX	1.173	6.648	13.44	1.584
RDX	1.216	6.609	14.89	1.690

(Continued)

Table 20.1   *(Continued)*

Explosive	ρ_0 (g/cm³)	Detonation Velocity (km/s)	P_{CJ} (GPa)	ρ_{CJ} (g/cm³)
RDX	1.29	7	16.4	1.742
RDX	1.46	7.6	20.8	1.938
RDX	1.6	8.13	26	2.122
RDX	1.72	8.46	30.85	2.295
RDX	1.762	8.622	32.5	2.343
RDX	1.8	8.754	34.1	2.391
RDX	1.8	8.754	34.7	2.405
RDX	1.8	8.59	34.1	2.422
RDX/TNT (65/35)	1.715	8.036	28.9	2.321
RDX/TNT (78/22)	1.755	8.306	31.7	2.377
TNT	0.624	3.82	2.62	0.876
TNT	0.81	4.4	4.213	1.108
TNT	0.866	4.444	5.889	1.321
TNT	0.91	4.555	5.384	1.273
TNT	0.96	4.243	5.74	1.437
TNT	1.001	4.673	7.096	1.482
TNT	1.58	6.88	18.4	2.096
TNT	1.58	6.88	17.7	2.070
TNT	1.583	6.79	18.3	2.113
TNT	1.595	6.7	18.9	2.167
TNT	1.63	6.94	21	2.225
TNT	1.63	6.86	19.44	2.183
TNT	1.632	6.94	19	2.152
TNT	1.636	6.932	18.84	2.152
TNT	1.638	6.92	19.8	2.191
TNT	1.64	6.95	19	2.157
TNT	1.64	6.95	17.7	2.330

CHNO liquids

NM	1.128	6.29	13.3	1.607
NM	1.13	6.37	12.5	1.554
NM	1.133	6.299	13.4	1.614
NM (23(C)	1.1354	6.35	13.4	1.605
NM (4(C)	1.162	6.42	14.2	1.652
NM/TNM (1/.071 molar)	1.197	6.57	13.8	1.633
NM/TNM (1/.25 molar)	1.31	6.88	15.6	1.750
TNM	1.638	6.36	15.9	2.155
TNT (molten)	1.447	6.58	17	1.986

CHNO gases

Etylene/air (stoichometric)	1.28E-03	1.790	1.97E-03	2.46E-03

CHNO Al solids

ALEX-20	1.801	7.53	23	2.325
ALEX-32	1.88	7.3	21.5	2.394

Table 20.1   *(Continued)*

Explosive	ρ_0 (g/cm^3)	Detonation Velocity (km/s)	P_{CJ} (GPa)	ρ_{CJ} (g/cm^3)
AN/ADNT/Al(2/1/2.66molar)	1.734	7.844	26.3	2.301
AN/ADNT/RDX/Al(5/1/1/ 3.3molar)	1.752	7.739	25	2.300
EARL-1	1.595	7.2	23	2.210
EARL-1	1.665	7.27	24	2.289
EARL-2	1.709	7.13	23	2.324
HBX-1	1.712	7.307	22.04	2.256
HBX-1	1.75	7.16	20.86	2.280
CHNO B solids				
$B_{2.159}H_{11.0226}C_{4.6477}N_8O_8$	1.665	8	27.5	2.244
CHNO B liquids				
$B_{10}H_{18}C_{5.75}N_{15}O_{30}$	1.4	6.74	17.2	1.919
$B_{10}H_{18}C_{6.45}N_{17.8}O_{35.6}$	1.427	6.82	16.7	1.907
ET	1.4	6.74	17.2	1.919
CHNO Ba solids				
Baratol(65/35)	2.35	5.15	13.49	2.999
Baratol(72/28)	2.452	5	15.37	3.273
CHNO B F liquids				
$B_{10}H_{18}C_{17}F_{30}O_{30}N_{15}$	1.467	6.91	20.6	2.078
CHNO Cl solids				
HMX/AP/EDNP (51/20/29)	1.67	7.19	23	2.276
HMX/AP/PB (57/29/14)	1.67	7.76	26	2.252
HMX/AP/PB (69/17/14)	1.67	8.05	27.5	2.239
HMX/AP/PB (80.3/5.9/13.8)	1.66	8.19	27.5	2.204
CHNO Cl K solids				
HMX/KClO4	1.876	6.25	18.93	2.529
HMX/KP/EDNP (31/45/24)	1.87	6.66	23.5	2.609
HMX/KP/PB (33.4/53.4/13.2)	1.88	6.18	17.25	2.474
HMX/KP/PB (51/35/14)	1.78	7.15	22	2.348
HMX/KP/PB (52.6/34.7/12.7)	1.82	7.13	25	2.494
HMX/KP/PE (52/43/5)	1.985	7.63	32.5	2.762
HMX/KP/PE (52/43/5)	1.992	7.54	30.5	2.726
HMX/KP/PE (52/43/5)	1.994	7.76	35	2.814
CHNO Cl Na solids				
RDX/NaCl (80/20)	1.3	6.062	12.69	1.770

(Continued)

Table 20.1 *(Continued)*

Explosive	ρ_0 (g/cm^3)	Detonation Velocity (km/s)	P_{CJ} (GPa)	ρ_{CJ} (g/cm^3)
CHNO F solids				
HMX/Viton (85/15)	1.866	8.47	35	2.527
RDX/TFNA	1.754	8.22	32.4	2.414
TFNA	1.692	7.4	24.9	2.314
CHNO F liquids				
TFENA	1.523	6.65	17.4	2.054
CHNO Pb solids				
HMX/Exon/Pb (60/10/30 volume)	4.6	5	24.8	5.865
CHNO Si liquids				
NM/PMMA/GMb[a] (87.3/2.7/ 10w%)	7.80E-01	4.140	4.30E+00	1.15E+00
NM/PMMA/GMb (87.3/2.7/10)	7.80E-01	4.140	3.74E+00	1.08E+00
NM/PMMA/GMb (82.5/2.5/15)	6.77E-01	3.550	2.45E+00	9.50E-01
NM/PMMA/GMb (82.5/2.5/15)	6.77E-01	3.550	2.60E+00	9.74E-01
NM/PMMA/GMb (77.6/2.4/20)	5.75E-01	3.000	1.50E+00	8.10E-01
NM/PMMA/GMb (67.9/2.1/30)	3.90E-01	2.080	6.00E-01	6.05E-01
NM/PMMA/GMb (58.2/1.8/40)	2.58E-01	1.500	4.00E-01	8.30E-01
CHNO W solids				
HMX/W/binder (13.22/85.48/1.3)	7.41	4.64	29.7	9.105
HMX/W/binder (13.22/85.48/1.3)	7.47	4.54	29.7	9.255
H O gases				
Hydrogen/oxygen (8/1 molar)	2.39E-04	3.532	1.44E-03	4.62E-04
Hydrogen/oxygen (8/1)	2.39E-04	3.532	1.30E-03	4.23E-04
Hydrogen/oxygen (4/1)	3.57E-04	3.273	1.75E-03	6.58E-04
Hydrogen/oxygen (4/1)	3.57E-04	3.273	1.76E-03	6.60E-04
Hydrogen/oxygen (2/1)	5.36E-04	2.820	1.77E-03	9.18E-04
Hydrogen/oxygen 2/1)	5.36E-04	2.820	1.83E-03	9.39E-04
Hydrogen/oxygen 2/1)	5.36E-04	2.820	1.83E-03	9.40E-04
Hydrogen/oxygen (1/1)	7.59E-04	2.314	1.76E-03	1.34E-03
Hydrogen/oxygen (1/1)	7.59E-04	2.314	1.73E-03	1.32E-03
Hydrogen/oxygen (1/2)	8.42E-04	1.922	1.55E-03	1.68E-03
Hydrogen/oxygen (1/2)	8.42E-04	1.922	1.55E-03	1.68E-03
Hydrogen/oxygen (1/3)	1.13E-03	1.707	1.46E-03	2.03E-03
Hydrogen/oxygen (1/3)	1.13E-03	1.707	1.43E-03	2.00E-03
Hydrogen/oxygen (1/3)	1.13E-03	1.707	1.40E-03	1.97E-03
H2/O2 (ER=1.00, IT=300, IP=86)[b]	4.00E-04	2.819	1.66E-03	8.37E-04

Table 20.1   *(Continued)*

Explosive	ρ_0 (g/cm^3)	Detonation Velocity (km/s)	P_{CJ} (GPa)	ρ_{CJ} (g/cm^3)
H2/O2 (ER=1.00, IT=100, IP=101)	1.50E-03	3.175	5.10E-03	2.26E-03
H2/O2 (ER=1.00, IT=100, IP=188)	2.60E-03	3.115	9.50E-03	4.17E-03
H2/O2 (ER=0.33, IT=100, IP=405)	9.80E-03	2.028	2.16E-02	2.11E-02
H2/O2 (ER=0.75, IT=100, IP=405)	6.80E-03	3.018	2.28E-02	1.08E-02
H2/O2 (ER=1.00, IT=100, IP=405)	5.80E-03	3.008	2.40E-02	1.07E-02
H2/O2 (ER=2.00, IT=100, IP=405)	3.90E-03	3.636	1.92E-02	6.21E-03
H2/O2 (ER=3.07, IT=100, IP=405)	3.00E-03	3.688	1.87E-02	5.54E-03
N O liquids				
NO	1.294	5.62	10	1.713
N Pb solids				
Lead azide	3.18	4.03	12.4	4.185
Lead azide	3.23	4.06	12.6	4.231
Lead azide	3.66	4.42	15.5	4.673
Lead azide	3.7	4.48	15.8	4.700

References 1 and 2.
[a] GMb, glass microballoons, SiO$_2$.
[b] ER, equivalence ratio; IT, initial temperature (K); IP, initial pressure (kPa).

derived data for a large number of explosives and explosive mixtures for which the density, detonation velocity, and detonation pressure were all measured independently in the same experiment for each set of data.

20.2.1   Relationship of Initial Density to CJ Density

If we plot the initial density of unreacted explosive versus the density at the CJ state for all the explosives in Table 20.1, we find that the data fit a straight line on a log plot (Figure 20.5). Specifically, the equation of this line (Ref. 2) is

$$\rho_{CJ} = 1.386 \, \rho_0^{0.96}$$

This correlation will become very handy as an estimating tool. In Section I we had shown an estimate that

$$P_{CJ} = \rho_0 D^2/4$$

This implies (when combined with the mass and momentum equations) that

$$\rho_{CJ} = (4/3)\rho_0$$

When we plot that equation on the same graph (Figure 20.5), we see that it is close but underestimates ρ_{CJ} at the lower densities.

20.2.2 Estimating the CJ Pressure from ρ_0 and D

If we look back at the mass and momentum equations for the jump from unreacted explosive to the CJ state,

$$\frac{\rho_{CJ}}{\rho_0} = \frac{D}{D - u_{CJ}}$$

$$P_{CJ} = \rho_0 u_{CJ} D$$

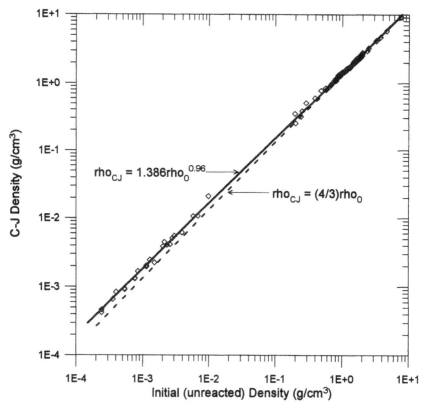

Figure 20.5 Correlation of CJ density to initial unreacted explosive density (Refs. 1 and 2).

We can rearrange the terms and get

$$P_{CJ} = \rho_0 D^2 \left(1 - \frac{\rho_0}{\rho_{CJ}}\right)$$

If we now combine this with the $\rho_{CJ} - \rho_0$ relationship derived from the experimental data in Table 20.1, we get

$$P_{CJ} = \rho_0 D^2 (1 - 0.7125\rho_0^{0.04})$$

This equation, knowing only ρ_0 and D for any explosive at any density, estimates P_{CJ} within 5% of the experimentally measured values. This is slightly better than estimates from computer codes using various nonlinear equations of state, which agree with the experimental data within 7 to 9%.

Example 20.1 Suppose we have a new and, as yet, not well-known explosive. A sample is measured and weighed and its density is found to be 1.43 g/cm³. Measurement at this density is made of its detonation velocity, which is found to be 6.95 km/s. What do you estimate its CJ pressure to be?

$$P_{CJ} = \rho_0 D^2 (1 - 0.7125\rho_0^{0.04})$$

Solution

$$P_{CJ} = (1.43)(6.95)^2 [1 - (0.7125)(1.43)^{0.04}]$$

$$= 19.15 \text{ GPa}$$

20.2.3 Estimating the *P-u* Hugoniot of Detonation Products

Similar to what we saw with shock-wave interactions between inert materials, we will need to know the properties of detonation reaction products at the shock states created when an explosive detonates in contact with another material. As we shall see later, if the adjacent material has a shock impedance greater than that of the detonation reaction products at the CJ state, then the resulting pressure at the interface will be greater than the CJ pressure. These pressures lie along the shock adiabat of the detonation reaction products (states adiabatically shocked up from the CJ state). Conversely, if the adjacent material has a shock impedance lower than that of the detonation reaction products at the CJ state, then the resulting pressure at the interface will be lower than the CJ pressure. These pressures lie along the expansion isentrope of the detonation reaction products. Although, strictly speaking, only the adiabat is usually called the Hugoniot, for our purposes here, the Hugoniot is the combination and continuum of these two regimes, joined at the CJ state.

This Hugoniot can be estimated by computer codes that utilize estimated product composition equilibria along with nonlinear empirical EOSs for the gases. While these codes are quite good at estimating the values along the Hugoniot, they are not readily available to most engineers, nor are the large computers that are required to run them.

Some experimental data are available in the open literature, but are limited to relatively few explosives and only a few initial densities of those explosives. Therefore, a simple means is needed to estimate the values along the Hugoniot for any explosive and at any density, based only upon easily obtained or estimated parameters.

When experiments are conducted for a given explosive with a variety of targets spanning the range from low- to high-target shock impedance, the Hugoniot of the detonation reaction products of that particular explosive can be constructed. One such Hugoniot is shown in Figure 20.6 for a plastic-bonded explosive consisting of TATB and a binder. The target materials used in these experiments were copper, aluminum, magnesium, transacryl (a polymer), water, and argon gas at various initial pressures ranging from 5 to 705 bar (Ref. 3). We see that this Hugoniot in Figure 20.6 is similar to those for the detonation reaction product Hugoniots spread across the Jones plots we saw earlier as Figure 18.4.

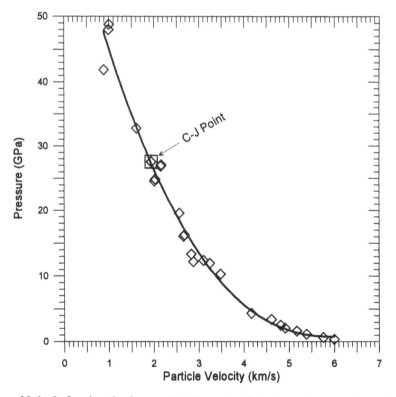

Figure 20.6   Left-going shock wave P-u Hugoniot of the detonation reaction products of TATB/T2, developed from experimental data (Ref. 3).

Experimental data similar to those shown in Figure 20.6 were found for a number of other explosives and reduced to the form P/P_{CJ} and u/u_{CJ}. When these parameters were plotted against each other, they fell into a narrow band that could be approximated by a simple correlation, shown in Figure 20.7, where P/P_{CJ} is plotted versus u/u_{CJ} on log axes.

The plotted data fall into two regions. For reduced pressure above 0.08, the data are correlated (with a correlation coefficient = 0.987) by:

$$\frac{P}{P_{CJ}} = 2.412 - 1.7315\left(\frac{u}{u_{CJ}}\right) + 0.3195\left(\frac{u}{u_{CJ}}\right)^2$$

or

$$P = (2.412 P_{CJ}) - \left(\frac{1.7315 P_{CJ}}{u_{CJ}}\right) u + \left(\frac{0.3195 P_{CJ}}{u_{CJ}^2}\right) u^2$$

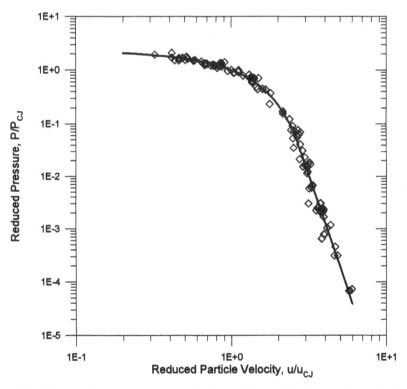

Figure 20.7 Reduced pressure versus reduced particle velocity for experimental data in Table 13.1 (Ref. 4).

For reduced pressure below 0.08, the data are correlated (with a correlation coefficient = 0.898) by:

$$\frac{P}{P_{CJ}} = 235\left(\frac{u}{u_{CJ}}\right)^{-8.71}$$

or

$$P = (235P_{CJ}u_{CJ}^{8.71})u^{-8.71}$$

Example 20.2   In Example 20.1, we saw a hypothetical explosive that had the CJ state properties: $\rho_0 = 1.43$ g/cm^3, $D = 6.95$ km/s, and $P_{CJ} = 19.15$ GPa. What is the equation of the left-going shock wave P-u Hugoniot of its detonation reaction products?

Solution   We will need the CJ particle velocity and pressure. The momentum equation can be rearranged to give

$$u_{CJ} = P_{CJ}/\rho_0 D$$
$$=(19.15)/(1.43)(6.95)$$
$$=1.93 \text{ km/s}$$

The detonation product Hugoniot was given by

$$P = (2.412P_{CJ}) - (1.7315P_{CJ}/u_{CJ})u + (0.3195P_{CJ}/u_{CJ}^2)u^2$$
$$= (2.412)(19.15) - [(1.7315)(19.15)/(1.93)]u + [(0.3195)(19.15)/(1.93)^2]u^2$$
$$P = 46.2 - 17.2u + 1.64u^2$$

20.3 Detonation Interactions

Just as with shock waves (nonreacting), the detonation is a jump process and is handled the same as shock waves are on the P-u plane. The detonation jump condition from the state of an unreacted explosive to the CJ state is the straight line joining those two states. The difference here is that state zero is the solid HE, and the CJ state is on the product Hugoniot. The initial unreacted state, if we assume $u_0 = 0$, is at the $P = 0$, $u = 0$ origin on the P-u plane. Figure 20.8 shows the detonation jump condition. Where the slope of the jump line for nonreactive shocks was $\rho_0 U$, the slope of the jump line for a detonation is $\rho_0 D$.

The major interaction of interest with explosives is the case where an explosive is in contact with another material, and the detonation wave interacts at that interface. As with the analogous nonreactive shock, it is important to know the relative shock impedance of the material and the explosive reaction products. The shock impedance of the products is $Z_{det} = \rho_0 D$.

20.3.1 Case 1, $Z_{material} > Z_{det}$

This case deals with a detonation causing a shock into an abutted material whose impedance is higher than that of the detonation product at the CJ state. Figures 20.9 and 20.10 show the P-x diagram and the x-t plane of this interaction.

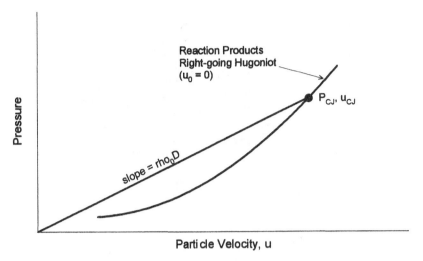

Figure 20.8   The reaction product Hugoniot (P-u).

The interaction will produce a right-going shock wave in material B coming from $P = 0$, $u = 0$, and a left-going shock wave back into the detonation product gases coming from P_{CJ}, u_{CJ}. The intersection of the Hugoniots from these two waves is the solution to the interaction problem. The left-going wave Hugoniot of the products is the mirror image of the Hugoniot rotated around the CJ state.

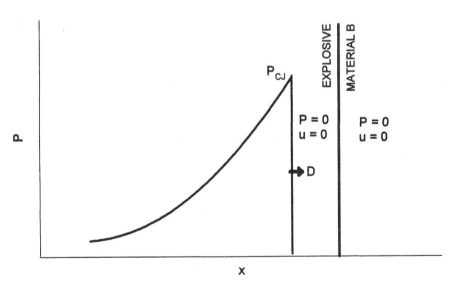

Figure 20.9 P-x diagram, detonation approaching material B.

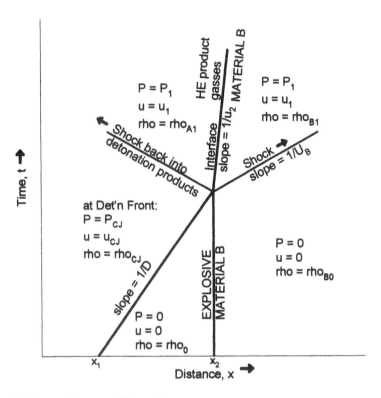

Figure 20.10   *x-t* diagram of interaction.

The interaction, shown in Figure 20.11, produces a shock pressure in *B* greater than the CJ pressure of the explosive.

Example 20.3   We have a slab of Composition B explosive ($\rho_0 = 1.733$) in contact with a brass plate. When the Composition *B* detonates, what shock pressure will be created at the surface of the brass plate?

Solution   First, we will need the *U-u* Hugoniot values for the brass and the CJ state parameters for Composition *B*. These will be found in Tables 17.1 and 20.1, respectively.

Brass: $\rho_0 = 8.45$ g/cm³, $C_0 = 3.726$ km/s, s = 1.434

Composition B: $\rho_0 = 1.733$ g/cm³, $P_{CJ} = 30$ GPa, $D = 8.0$ km/s

$\qquad u_{CJ} = P_{CJ}/\rho_0 D = (30)/(1.733)(8) = 2.16$ km/s

From these data we can now construct both the resultant right-going shock Hugoniot in the brass as well as the left-going wave in the detonation products.

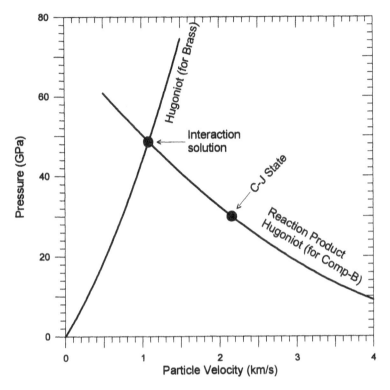

Figure 20.11   Interaction on P-u plane $Z_B > Z_{det}$.

Brass: $P = \rho_0 C_0 u + \rho_0 s u^2$

$= (8.45)(3.726)u + (8.45)(1.434)u^2$

$P = (2.412 P_{CJ}) - (1.7315 P_{CJ}/u_{CJ})u + (0.3195 P_{CJ}/u_{CJ}^2)u^2$

Composition B: $= (2.412)(30) - [(1.7315)(30)/(2.16)]u$

$+ [(0.3195)(30)/(2.16)^2]u^2$

Equating these and solving for u yields: $u = 1.09$ km/s, and using this in either of two Hugoniots, then yields: $P = 48.6$ GPa.

20.3.2 Case 2, $Z_{material} < Z_{det}$

This case is the same as the previous one, except the target material B now has a lower impedance than the detonation products at the CJ state, as seen in Figure 20.12. This results in a shock pressure at the interface lower than P_{CJ}, and a partial rarefaction reflected back into the compressed reaction product gases.

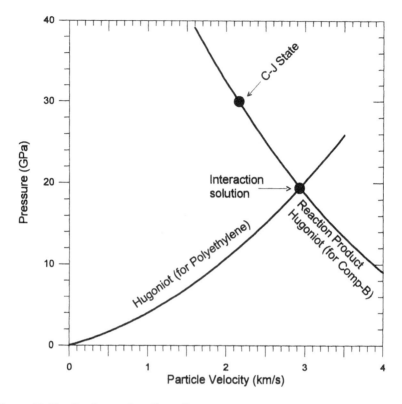

Figure 20.12 $P\text{-}u$ interaction, $Z_B < Z_{det}$.

Example 20.4   Cast TNT with initial unreacted density of 1.638 is in contact with a slab of polyethylene. Upon detonation, what pressure is developed at the interface?

Solution   From Table 17.1 we have

Polyethylene: $\rho_0 = 0.915$ g/cm³, $C_0 = 2.901$ km/s, $s = 1.481$

From Table 14.1 we have

TNT: $\rho_0 = 1.638$ g/cm³, $D = 6.92$ km/s, $P_{CJ} = 19.8$ GPa

From $P_{CJ} = \rho_0 u_{CJ} D$, we get

$$u_{CJ} = P_{CJ}/\rho_0 D$$
$$= (19.8)/(1.638)(6.92)$$
$$= 1.75 \text{ km/s}$$

This interaction will form a right-going wave in the polyethylene and a left-going wave in the detonation products. The $P\text{-}u$ Hugoniots for these two waves are

Polyethylene: $P = \rho_0 C_0 u + \rho_0 s u^2$

$= (0.915)(2.901)u + (0.915)(1.481)u^2$

TNT: $P = (2.412 P_{CJ}) - (1.7315 P_{CJ}/u_{CJ})u + (0.3195 P_{CJ}/u_{CJ}^2)u^2$

$= (2.412)(19.8) - [(1.7315)(19.8)/(1.75)]u$

$+ [(0.3195)(19.8)/(1.75)^2]u^2$

Equating these and solving for u yields

$u = 2.32$ GPa

and solving for P by using this particle velocity in either of the Hugoniots yields

$P = 13.4$ GPa

20.4 Summary

In this session we saw that the detonation wave is, and is treated as, a shock with the special feature that its pressure and shape, as well as its velocity, remain fixed with time.

We saw how the initial density of the explosive affects both the detonation velocity and pressure.

We learned that we can easily estimate the density of the detonation product gases at the detonation front, and using this can then estimate the detonation pressure.

We learned, knowing the CJ parameters, how to estimate the P-u Hugoniot of the detonation product gases.

We saw how to calculate the interaction of a detonation with another material, using the P-u Hugoniots.

21

Real Effects in Explosives

Up to this point, both for detonation as well as nonreactive shock waves, we have examined special ideal conditions such as uniaxial phenomena (one dimensional without edge effects) and only at ideal detonation conditions in explosives at unspecified lengths. In this session we will explore the phenomena that exist outside these limits. These phenomena, the effects of physical dimensions and temperature, are very complex, and so we will treat them only on the empirical level. We will also look somewhat into methods of estimating or scaling these effects where possible.

21.1 The Reaction Zone

In the simple model of detonation we treated the reaction zone as if it did not exist, had zero length. We know, of course, that this is not really the case. Figure 21.1 shows an idealized detonation wave where the structure and size of the reaction zone are indicated. Notice that although finite, its length really is tiny compared to that of a Taylor wave. The reaction-zone length appears to increase with decreasing density (Figure 21.2), and to decrease with increasing initial ambient temperature of the explosive.

Reaction-zone lengths cannot conveniently be measured directly. Their dimensions are deduced from their effects on other parameters, such as initial free-surface velocity of very thin foil flyers. Lengths vary from as little as a hundredth of a millimeter for some high-density high explosives and some liquids, up to several centimeters for some blasting agents. Table 21.1 lists some reaction-zone length data.

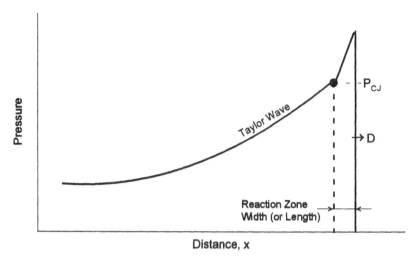

Figure 21.1 Idealized *P-x* diagram of a detonation.

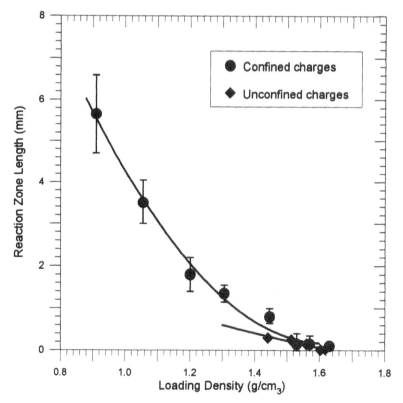

Figure 21.2 Detonation reaction zone length for TNT as a function of loading density (Ref. 5).

We are interested in reaction-zone length because it appears to be the major parameter controlling detonation velocity in the nonideal detonation region. It appears that explosives with thick reaction zones have a larger effect on detonation-velocity/diameter and failure diameters than explosives with thin reaction zones.

21.2 Diameter Effects

If we detonate a cylindrical column of explosive, and measure the detonation velocity, we will find that the velocity changes if we change the diameter of the column. The velocity decreases as the diameter of the column decreases. This

Table 21.1 Reaction-Zone Length

Explosive	Density (g/cm³)	Approximate Length (mm)	Conditions
Amotal 80/20	1.67	4	
AP (10m)	1.00	6.3	203 mm long
	1.10	6.7	"
	1.20	8.0	"
	1.26	10.0	"
CompB	1.67	0.13	(Al plate) 140 × 140 × 76 mm
HBX-1	1.60	0.198	
NG		0.21	
NM	1.128	0.3 − 0.6	0.25-mm thick-walled paper tube
	1.128	0.03	Pyrex cylinder
		0.08 at −5°C	25.4-mm-O.D. brass tube
		0.27 at 33°C	"
NM/acetone (75/25)		0.21	
	1.05	0.80	55 mm dia.
PBX-9502	1.895	3.3	200 mm dia.
Picric acid		2.2	glass cylinder
RDX		0.826	
microporous	1.30	1.82	
single crystal	1.80	2.90	
TNT		0.36	steel cylinder
	1.00	0.32	(Mg plate) 40 × 90 mm long
	1.55	0.18	(Al plate) 40 × 90 mm long
		0.13	(Cu plate) 40 × 90 mm long
		0.21	(Mg plate) 40 × 90 mm long
	1.59	0.70	60 mm long
pressed	1.63	0.3	90 mm long
cast	1.615	0.42 at 291 K	
	1.70	0.55 at 77.4 K	
	1.71	0.62 at 20.4 K	paper cylinder
liquid		0.9 at 100°C	glass cylinder
		1.1 at 100°C	Dural cylinder

Reference 5.

Table 21.2   Composition B Detonation Velocity Versus Rate-Stick Diameter

Diameter (mm)	Detonation Vel. (mm/μs)[a]	Density (g/cm^3)	Length/ Diameter
25.5	7.868	1.706	2
25.5	7.887	1.706	2
24.8	7.869	1.704	5.2
24.8	7.864	1.702	5.2
24.8	7.847	1.698	5.2
12.7	7.816	1.704	4
12.7	7.819	1.703	4
10.0	7.787	1.703	5
10.0	7.792	1.701	5
10.0	7.755	1.701	5
8.48	7.738	1.704	6.3
8.47	7.742	1.708	6
7.95	7.738	1.704	6.4
7.95	7.725	1.704	6.4
7.96	7.746	1.704	6.4
6.36	7.648	1.703	10.4
6.35	7.650	1.700	8
5.61	7.572	1.706	
5.61	7.561	1.706	
5.10	7.476	1.705	9.2
5.08	7.476	1.705	9.9
4.64	7.326	1.703	10.9
4.60	7.308	1.706	11.0
4.45	7.092	1.701	11.4
4.43	7.066	1.703	11.5
4.28	6.709	1.704	7.9
4.27	Failed	1.700	11.8

Reference 6.

[a] Average velocity through the stick.

effect is caused by energy losses to the side of the column. When the diameter is large the losses are small relative to the energy production at the wave front. When the column diameter is small the energy losses are larger relative to the energy generated at the wave front. The decrease in velocity continues until a diameter is reached where the energy losses are so great relative to the energy production that the detonation fails to propagate at all. In the following material, we will see quantitatively what these changes are and how they are related to each other as well as to other properties of explosives.

21.2.1 Effect of Diameter on Detonation Velocity

Measurements of this effect are fairly easy to make. We can take long cylinders of explosive of different diameters and detonate them from one end. We then measure the rate of detonation along the length of the cylinder by any one of several techniques, including high-speed framing cameras, streak photography, or ionization switches. The data can then be compiled and graphed. Table 21.2 gives typical data from "rate-stick experiment" for Composition *B*.

Although this type of experiment is fairly easy to conduct, they are rather expensive and time consuming. Therefore, we find a limited amount of such data available for explosives. The data in Table 21.2 are presented in graphical form in Figure 21.3.

This type of presentation is interesting, but does not tell us anything about the relative effect in a particular explosive or among different explosives. Notice, however, that the detonation velocity asymptotically approaches a constant value as the diameter becomes larger and larger. If we plot these same data in the form of detonation velocity versus reciprocal diameter, $1/d$, we see that the relationship becomes linear in $1/d$ as diameter increases (as $1/d$ approaches zero). This is shown in Figure 21.4.

When the linear portion of the relationship is extrapolated to $1/d = 0$, we find the value of the detonation velocity for an infinite diameter (uniaxial) charge.

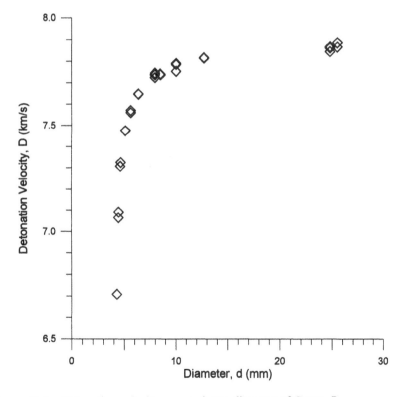

Figure 21.3 Detonation velocity versus charge diameter of Comp. *B*.

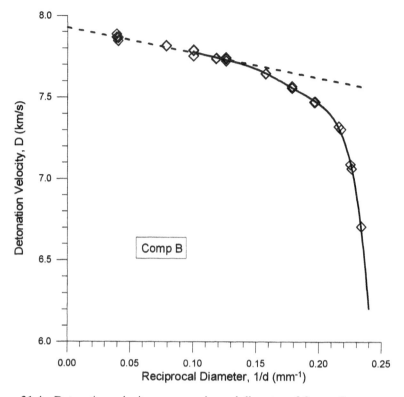

Figure 21.4   Detonation velocity versus reciprocal diameter of Comp. *B*.

This veolcity is called the ideal detonation velocity or the infinite diameter detonation velocity, and is designated as either D_i or D_∞. Knowing D_i, we can plot the same data in reduced (or relative) terms in the form of D/D_i versus $1/d$, as seen in Figure 21.5.

The linear portion of this relationship is expressed as

$$\frac{D}{D_i} = 1 - a\,\frac{1}{d}$$

The term a, the slope of the linear portion, a constant, is different for different explosives and is different for the same explosive at different initial conditions, which include density, temperature, and particle size. Although there are few data, the constant a appears to be proportional to reaction-zone length. Table 21.3 gives data for several different explosives as well as for several densities of one explosive, cyclotol (60/40).

A somewhat different treatment is given to rate-stick data in Ref. 6. There the data are correlated by

$$\frac{D}{D_i} = 1 - A\,\frac{1}{R - R_C}$$

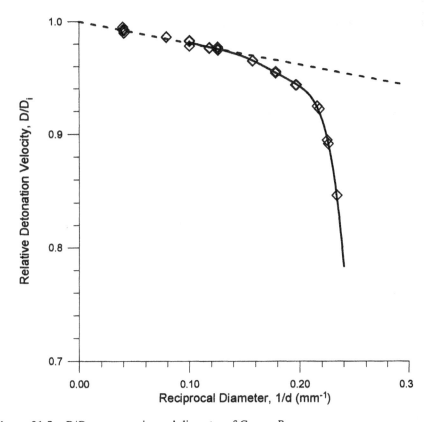

Figure 21.5 D/D_i versus reciprocal diameter of Comp. *B*.

where R is the radius of the charge and A and R_C are fitting constants. Data for A and R_C are given in Table 21.4 for several explosives.

Example 21.1   Cyclotol (60/40) is to be loaded into a long paper tube with an inside diameter of one-quarter inch. It will be pressed into the tube at a density of 1.10 g/cm³. What will the detonation velocity be in this configuration?

$$\frac{D}{D_i} = 1 - a\frac{1}{d}$$

From Table 21.3 we find that $D_i = 6.2$ km/s and $a = 1.02$ mm. The diameter $d = 0.25$ in. or 6.35 mm. The detonation velocity will be

$$D = D_i - aD_i/d$$

$$= (6.2) - (1.02)(6.2)/(6.35)$$

$$= 5.2 \text{ km/s}$$

The addition of confinement to an explosive charge, such as housing the charge column in a metal sleeve, helps to increase the detonation velocity or bring it

Table 21.3 Effect of Diameter on Detonation Velocity of Cylindrical Charges Fired in Air

Explosive	Density (g/cm³)	D_∞ (km/s)	a (mm)	Ref.
Baratol(73/27)	~2.60	4.96	1.83	4
CompB(GradeA)	1.715	7.99	0.189	4
Cyclotol(75/35)	1.755	8.298	0.139	4
Cyclotol(60/40)	0.50	4.26	3.19	17
	0.74	5.10	1.96	17
	0.90	5.60	1.55	17
	1.10	6.20	1.02	17
	1.40	7.15	0.49	17
DATB	1.788	7.52	0.14	4
LX-02	1.44	7.44	0.–116	4
LX-04	1.86	8.46	0.0568	4
Octol(75/25)	1.814	8.48	0.153	4
PBX-9010	1.781	8.371	0.0243	4
PBX-9404	1.84	8.80	0.0548	4
TATB	1.876	7.79	0.0431	4
XTX-8003	1.53	7.26	0.00832	4

References 5 and 7.

Table 21.4 Parameters of the Diameter-Effect Curve

Explosive	Data Points/Dia.[b]	Density/TMD[a] (g/cm³)	D_∞ (km/s)	R_C (mm)	A (mm)	Experiment Failure Radius (mm)[c]
Nitromethane (liquid)	9/5	1.128/1.128	6.213	−0.4	0.26	1.42
Amatex 80/20	4/4	1.613/1.710	7.030	4.4	5.9	8.5
Baratol 76/24	3/3	2.619/2.63	4.874	4.36	10.2	21.6
Comp. A	5/5	1.687/1.704	8.274	1.2	0.139	<1.1
Comp. B	26/12	1.700/1.742	7.859	1.94	0.284	2.14
Cyclotol 77/23	8/8	1.740/1.755	8.210	2.44	0.489	3.0
Dextex	7/4	1.696/1.722	6.816	0.0	5.94	14.3
Octol	8/6	1.814/1.843	8.481	1.34	0.69	<3.2
PBX 9404	15/13	1.846/1.865	8.773	0.553	0.089	0.59
PHX 501	7/5	1.832/1.855	8.802	0.48	0.19	<0.76
X-0219	8/6	1.915/1.946	7.627	0.0	2.69	7.5
X-0290	5/5	1.895/1.942	7.706	0.0	1.94	4.5
XTX 8003	162/4	1.53/1.556	7.264	0.113	0.0018	0.18

Reference 6.

[a] TMD, theoretical maximum density.

[b] Number of shots that propagated a steady wave/number of distinct diameters at which observations were made.

[c] R is the average of the radii from two go/no go shots (all shots fired in air except NM, which was in brass tubes with 3.18 mm thick walls).

closer to ideal performance. For explosives pressed into steel sleeves, the following relationship was found

$$\frac{D}{D_i} = 1 - 8.7\left(\frac{W_e}{W_c}\right)\left(\frac{a}{d}\right)^2$$

where (W_e/W_c) is the weight ratio of explosive to casing per unit length. Data supporting this relationship are shown in Figure 21.6.

Example 21.2 If the cyclotol in the previous example were loaded into a steel tube instead of a paper tube, what would the detonation velocity be if the tube wall was an eighth inch thick?

$$\frac{D}{D_i} = 1 - 8.7\left(\frac{W_e}{W_c}\right)\left(\frac{a}{d}\right)^2$$

The weight of explosive to casing ratio is easily found per unit length from the volumes and densities of the materials. In this case $(W_e/W_c) = 0.0465$; therefore

$$D = 6.2[1 - 8.7 \times 0.0465 \, (1.02/6.35)^2]$$

$D = 6.14$ km/s

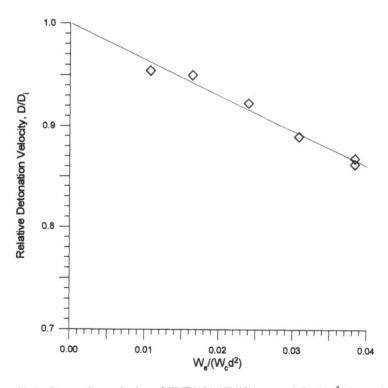

Figure 21.6 Detonation velocity of TNT/AN (60/40), $\rho_0 = 1.5$ g/cm³, in steel tube versus reciprocal diameter squared times weight ratio of charge to casing (Ref. 7).

21.2.2 Detonation Failure Diameter

The same mechanism, side losses that cause steady-state detonation velocity to decrease in the nonideal region, eventually become so dominant with decreasing diameter that a point is reached where steady-state detonation cannot be maintained. At this point detonation fails; it either suddenly slows down to below the sound speed in the unreacted explosive or stops altogether. This point is called the *failure diameter, D_f*, it is also called the *critical diameter, D_{crit}*. Failure diameter is strongly affected by confinement, particle size, initial density, and ambient temperature of the unreacted explosive. Failure diameter can be roughly correlated to the velocity-diameter constant a, as seen in Figure 21.7.

The effect of temperature is shown in Figure 21.8, where we see that increasing the initial or ambient temperature of the explosive decreases D_f. This effect is true for all explosives tested.

Decreasing particle or grain size also decreases D_f. This effect is seen in both Figure 21.9 and 21.10. The effect of initial density on D_f is not the same in all explosives, nor at all densities. For most single-component HEs, such as RDX, HMX, TNT, etc., the D_f decreases with increasing density. This trend continues

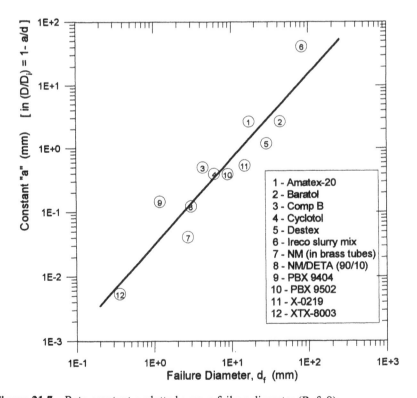

Figure 21.7   Rate constant a plotted versus failure diameter (Ref. 8).

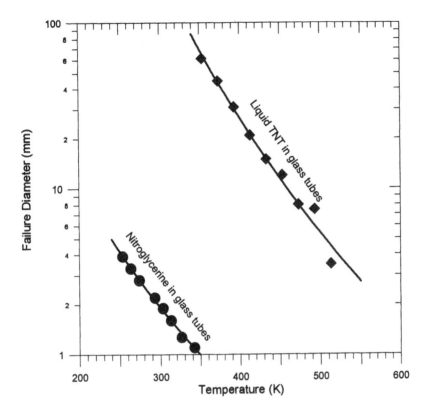

Figure 21.8   Failure diameter of nitroglycerine and liquid TNT as a function of temperature (Ref. 5).

until the maximum theoretical density (TMD) or crystal density is approached. For those explosives in this group that tend to fuse at or near TMD, the D_f suddenly jumps to a much higher value. This phenomenon is indicated by the data shown in Figure 21.9, where we see D_f versus ρ_0 for two different particle-size conditions of TNT.

Notice also on Figure 21.9 that the casting technique also affected D_f. The casting technique that leads to a final material with little or no fine voids or pronounced grain boundaries yields the highest D_f (poured clear), while that with the most fine bubbles and grain boundaries (creamed and powder) yields the lowest D_f. Perhaps the sudden change in D_f is due to a sudden shift in initiation mechanism, as affected by porosity. A second group of explosives, primarily those that contain large amounts of ammonium nitrate (AN), or ammonium perchlorate (AP), behave exactly the opposite in respect to density. With these explosives, D_f increases with increasing density. This is seen in Figures 21.10 and 21.11. In Figure 21.10 the trend of decreasing D_f with decreasing particle size is still the same for all explosives.

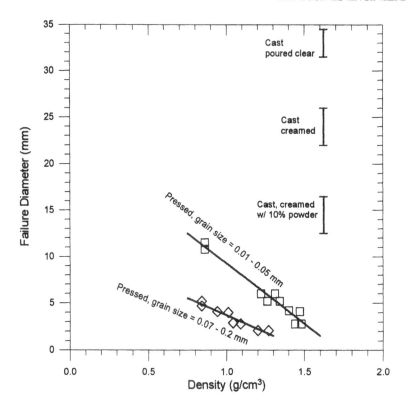

Figure 21.9   Failure diameter of TNT as a function of initial density (Ref. 7). 1. Pressed or powder: Grains ize: (1) 0.01–0.05 mm; (b) 0.07–0.2 mm. 2. Cast: (a) Poured clear; (b) creamed; (c) creamed with 10% powder added.

A very interesting set of data is seen in Figure 21.12, where an explosive from each group is mixed into identical charges. Here, for TNT/AN (50/50), the effect of density for the second group (ANs and APs), where D_f increases with increasing density, predominates the effect in the lower-density region. The effect of density of the first group (single-molecule organics), where D_f decreases with increasing density, dominates in the higher-density region.

Data for both groups of explosives are given in Table 21.5, which includes notes explaining the particular confinement and/or temperature conditions at which the data were obtained. Some failure diameters were shown previously in Table 21.4.

Similar to D_f, but for plates or slabs of explosive, there is a failure thickness that can and has been measured. These experiments, to determine failure thickness, are run on tapered explosive wedges initiated at the thicker end. The tests are conducted using a brass witness plate to indicate where failure occurred.

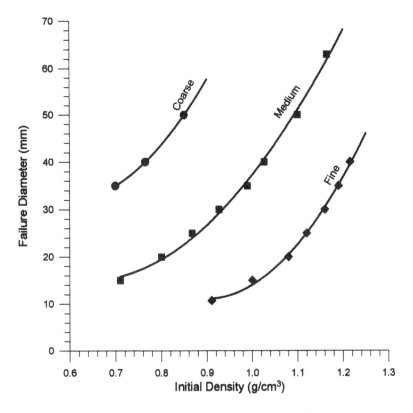

Figure 21.10   Failure diameter versus initial density for AN-fuel (peat) explosives at different grain sizes (Ref. 7).

Since the brass affords heavy confinement on one side of the explosive, and steel bars confine the sides, the failure thickness measured is most likely less than that for an unconfined explosive wedge. Figure 21.13 shows the test setup and Table 21.6 the test results for several explosives.

Data for, and estimates of, D_f and failure thickness are obviously important to the designer or analyst in spotting potential detonation reliability problems where explosive charges or systems must be minimized in size and weight.

21.3 Density Effects

From the previous chapter on ideal detonation, we know that both P_{CJ}, the detonation pressure, and D, the detonation velocity, are dependent upon the initial density of the unreacted explosive. We will recall that at the CJ point, the product

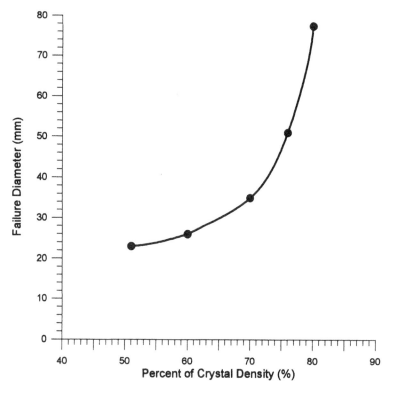

Figure 21.11   Failure diameter for detonation of AP as a function of initial density. Average grain size, 10 m (Ref. 7).

Hugoniot is tangent to the Rayleigh line, which connects the CJ point to v_0, the specific volume of the unreacted explosive. Therefore, we can say that at the CJ point, the slope of the Hugoniot equals the slope of the Rayleigh line (condition of tangency). We also know that the slope of the Rayleigh line equals $-(D/v_0)^2$ or $-D^2\rho_0^2$. So, at the CJ point

$$\left(\frac{dP}{dv}\right)_{CJ} = -D^2\rho_0^2$$

Since we are not really sure of, or do not wish to specify, the equation of the Hugoniot, we can just say that for any particular explosive

$$P_{CJ} = f_1(v_{CJ})$$

and the slope of the Hugoniot at the CJ condition is

$$(dP/dv)_{CJ} = f_1'(v_{CJ})$$

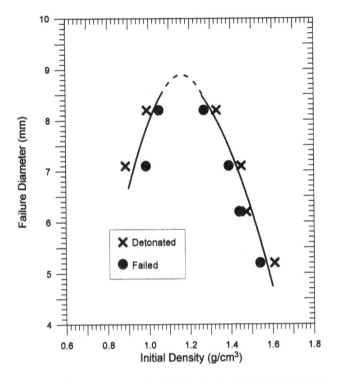

Figure 21.12   Failure diameter versus initial density for TNT/AN 50/50 (Ref. 7).

We now know two other facts, one, that $v = 1/\rho$, and therefore

$$(dP/dv)_{CJ} = f'_2(\rho_{CJ})$$

and, two, $\rho_{CJ} = a\rho_0^b$, and therefore

$$(dP/dv)_{CJ} = f'_3(\rho_0)$$

So we can equate the slope of both the Hugoniot and Rayleigh line thus

$$-D^2\rho_0^2 = f'_3(\rho_0)$$

$$D = \left(\frac{f'_3(\rho_0)}{\rho_0^2}\right)^{1/2}$$

The only way D versus ρ_0 can be linear is that the Hugoniot be a perfect poly-nomial in $(1/v_{CJ})$. Nature is certainly not that easy. The whole purpose of this little exercise was to demonstrate that if we plot D versus ρ_0 for some given explosive, we should not expect to get a straight line. The data in Figure 21.14 for HBX (a mixture of RDX, TNT, and aluminum) bear this out.

However, for most explosives over reasonable ranges of density, D versus ρ_0

Table 21.5   Critical Diameter (D_c)

Explosive	Density, ρ (g/cm^3)	Critical Diameter (D_c) (mm)	Conditions
Amatol 80.20 (cast)	—	80	—
AN	low-density	~100	Confined in steel tube
	~0.95	~12.7	Encased in paper tube, poor reproducibility
pressed	1.4	no detonation	100-mm-diameter charge confined in glass tubing
	1.61	no detonation	36.5-mm-diameter charge confined in 11-mm-thick steel tube
AP (particle	0.8–1.0	14	
size 5m; sifted	1.1	23 at 20°C	Charge length is 8–10 times the diameter
through nylon			
mesh having	1.1	12 at 200°C	Charge length is 8–10 times the diameter
70+ 10M openings)			
	1.2	~28 at 20°C	In cellophane tube
poured, 200m	1.29	>76.2	203-mm-long charge
pressed, 10m	1.56	76.2	203-mm-long charge
Baratol (cast)	2.619	43.2	Unconfined
Black powder (low density)		~100	Confined in steel tube
Comp. A-3 (pressed)	1.63	2.2	—
Comp. B 36/63/1 (cast)	1.70	4.28	Unconfined
Comp. B-3		3.73–4.24	Unconfined
cast		~3.18	Encased in Plexiglas tube
		~2.54	Confined in steel tube
Comp			
C-4	1.53–1.56	$3.81 < D_c < 5.08$	Confined
Cyclotol 77/23 (cast)	1.740	6.0	Unconfined
DATB	1.800	5.3	Unconfined
Explosive D	1.65	<25.4	Unconfined
FEFO		<3.43	Confined in 3.18-mm-thick 102-mm-long steel tube
HXB-1, pressed	1.72	6.35	Unconfined
cast	1.72	>6.35	
HMX/Wax 90/10	1.10	$6.0 < D_c 7.0$	—
78/22	1.28	$7.0 < D_c < 8.0$	—
70.30	1.42	$8.0 < D_c < 9.0$	—
Lead azide	3.14	0.4–0.6	—
NM	1.128	2.86	Encased in 3.18-mm-thick brass tube
	1.127	<3	Encased in 12.7-mm-diameter 6.4-mm-long pellet
	1.127	>11.76	Unconfined at ~25°C
	1.128	16.2	Encased in 22-mm-I.D. Pyrex tube at 24.5°C
	1.128	36	Encased in 16.3-mm-I.D. glass tube at 24.5°C
	1.128	28	Encased in 16.3-mm-I.D. glass tube at −8°C

Table 21.5   *(Continued)*

Explosive	Density, ρ (g/cm^3)	Critical Diameter (D_c) (mm)	Conditions
	1.128	20	Encased in 16.3-mm-I.D. glass tube at 12°C
	1.128	14	Encased in 16.3-mm-I.D. glass tube at 34°C
	1.128	27	Encased in 0.25-mm-thick-walled paper tube at 18–22°C
NQ	1.52	$1.27 < D_c < 1.43$	—
Octol 75/25 (cast)	1.814	< 6.4	Unconfined
PBX-9404	—	~ 1.02	Encased in Plexiglas or steel tubes
	1.846	~ 1.18	Unconfined
PBX-9501	1.832	< 1.52	Unconfined
PBX-9502	1.893	$10 < D_c < 12$	At −55°C
	1.894	$8 < D_c < 10$	At 24°C
	1.895	9	Unconfined
PBX	1.897	$4 < D_c < 6$	—
Pentolite 50/50 (cast)	—	6.7	Unconfined
PETN, powder	0.4–0.7	> 0.3	Encased in 0.05-mm-thick cellophane casing
single crystal	—	> 8.38	6.4 × 11.1-mm rod
Picric acid	0.9	5.20	—
RDX	0.9	5.20	—
RDX/TNT 100/0		3	—
90/10	1.0	3.5	—
80/20	1.0	3.75	—
70/30	1.0	4.25	—
50/50	1.0	5.25	—
40/60	1.0	5.75	—
20/80	1.0	7.0	—
10/90	1.0	7.5	—
0/100	1.0	7.5	—
RDX/Wax 95/5	1.05	$4.0 < D_c < 5.0$	
90/10	1.10	$4.0 < D_c < 5.0$	Encased in cellophane
80/20	1.25	$3.8 < D_c < 5.0$	shells with $D{:}L = 1{:}{>}10$
72/28	1.39	$3.8 < D_c < 5.0$	
TACOT	1.45	3	Unconfined
TATB	1.7	6.35	
TNT, cast	1.70	9	Encased in 0.2-mm paper at 77.4 K
	1.71	11	Encased in 0.2-mm paper at 20.4 K
	1.61	7	Encased in 0.2-mm paper at 290 K
powder	0.5–0.8	7.5	Encased in 0.05-mm-thick cellophane casing
	1.0	6	Encased in glass tube at 20°C

(Continued)

Table 21.5 *(Continued)*

Explosive	Density, ρ (g/cm³)	Critical Diameter (D_c) (mm)	Conditions
84% 0.5 mm, 16% 0.1 mm	0.95	22.52	
cast	1.6	27.43	
cast, poured cloudy	1.615	$22.0 < D_c < 25.4$	Unconfined
cast, creamed	1.615	$12.6 < D_c < 16.6$	Unconfined
cast	1.62	14.5	Unconfined
	1.615	15	Encased in 0.2-mm paper at 291 K
cast, poured clear	1.625	<3.7	Unconfined
liquid	1.443	62.6	Encased in 2.54-mm-thick glass tube
		$30 < D_c < 32.5$	Encased in 70-mm-diameter by 510-mm-long Pyrex tube at 100°C
XTX-8003	1.53	0.36	Encased in polycarbonate at 2-mm diameter
	~1.53	<0.39	
XTX-8004	~1.53	~1.4	Encased in polycarbonate at 2-mm diameter
	1.553	>1.78	

Reference 5.

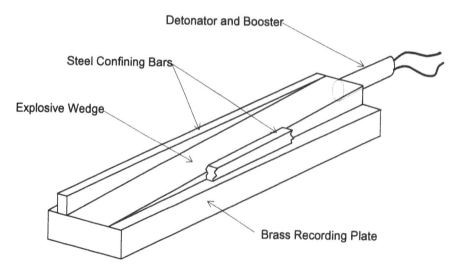

Figure 21.13   Minimum failure thickness test assembly (Ref. 6).

Table 21.6 Detonation Failure Thickness

Explosive	Density (g/cm^3)	Failure Thickness (mm)
Pure explosives		
Ammonium picrate	1.64	3.29
TNT	1.61	1.91[a]
Castable mixtures		
Comp. A-3	1.63	0.57
Comp. B-3	1.72	0.94
Cyclotol 75/25	1.75	1.51
Octol 75/25	1.79	1.43
Pentolite	1.70	1.39[b]
Plastic-bonded explosives		
HMX-based		
PBX 9011	1.77	0.61
PBX 9404	1.83	0.46
X-0204	1.922	0.41
RDX-based		
PBX 9010	1.78	0.52
PBX 9205	1.69	0.57
PBX 9407	1.77	0.30

Reference 6.
[a] Pressed 65°C.
[b] Cast 50-mm wedge.

is very close to linear, as seen in the data for PETN and for TNT shown in Figure 21.15. It is wrong to assume that you will find this in all explosives. If you are absolutely without density data, then you will have to assume that the relationship is linear in order to extrapolate some known D, ρ_0 condition to the one in which you are interested. If you do this you can also assume that the slope of the linear relationship is 3, which is a fair average over a short density range for many explosives, or use the Urizer estmate (given in Chapter 5), which is also shown for those same explosives in Figure 21.15. Table 21.7 gives D versus ρ_0 relationships for a number of explosives.

21.4 Temperature Effects

If we raise the initial temperature of an explosive, we might expect that we should raise the detonation velocity and also the detonation pressure. Quite the opposite is what actually happens. How can that be? It turns out that in raising the temperature we are expanding the explosive, thereby lowering the density. By lowering density, we are lowering P_{CJ} and also D.

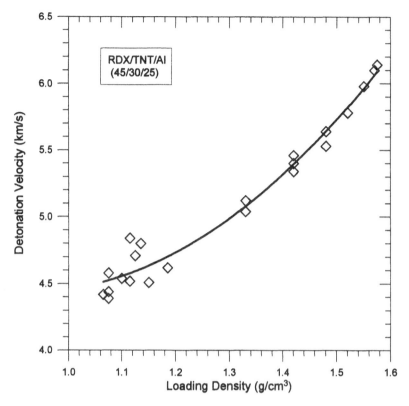

Figure 21.14 Detonation velocity versus initial density for HBX (RDX/TNT/A1, 45/30/25) (Ref. 9).

Although the effect is small, raising the initial temperature of an explosive decreases its detonation velocity and vice versa. Data for some explosives have been measured and are presented in Table 21.8. There are sufficient data on such properties as C_p, σ (linear temperature expansion coefficient), ΔH_d^0 (heat of detonation), ρ_0, P, and D to calculate this effect using the jump equations and simple thermophysics (Ref. 5 lists such data) for many explosives. Typically, $(\Delta D/\Delta T)$, the change in detonation velocity per unit change in temperature, will be found to be in the range of from -0.4×10^{-3} to -4×10^{-3} (mm/μs)/(°C).

21.5 Geometry Effects (L/D)

If we initiate the end of a cylinder of explosives by placing a detonator on the centerline, we would expect the detonation to grow out spherically from this

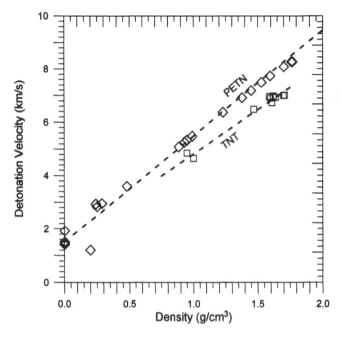

Figure 21.15 Detonation velocity versus initial density for PETN and TNT (Refs. 1, 2, and 5).

Table 21.7 Effect of Initial Density on Detonation Velocity

Explosive	Const. j	Const. k	Range of Applicability
Ammonium	1.146	2.276	$0.55 < \rho < 1.0$
perchlorate	−0.45	4.19	$1.0 < \rho < 1.26$
BTF	4.265	2.27	
DATB	2.495	2.834	
HBX-1	0.063	4.305	
LX-04-1	1.733	3.62	
Nitroquanidine	1.44	4.015	$0.4 < \rho < 1.63$
PBX-9010	2.843	3.1	
PBX-9404	2.176	3.6	
PETN	2.14	2.84	$\rho < 0.37$
PETN	1.82	3.7	$.37 < \rho < 1.65$
PETN	2.89	3.05	$1.65 < \rho$
Picric acid	2.21	3.045	
RDX	2.56	3.47	$1.0 < \rho$
TATB	0.343	3.94	$1.2 < \rho$
TNT	1.67	3.342	

Data from Ref. 5.
$D = j + k\rho_0$, D in km/s, ρ_0 in g/cm^3.

Table 21.8 Effect of Temperature on Detonation Velocity

Explosive	$\Delta D/\Delta T$ (km/s)/(°C)	Temperature Range (°C)
Comp. B (Grade A)	-0.5×10^{-3}	
LX-01-0	-3.8×10^{-3}	
LX-04-1	-1.55×10^{-3}	-54 to $+74$
LX-07	-1.55×10^{-3}	-54 to $+74$
LX-08	-3.56×10^{-3}	-36 to $+23$
LX-09	-3.31×10^{-3}	-5.4 to $+74$
Nitromethane	-3.7×10^{-3}	-20 to $+70$
PBX 9404	-1.165×10^{-3}	-54 to $+74$
Pentolite (50/50)	-0.4×10^{-3}	
XTX 8003 (LX-13)	-2.34×10^{-3}	-54 to $+74$

Reference 5.

point, and indeed at first it does. But after going some fixed distance, the detonation ceases to grow spherically and maintains a constant radius of curvature at the detonation front. If you make the charge larger, this point moves out further, but the ratio of radius of curvature to charge diameter remains constant. The value of this ratio changes from explosive to explosive (probably a function of reaction-zone length among others). Figure 21.16 shows this effect for several explosives. In this figure R is the radius of curvature of the front, d is the charge diameter, and L is the charge length.

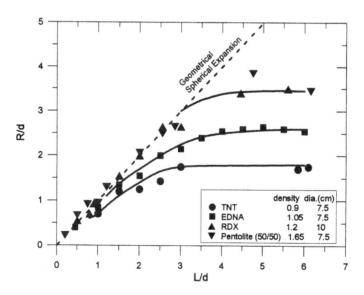

Figure 21.16 Variation of wave shape with charge length in ideal detonation (Ref. 9).

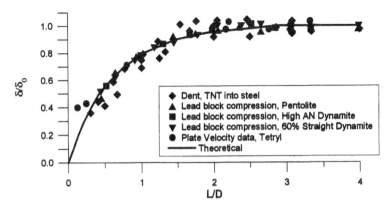

Figure 21.17 Variation of end effect with charge length (Refs. 6 and 9).

One of the manifestations of this effect is that the work done by the pressure at the end of the charge is a function not only of the pressure, etc., but also of the L/d of the charge. If we define some work function as δ (this could be depth of dent in a witness plate, degree of crushing of an adjacent material even properties of a jet in a shaped charge), then as L increases, δ also increases, but only up to some maximum L. This L we can call L_{max}; beyond this, δ remains constant no matter how long we make the charge. At this L_{max} then, we have a δ_{max} or δ_0. If we plot (δ/δ_0), (the reduced work function) versus (L/d) (the reduced, or scaled length), we find that all these parameters do indeed scale. Figure 21.17 shows this "end effect" or "length effect" for explosively driven flyer-plates, compression of lead blocks, and dent depth in steel witness plates.

In this figure we see that δ/δ_0 is approaching a constant for values of $L/d > 2$.

21.6 Summary

In this chapter we investigated some of the real effects in explosives. These included, in the nonideal region of detonation, the effects of diameter in lowering the detonation velocity and eventually causing failure in detonation. In ideal detonation we examined the effects of temperature, density, and geometry.

References

1. Cooper, P. W., *Estimation of the C-J Pressure of Explosives*, Proceedings of the 14th International Pyrotechnics Symposium, Jersey, United Kingdon, September 1989.

2. Cooper, P. W., *Extending Estimation of C-J Pressure of Explosives to the Very Low Density Region*, Proceedings of 18th International Pyrotechnic Symposium, Breckenridge, Colorado, July 1992.

3. Pinegree, M., Aveille, Leroy, J., Leroy, M., Protat, J. C., Cheret, R., and Camarcat, N., *Expansion Isentropes of TATB Compositions Released into Argon*, in Proceedings of the Eighth Symposium (International) on Detonation, Albuquerque, New Mexico, July 1985.

4. Cooper, P. W., *Shock Behavior of Explosives about the C-J Point*, Proceedings of the Ninth Symposium (International) on Detonation, Portland, Oregon, August 1989.

5. Dobratz, B. M., *LLNL Handbook of Explosives*, UCRL-52997, Lawrence Livermore National Laboratory, March 1981 (updated Jan. 1985).

6. Gibbs, T. R., and Popolato, A., *LASL Explosive Property Data*, University of California Press, Berkeley, 1980.

7. Johansson, C. H., and Persson, P. A., *Detonics of High Explosives*, Academic Press, London, 1970.

8. Cooper, P. W., *A New Look at the Run Distance Correlation and its Relationship to Other Non-Steady-State Phenomena*, Proceedings of the Tenth Symposium (International) on Detonation, Boston, 1993.

9. Cook, M. A., *The Science of High Explosives*, Reinhold Publishing Corp., New York, 1971.

INITIATION AND INITIATORS

In this section we will examine the phenomenon of initiation of explosives. Starting with the extant theories, we will see how all methods of initiation are basically thermal in nature. We will then examine the common types of initiating devices and see how the various modes of initiation and the theories that explain them are applied to both design and performance analysis of these devices. We will also examine the interplay of electrical initiator design and electrical firing circuits.

22

Theories of Initiation

In this chapter we will examine two major modes of initiation, the initiation of burning or deflagration, and the initiation of detonation. In the former, we will see that initiation is entirely a thermal phenomena; in the latter, we will see that the thermal effects must be coupled with hydrodynamic effects.

22.1 Initiation of Deflagration

As explosives decompose, they generate heat that can accelerate the rate of decomposition leading to a runaway or thermal ignition condition. This decomposition can be started merely by allowing an explosive to be exposed to high ambient temperatures, or by generating heat mechanically within the explosive.

22.1.1 Thermal Decomposition

When an explosive slowly decomposes, the reaction products are not necessarily formed at the maximum oxidation state. The various nitro, nitrate, nitramine, acid, etc., groups in an explosive molecule can slowly break down. This is due to low-temperature kinetics as well as the influence of light, infrared, and ultraviolet radiation, and any other mechanism that can feed energy into the molecule. Upon decomposition, products such as NO, NO_2, H_2O, N_2, acids, aldehydes, ketones, etc. are formed. Large radicals of the parent explosive molecule are left, and these react with their neighbors. As long as the explosive is at a temperature above absolute zero, decomposition occurs. At lower temperatures the rate of decomposition is infinitesimally small. As temperature increases, the decomposition rate increases. Although we do not always, and in fact seldom do, know the exact chemical mechanism, we do know that most explosives in the common use range of temperatures decompose at a zero-order reaction rate. This means that the rate of decomposition is usually independent of the composition of, or the presence of, the reaction products. The rate depends only upon temperature.

The reaction rate can be determined experimentally by holding the explosive at some constant temperature and measuring its weight loss as a function of time. When we do this we find that the rate of weight loss relative to the starting weight is a constant at constant temperature. This is expressed as

$$K = \frac{d(A'/A)}{dt} \tag{22.1}$$

where K is the reaction rate; A', the weight remaining at any given time; A, the initial weight; and t, the time. We can repeat this experiment at a number of different temperatures, obtaining values of K at each temperature. If we then plot these data on a graph of log K versus reciprocal temperature, $1/T$, we find that the data points form a straight line (Figure 22.1)

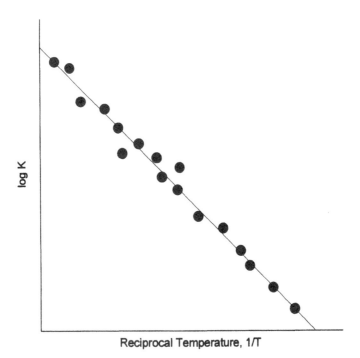

Figure 22.1. Typical plot of log of reaction rate versus reciprocal temperature.

The equation of a straight line on this type of plot is

$$\ln K = \ln Z - a \frac{1}{T} \tag{22.2}$$

where $\ln Z$ is the y-axis intercept and a is the slope. In the antilog form, this equation becomes

$$K = Ze^{-a/T} \tag{22.3}$$

The term a, the slope, has a commonality to all chemical reactions if we put it in terms

$$a = E_a/R \tag{22.4}$$

where E_a is the reaction activation energy and R is the universal gas constant. Thus, replacing a in Eq. (22.3), we obtain

$$K = Ze^{-E_a/RT} \tag{22.5}$$

This equation is known as the Arrhenius equation. It is in this manner that we determine the activation energy for a particular chemical reaction. The term Z is a constant unique to that particular reaction and is often called the "pre-exponential factor."

As the decomposition reaction progresses, it produces energy in the form of heat. For a given quantity of explosive, the faster the rate, the greater the rate of heat evolved. Combining the thermochemical characteristics of the reaction with the rate of reaction yields an expression for the rate of energy, or heat, produced.

$$Q = \rho \, \Delta H \, Ze^{-E_a/RT} \tag{22.6}$$

where Q is the rate of heat evolved per unit volume; ρ, the density; and ΔH, the heat of reaction.

The heat produced in this manner is transferred to the surrounding explosive material. The heat transfer rate is dependent upon temperature as well as thermal conductivity, heat capacity, and density. One of the classical three-dimensional heat transfer equations that relates the rate of heat production to the rate of temperature rise of the reacting material and to its surroundings is the Frank-Kamenetskii (FK) equation (Ref. 1).

$$-\lambda \, \nabla^2 T + \rho \, C\left(\frac{dT}{dt}\right) = \rho \, \Delta H \, Ze^{-E_a/RT} \tag{22.7}$$

where λ is the thermal conductivity and C is the heat capacity.

In essence, this equation says that if heat is evolved by the reaction faster than it can be transferred away, then the temperature of the reacting material must increase. Increasing temperature, as we have seen, increases the reaction rate and hence increases the heat production rate. Thus we can envision a situation where if the heat transfer rate cannot keep pace with the rate of heat produced, then the temperature will continue to rise at a greater and greater rate. At higher temperatures, the reaction mechanism changes toward higher oxidation states of the products, increasing ΔH as well as the relative amounts of gaseous products. Convective heat transfer then also increases along with increasing pressure and the result is an explosion.

The heat transfer rate is also a function of the thickness of the material through which it is being conducted; that is, the thicker the material, or the longer the heat transfer path, the lower the heat transfer rate. Thus a large sample or charge of explosive conducts internal heat away slower than a smaller one. From the above, we can see that for a given size and shape of a given explosive material, there must be some maximum *initial* temperature, which if exceeded, will lead to a runaway reaction or explosion. This temperature is called the "critical temperature," T_C.

22.1.2 Critical Temperature

When the *FK* equation is solved for critical temperature at the asymptotic, or steady-state, condition where time approaches infinity, the following expression is obtained.

$$\frac{E_a}{T_C} = R \ln\left(\frac{r^2 \rho \, \Delta H \, ZE_a}{T_C^2 \lambda \, \delta \, R}\right) \tag{22.8}$$

where r is the radius of a sphere, cylinder, or half-thickness of a slab in cm; ρ, the density in g/cm^3; ΔH, the heat of decomposition reaction in cal/mole; Z, the pre-exponential factor in s^{-1}; E_a, the activation energy in cal/mole; T_C, the critical temperature in K; R, the universal gas constant (1.9872 cal/mole K); λ, the thermal conductivity in cal/cm sec K; and δ, the shape factor; 0.88 for an infinite slab; 2.00 for an infinite-length cylinder; 3.32 for a sphere.

This solution of the *FK* equation represents the condition where the rate of heat evolved exactly equals the rate at which it is transferred away for a given size and shape of an explosive charge. Any increase in ambient temperature above T_C would lead to a runaway reaction or explosion within a finite time.

The validity of Eq. (22.8) is demonstrated by the results of small-scale tests where a slab of explosive of known properties is held between the two heated anvils in a manner that seals in all evolved gases. The anvils are electrically heated and held at a constant temperature. The time it takes to explosion is measured. The test is repeated over a range of temperature, and the results are plotted as reciprocal anvil temperature versus log of time to explosion. The temperature at which the relationship become asymptotic (time approaches infinity) is defined as the critical temperature, T_C, corresponding to the conditions of Eq. (22.8).

Table 22.1 gives the measured values of T_C obtained in the above manner for several explosives, along with the values calculated from Eq. (22.8). Let us apply

Table 22.1 Critical Explosion Temperatures

HE	Sample Thickness d (mm)	T_C(°C) Exp.[a]	Calc.	a (cm)	ρ (g/cm³)	Q (cal/g)	Z (s⁻¹)	E (kcal/mol)	λ (cal/cm s K)
BTF	0.66	248–251	275	0.033	1.81	600	4.11×10^{12}	37.2	0.0005
Comp. B	0.80	216	215	0.040	1.58	758	4.62×10^{16}	43.1	0.00047
DATB	0.70	320–323	323	0.035	1.74	300	1.17×10^{15}	46.3	0.0006
HMX	0.80	258	253	0.033	1.81	500	5×10^{19}	52.7	0.0007
HNS	0.74	320–321	316	0.037	1.65	500	1.53×10^{9}	30.3	0.0005
NQ	0.78	200–204	204	0.039	1.63	500	2.84×10^{7}	20.9	0.0005
PETN	0.80	197	196	0.034	1.74	300	6.3×10^{19}	47.0	0.0006
RDX	0.80	214	217	0.035	1.72	500	2.02×10^{18}	47.1	0.00025
TATB	0.70	353	334	0.033	1.84	600	3.18×10^{19}	59.9	0.0010
TNT	0.80	286	291	0.038	1.57	300	2.51×10^{11}	34.4	0.0005

Reference 1.
[a] All experimental critical temperatures (T_C) are for the stated sample thickness d.

this equation to large explosive charges. Given a particular storage temperature, we can solve the maximum-size charge that can be safely held at that temperature. Using Eq. (22.8) to solve for critical temperature versus the radius of a sphere of a given explosive, we obtain the results shown plotted in Figure 22.2.

The decomposition reactions dealt with in these equations are strongly a function of the characteristics of the particular explosive charge. Particle size and surface area, the presence of chemical impurities, and other often uncontrollable factors all affect the decomposition reaction mechanism and hence its rate and thermochemical characteristics. Values of E_a, ΔH, and Z are not readily available in the literature, and often must be experimentally determined for the particular batch of explosives of interest.

Although values for thermal conductivity, λ, may be found, they may not be for the particular conditions of interest. Thermal conductivity changes with density, as seen in Figure 22.3, where data for TATB (Ref. 1) are plotted. The slope, $d\lambda/d\rho$ for these data is 0.00235 (cal/cm s K)/(g/cm³). This is probably in the ball park for most explosive powders and can be used to estimate value of λ for other densities of other explosives.

Figure 22.4 shows the temperature dependence of thermal conductivity for AP along with two different compositions of HMX/VITON. The slopes of these plots are probably representative of most explosives. Table 22.2 gives thermal conductivity data for a number of different explosives.

22.1.3 Heating by Impact at Low Velocity

We know that under certain conditions explosives will be initiated if they are subjected to impact. Low-velocity impact has caused many accidental explo-

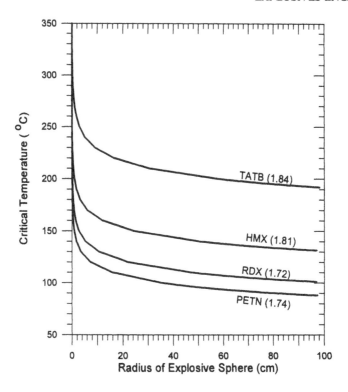

Figure 22.2. Critical temperature versus radius for several explosives.

sions, and we use low-velocity impact machines to test ''sensitiveness'' of explosives.

The phenomenon of impact or mechanical initiation is really thermal in nature. This was first proposed by Bowden and Yoffe (Refs. 2 and 3), who postulated that any of several mechanical mechanisms could produce heat at tiny local areas and thus raise the local temperature to the ignition point of an explosive. They referred to these tiny locally heated areas as ''hot spots.'' The mechanisms that they proposed were:

1. Adiabatic compression of air or vapor bubbles included in the explosive;
2. Intercrystalline friction;
3. Friction of impacting surfaces;
4. Plastic deformation of a sharply pointed impacting surface; and
5. Viscous heating of impacted material as it flows past the edges of the impacting surfaces.

Further studies by Kholevo (Ref. 4) and Andreev (Ref. 5) indicated that these mechanisms would not produce sufficiently high temperatures; so they proposed that microjetting in bubbles and particle interstices and/or inelastic compression

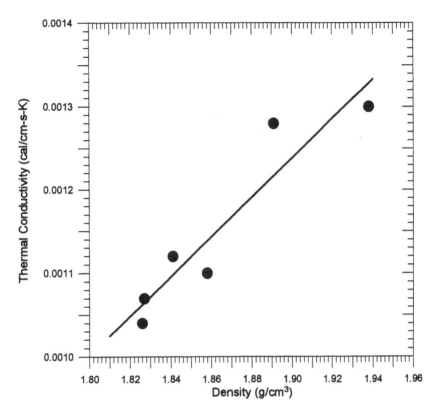

Figure 22.3. λ versus density for TATB.

of the solid particles was at the heart of the matter. A more believable mechanism was later proposed by Afanas'ev and Bobolev (Ref. 6), which is today the leading contender for acceptance. They proposed that the inelastic flow of the explosive under impact would produce the required temperatures. Because the explosives (solids) are particulates, local anisotropic behavior can create localized high shear stresses, where the solids in that area can suddenly break down into plastic flow relative to the surrounding (or included) solid particles. Flow in these local "shear bands" converts the quasistatic stresses on the entire bed into heat by viscous effects. They further point out that temperature in these areas is not limited by the normal melting point of the explosive because these local areas are at high stress or pressure, and therefore the melting point is raised.

$$T_m = T_m^0 + \alpha P \tag{22.9}$$

where T_m is the melting point; T_m^0, the normal melting point at 1 atm; α, the melting point pressure coefficient (for most CHNO explosives, α is approximately 0.02 °C/atm); and P, the pressure.

Further, we define a critical temperature, T_C', in a similar manner to the critical

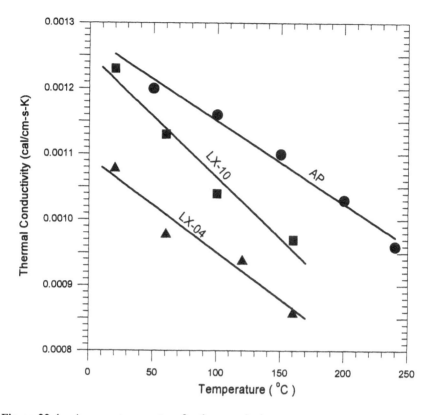

Figure 22.4. λ versus temperature for three explosives.

temperature we just dealt with in the previous section. T_C', however, is calculated not for infinite time to explosion, but for times less than 10^{-5} (which is the ignition delay time derived from observations in impact-machine experiments). The local critical stress to cause ignition was then stated (by Afanas'ev and Bobolev) to be

$$P_{cr} = (T_C' - T_m)/\alpha \tag{22.10}$$

where P_{cr} is the critical stress.

Thus, the higher the local pressure, the higher the local melting point, and the lower the critical stress to produce ignition. They then go on (in Ref. 6) to analyze the mechanical conditions and pressure or stress distributions in both thick and thin explosive charges undergoing inelastic flow due to low-velocity impact. The comparative database they use in these analyses is from impact machine tests. They find that for most explosives the critical temperatures, T_C', are between 400 and 600°C, and that the critical diameter of a "hot spot" is between 10^{-5} and 10^{-3} cm.

Table 22.2 Thermal Conductivities (λ) of Explosives and Binders

MATERIAL	Density, ρ (g/cm^3)	Thermal Conductivity $\lambda \times 10^4$ (cal/cm s K)	Temperature (°C)
AN		2.9–3.9	
AP[a]		12.0	50
		11.6	100
		11.0	150
		10.3	200
		9.6	240
Baratol		11.84	18–75
Comp. B	1.70	5.4	25
Comp. B-3		6.27	18–75
	1.73	5.23	46
Comp. C-4		6.22	
Cyclotol (75/25)	1.760	5.41	46
DATB	1.834	6.00	
Estane 5702		3.48	
Estane 5703	1.18	3.53	41.4
H-6		11.01	35
HBX-1		9.7	35
HBX-3		17.0	35
HMX		12.2	24
		13.3	
	1.91	9.83	
		10.13	
HNS-I	1.646	2.04	20
HNS-II	1.646	1.91	20
Kel-F 800	1.90	1.26	41.4
Lead azide	4.1	4.2	
	3.6	6.61	72–130
	0.88	1.55	
LX-04	1.87	10.7	20
LX-07	1.87	12.0	20
LX-09	1.84	12.3	20
LX-10	1.86	12.3	20
LX-14-0	1.83	10.42	20
LX-17-0	1.88	19.1	20
	1.89	12.1	40
Minol-2	1.74	16.5	
NC (12.7% N)		5.5	
	1.5	2.15	
NQ	1.651	10.14	41.3
NQ	1.689	9.85	41.3
PBX9010	1.875	5.14	48.8
PBX9011		10.3	21.1
	1.772	9.08	43.4
PBX9404		10.3	21.1
	1.845	9.2	46.2
PBX9501	1.847	10.84	55
PBX9502	1.893	13.2	38

(Continued)

Table 22.2 *(Continued)*

MATERIAL	Density, ρ (g/cm³)	Thermal Conductivity $\lambda \times 10^4$ (cal/cm s K)	Temperature (°C)
Picric acid	1.60	2.4	
Polystyrene		2.51	0
		2.78	50
		3.06	100
RDX	1.806	2.53	
	1.66	1.75	20
	1.81	2.53	
Sylgard 182 (cured)		3.5	
TATB	1.938	13	
	1.891	12.8	38
	1.841	11.2	
	1.858	11.0	
	1.827	10.7	
	1.826	10.4	
Tetryl	1.53	2.48	
	1.53	6.83	
	1.7	2.3	
	0.767	2.0	
TNT	1.654	6.22	18–45
	1.63	7.1	90–100
	1.56	4.8	
	0.846	3.5	
	1.65	3.1–6.2	
Viton A	1.815	5.4	
XTX-8003	1.54	3.42	
XTX-8004	1.540	3.42	40

[a] 43–61 mm particle size.

22.2 Initiation of Detonation

We have seen in the previous sections how burning or deflagration can be initiated in an explosive. If the decomposition reaction is completed at shock velocities in the explosive, that is called a detonation. The initiation of chemical reaction in a detonation is similar to what we saw with low-velocity impact. The shock front compresses the unreacted explosive material, causing local shear failure and inelastic flow (Ref. 7). These processes create "hot spots" that grow into complete reaction. The difference in the case of detonation is that the ensuing reaction is completed at a much higher rate.

22.2.1 Critical Energy Fluence

Let us consider that we shock an explosive with a square-wave pulse shock wave. This shock pulse has an amplitude P, the shock pressure, and a duration

t, the shock width (in time). The particle velocity behind the shock front is u. The rate at which work is done, per unit area, on the explosive being compressed by the shock is

$$\frac{\text{rate of work}}{\text{unit area}} = Pu \tag{22.11}$$

Since the work is being applied over a time t, the amount of energy, per unit area, deposited in the explosive is

$$E = Put \tag{22.12}$$

where E is the energy per unit area.

The term *energy per unit area* is referred to as the *energy fluence* (Ref. 8). Recalling the Rankin-Hugoniot jump conditions, specifically the mass and momentum equations for a shock, we had derived that

$$P = \rho_0 u U \tag{22.13}$$

where ρ_0 is the density of unshocked material and U is the shock velocity.

We can rearrange this slightly to

$$u = P/\rho_0 U$$

and replace this for u in Eq. (22.12), giving us

$$E = P^2 t/\rho_0 U \tag{22.14}$$

A number of experiments have been conducted in which explosives were subjected to square-wave shock pulses caused by the impact of flyer plates. The pulse duration was varied by changing the thickness of the flyer plates, and the shock pressure was varied by changing the impact velocity of the flyer plates. Each explosive was found to have a unique range of energy fluence above which prompt detonation was always obtained, and below which it was not. The average of this range is called the "critical energy fluence," E_C. Table 22.3 lists critical energy fluence for shock initiation for a number of various explosives.

The quantity $\rho_0 U$ [in Eq. (22.14)] is often called the shock impedance of a material. It increases very slowly with increasing pressure, but over the pressure ranges of general interest (in shock initiation) it can be considered to be nearly constant. Because of this, many workers in the explosives field combine this value into the critical energy fluence and use the term $P^2 t_{\text{crit}}$ instead as the critical value for initiation.

Example 22.1   In a previous section we gave an example (section 19.2) of the impact of a polyethylene flyer, 5 mm thick, impacting a piece of PBX9404 at 2.5 km/s. We saw that this formed in the unreacted explosive a square-wave shock pulse with pressure of 7.73 GPa and duration of 1.6 s. Will the PBX9404 detonate promptly from this input?

Solution   In the referenced example we also saw that the shock particle velocity in the explosive was $u = 0.898$ km/s. From Table 17.2 we find the U-u Hugoniot of PBX9404

Table 22.3 Critical Energies for Shock Initiation

Explosive	Density ρ (gm/cm³)	E_C (cal/cm²)
Comp. B	1.73	44
Comp. B-3	1.727	33
DATB	1.676	39
HNS-I	1.555	<34
Lead azide	4.93	0.03[a]
LX-04	1.865	26
LX-09	1.84	23
NM	1.13	404.7[a]
PBX-9404	1.84	15
	1.842	15
PETN	~1.0	~2
	1.0	2.7
	~1.6	~4
RDX	1.55	16[b]
TATB	1.93	226[a]
	1.762	72–88
Tetryl	1.655	10
TNT, cast	1.6	100[a]
pressed	1.620	32
	1.645	34

Reference 1.
[a] Values were estimated from data other than critical energy determinations and should be considered tentative.
[b] Constant-energy threshold not confirmed.

at $\rho_0 = 1.84$ g/cm³ is $U = 2.45 + 2.48u$. Therefore, U, the shock velocity in the unreacted explosive, is $U = (2.45) + (2.48)(0.898) = 4.68$ km/s, and the energy fluence is therefore:

$$E = \frac{P^2 t}{\rho_0 U}$$

$$E = \frac{(7.73)^2\,(1.6)}{(1.84)(4.68)}$$

$$E = 11.1\,\frac{(GPa)^2\,(\mu s)}{(g/cm^3)(km/s)} = 267 \text{ cal/cm}^2$$

From Table 22.3, we see that E_C for this explosive is 15 cal/cm²; so we can say definitely that it will detonate.

22.2.2 Run Distance Versus Pressure

The critical energy fluence is a necessary condition for shock initiation of detonation, but is not, by itself, sufficient to describe the whole process in engineering terms. When shocked, an explosive does not instantly attain full steady-state detonation. The shock must travel some finite distance into the explosive before steady-state detonation can be achieved. This "run distance" is not a constant, but varies with the peak input shock pressure. The higher the pressure, the shorter the run distance. When run distance versus input shock pressure data are plotted on a log-log format, the data fall onto approximately straight lines. This data representation is called a "Pop-plot" after Alfonse Popalato, formerly of the Los Alamos National Laboratory. From the Pop-plots, we are able to establish equations of input shock pressure as a function of run distance for each explosive tested. These are shown for a number of explosives in Table 22.4. The run distance data in this table are obtained for long shock pulse duration. That is, the peak shock pressure remains constant over the entire length of the run distance.

If a very thin flyer plate were impacted into an explosive, a short or "thin" pressure pulse would be formed. If the pulse is thin enough, as we shall soon see, the pressure will not be maintained as a constant over the run distance.

Table 22.4 Least-Squares Fits for Shock Initiation Data

Baratol	2.611	$\log P = 1.2352 - 0.3383 \log x$	$6.8 < P < 12$
Comp. B	1.72	$\log P = 1.5587 - 0.7614 \log x$	$3.7 < P < 12.6$
HMX	1.891	$\log P = 1.18 - 0.59 \log x$	$4.4 < P < 9.6$
LX-04-1	1.862	$\log P = 1.228 - 0.656 \log x$	$6.8 < P < 16.7$
LX-17	1.90	$\log P = 1.4925 - 0.5657 \log x$	$6 < P < 23.5$
NQ, large grain	1.66–1.72	$\log P = 1.44 - 0.15 \log x$	$13.4 < P < 26.3$
commercial	1.688	$\log P = 1.51 - 0.26 \log x$	$21.2 < P < 29.1$
PBX-9011-06	1.790	$\log P = 1.1835 - 0.6570 \log x$	$4.8 < P < 16$
PBX-9404	1.840	$\log P = 1.1192 - 0.6696 \log x$	$2 < P < 25$
	1.721	$\log P = 0.9597 - 0.7148 \log x$	$1.2 < P < 6.3$
PBX-9407	1.60	$\log P = 0.57 - 0.49 \log x$	$1.4 < P < 4.7$
PBX-9501-01	1.833	$\log P = 1.0999 - 0.5878 \log x$	$2.5 < P < 6.9$
	1.844	$\log P = 1.1029 - 0.5064 \log x$	$2.5 < P < 7.2$
PBX-9502	1.896	$\log P = 1.39 - 0.31 \log x$	$10.1 < P < 15$
PETN	1.75	$\log P = 0.57 - 0.41 \log x$	$1.7 < P < 2.6$
	1.72	$\log P = 0.6526 - 0.5959 \log x$	$2.0 < P < 4.2$
	1.60	$\log P = 0.3872 - 0.5038 \log x$	$1.2 < P < 2.0$
	1.0	$\log P = -0.3855 - 0.2916 \log x$	$0.2 < P < 0.5$
TATB	1.876	$\log P = 1.4170 - 0.4030 \log x$	$11 < P < 16$
superfine	1.81	$\log P = 1.31 - 0.43 \log x$	$10 < P < 28$
micronized	1.81	$\log P = 1.41 - 0.38 \log x$	$14.3 < P < 27.8$
Tetryl	1.70	$\log P = 0.79 - 0.42 \log x$	$2.2 < P < 8.5$
	1.30	$\log P = 0.87 - 1.11 \log x$	$0.37 < P < 6.9$
TNT, cast	1.635	$\log P = 1.40 - 0.32 \log x$	$9.2 < P < 17.1$
pressed	1.63	$\log P = 1.0792 - 0.3919 \log x$	$4 < P < 12$
XTX-8003	1.53	$\log P = 0.7957 - 0.463 \log x$	$3.0 < P < 5.0$

Reference 1.
[a] Where x is the run distance to detonation in mm; P, the initial shock pressure in GPa.

Example 22.2 In the previous example we saw that a thick slab of PBX9404 had been impacted by a polyethylene flyer. The impact created an input pressure pulse of 7.73 GPa with a duration of 1.6 μs. How far into the explosive will this input shock wave travel before steady-state detonation is established?

Solution From Table 22.4 we find the Pop-plot or run distance equation for PBX9404 at $\rho_0 = 1.84$ g/cm^3.

$$\log P = 1.1192 - 0.6696\log X$$

or

$$\log P = \log(13.158) - 0.6696\log X$$

Therefore,

$$P = 13.158X^{-0.6696}$$

$$X = \left(\frac{P}{13.158}\right)^{-1/0.6696}$$

$$X = \left(\frac{7.73}{13.158}\right)^{-1/0.6696}$$

$$X = 2.2 \text{ mm}$$

22.2.3 Thick- Versus Thin-Pulse Criteria

The pulse duration or width of the shock wave formed at impact is dependent upon flyer thickness and material as well as upon the particular target explosive. The pulse width is determined by the time it takes the impact shock in the flyer to reach the flyer's rear, unconfined, free surface and for the ensuing rarefaction wave to return to the flyer-explosive interface. The shock velocity U_f in the flyer material can be found from the particle velocity using the velocity Hugoniot relationship.

$$U = C_0 + su \tag{22.15}$$

where C_0 is the bulk sound speed, and S is the velocity Hugoniot coefficient.
    The rarefaction wave speed can be approximated by

$$R = C_0 + 2\,su \tag{22.16}$$

where R is the rarefaction wave velocity.
    Thus the pulse duration of the square shock wave formed by the impact of a flyer is

$$t = \frac{x_f}{U_f} + \frac{x_f}{R_f} = x_f\left(\frac{1}{U_f} + \frac{1}{R_f}\right) \tag{22.17}$$

where t is the pulse width (in time); x_f, the flyer thickness; U_f, the shock velocity in the flyer; and R_f, the rarefaction velocity in flyer. The shock- and rarefaction-wave velocities in the unreacted explosive are found in the same manner from the velocity Hugoniot values for the unreacted explosives.

At the time the flyer rarefaction reaches the flyer-explosive interface, the shock in the explosive has traveled some distance x_L. At this same time, the interface is relieved and a rarefaction wave begins to travel into the shocked region of the explosive. At some distance x, the rarefaction catches up with the shock front and then begins to attenuate the shock peak pressure. This race is shown in the x-t diagram in Figure 22.5.

From Figure 22.5 we see that in the same time it takes the rarefaction in the explosive to travel the distance x, the shock has traveled the distance x-x_L.

Since time is equal to distance divided by rate, we have

$$\frac{x}{R_{HE}} = \frac{x - x_L}{U_{HE}} \tag{22.18}$$

We also see that

$$x_L = U_{HE} \, t \tag{22.19}$$

Combining these two, we obtain

$$x = \frac{U_{HE} R_{HE} t}{R_{HE} - U_{HE}} \tag{22.20}$$

This is the distance over which the shock maintains constant peak pressure. Beyond x, the rarefaction, as stated earlier, continually attenuates the peak pres-

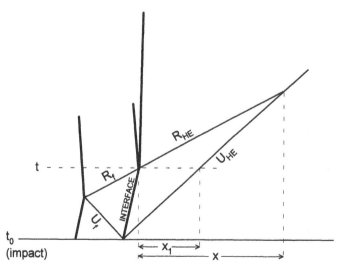

Figure 22.5. x-t diagram of shocks and rarefactions in both explosive and flyer after impact.

sure. This is shown in Figure 22.6. If this distance, x, is less than the run distance from the Pop-plot data (for this particular impact pressure), then this is called a "thin-pulse" condition. Since the peak pressure is not maintained over the entire "ideal" run distance, the actual run distance required to reach steady-state detonation will be greater. If the pulse is very thin, x is much shorter than the Pop-plot run distance, and the explosive will not detonate.

When designing explosive interfaces that utilize gaps and flyers to transfer detonation, these conditions must be taken into account.

Example 22.3   Again referring to the impact of a polyethylene flyer into PBX9404 as seen in the two earlier examples in this chapter, for these same conditions, what would the minimum flyer thickness be and still assure detonation?

Solution In Eq. (22.17), above, we saw that

$$t = x_f\left(\frac{1}{U_f} - \frac{1}{R_f}\right)$$

and in Eq. (22.20) we saw that

$$x_{\text{run}} = \frac{U_{HE}R_{HE}\,t}{R_{HE} - U_{HE}}$$

Manipulating the latter equation to solve for t and equating that to the former gives us

$$x_f = \frac{x_{\text{run}}(R_{HE} - U_{HE})}{U_{HE}R_{HE}\left(\dfrac{1}{U_f} - \dfrac{1}{R_f}\right)}$$

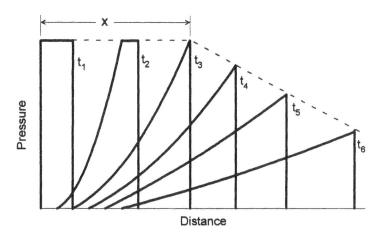

Figure 22.6.   Attenuation of the square impact shock pule with distance and time.

We know from the previous examples that

$$u = 0.898 \text{ and } U_f = 2.901 + 1.481u = 4.231 \text{ km/s}$$

$$R_f = 2.901 + 2 \times 1.481u = 5.561 \text{ km/s}$$

$$U_{HE} = 2.45 + 2.48u = 4.677 \text{ km/s}$$

$$R_{HE} = 2.45 + 2 \times 2.48u = 6.904 \text{ km/s}$$

Using these values in the equation above for x_f gives us $x_f = 2.68$ mm.

We have seen now that both critical energy fluence as well as pulse duration and magnitude are necessary conditions for shock initiation of detonation, but this is still not sufficient. One additional parameter must be taken into account, and that is the impact shock diameter.

22.2.4 Effects of Impact Shock Diameter

We saw that the rarefaction traveling axially into the rear of the shock pulse in an explosive can attenuate the peak shock pressure, and thereby cause longer than ideal run distance or even cause detonation failure. Rarefactions traveling radially into the sides or edges of the impact shock wave can do the same.

Picture a circular flyer impacting a slab of explosive. The shock wave generated at impact travels forward axially into the explosive. The edges of the shock are at ambient pressure; hence a rarefaction forms at the edge and propagates radially inward, relieving the shock pressure from the sides. This effect is shown in Figure 22.7.

In this manner, the shock is whittled down from the edges, forming a cone-shaped zone that defines the only location where the initial impact shock pressure can endure. The base angle of the cone is determined, approximately, by the ratio of the velocity of the radial rarefaction to the velocity of the axial shock. We saw earlier that the rarefaction velocity is greater than the shock velocity; therefore, this base angle must be less than 45°. If the apex distance of this cone is less than the run distance obtained from the Pop-plots, then the actual run

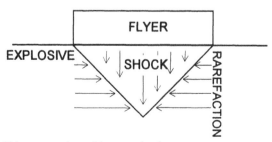

Figure 22.7.   Side attenuation of impact shock.

distance required will be greater than ideal. If the apex distance is small (close to half) compared to the ideal run distance, the explosive will fail to detonate.

This effect is seen quite dramatically in data obtained by both Moulard and Wenograd (Ref. 9). In both sets of tests reported, very long shock pulses were used (the flyers were actually cylinders). Therefore, the data are shown only for pressure, not energy fluence. Each data point represents the 50% pressure for detonation versus nondetonation for different diameter flyers. The explosive targets were Composition B, and the flyers were steel. These results are shown plotted in Figure 22.8.

Also shown in Figure 22.8 is the pop-plot run distance versus pressure. Note that the minimum diameter for which detonation could be achieved at any given pressure is approximately equal to the ideal run distance at that pressure. Good design practice is to make sure that detonation is always achieved within the constant pressure cone. Therefore, flyer diameter should always be equal to or greater than twice the run distance.

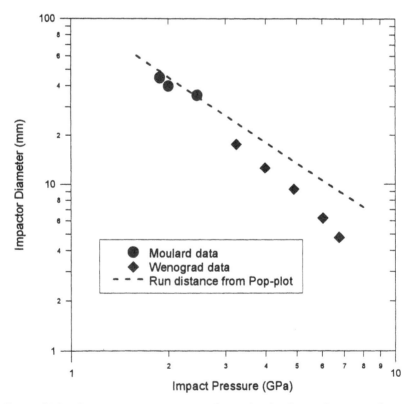

Figure 22.8. Impact pressure to cause detonation in Comp. B versus diameter of impacter.

Example 22.4   One more time, for that impact of polyethylene flyer into PBX9404: for the given impact condition, what is the minimum diameter flyer to assure detonation?

Solution   We saw in Example 22.2 that the run distance for this impact pressure is 2.2 mm. Good design practice is to ensure that the required run distance is equal to or less than the constant-pressure cone height. Since the cone is approximately 45° at the base, the diameter is about twice the height, so our flyer should be $2 \times 2.2 = 4.4$ mm diameter at the minimum.

22.3 Deflagration-to-Detonation Transition

We have now seen how deflagration can be initiated thermally, and also that detonation requires a shock for initiation. Under certain circumstances and conditions, a burning or deflagration reaction can grow into a full steady-state detonation.

If an explosive is ignited, it starts to deflagrate, and if it is confined such that the reaction product gases cannot escape, then the gas pressure in the deflagrating region builds up. Burning reaction rates are a function of pressure as well as temperature; therefore, the reaction rate increases as pressure increases. The high-pressure forces the hot gases into the surrounding material and the entire process accelerates. Pressure waves generated in the deflagrating region now can compact and compress the explosive material in the path of the waves. This causes greater confinement and hence even greater pressure buildup. The compressional waves will shock-up and, given sufficient time and distance, form the shock conditions to cause detonation.

We call this process the deflagration-to-detonation transition, or DDT. This process can occur accidentally in large explosive charges where the bulk of explosive itself provides the necessary confinement. DDT is utilized intentionally in the design of certain detonators where primary explosives cannot be used.

DDT distances for some secondary explosives under ideal confining conditions are as short as several millimeters, as for CP and PETN. DDT distances for some other explosives such as TNT are on the order of tens of centimeters.

The conditions necessary to achieve DDT depend upon such factors as confinement, particle size, particle surface area, packing density, charge diameter and length, heat transfer, and thermochemical characteristics of the particular explosive.

23

Nonelectric Initiators

The simplest initiators are the nonelectric initiators. They are simple in mechanical construction but not necessarily in terms of their mechanisms of initiation or of their chemical design or performance analysis. The nonelectric initiators can be broken down into four major categories according to the mechanism of initiation: flame or spark, friction, stab, and percussion.

23.1 Flame or Spark Initiators

These devices are usually detonators. The input end of the detonator has a charge of lead azide or other primary explosive that detonates instantly upon exposure to sparks, hot particles, or flame; and a secondary explosive as an output charge. The common nonelectric blasting cap is probably the detonator produced in highest volume (Figure 23.1).

The source of flame and sparks to this detonator come from the safety fuze that is crimped into the open end. The primary explosive charge has a lacquer seal over it to protect it from moisture. The "spit" of the safety fuze must be strong enough to break or burn through the lacquer seal in order to initiate the primary charge. The cap housing, or cup, is usually made of copper, gilding metal, or aluminum, but occasionally of extrudable steel alloys. The output charge is usually either PETN or RDX. The No. 8 blasting cap is the most popular size; it is loaded with 0.3-g lead azide, 1.2-g output high explosive, 7-mm outer diameter, and is 40-45 mm long.

Military fuze trains sometimes use flame-initiated detonators. In the military

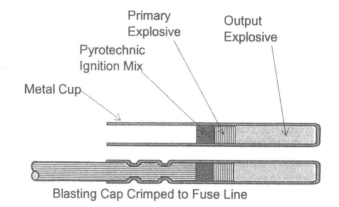

Figure 23.1. Commercial nonelectric blasting cap.

system they are called ''flash'' detonators. They are usually very small. Figure 23.2 shows two common military flash dets.

The flash dets are used in fuze trains preceded by some nondetonating burning element, such as a percussion cap or pyrotechnic delay element. They are used mainly in mechanical out-of-line safe-and-arm fuze trains and serve the purpose of both explosive relay and detonator.

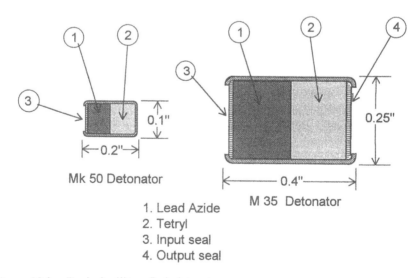

1. Lead Azide
2. Tetryl
3. Input seal
4. Output seal

Figure 23.2. Typical military flash detonators.

23.2 Friction-Initiated Devices

By far the most common ignitor in the world is the friction ignitor. Production in the United States of one type of inexpensive friction ignitor exceeds 500 billion units annually. This ignitor is called the *safety match*. The common safety, or book match, operates by bringing into intimate contact two chemcials that instantly react with each other. The potassium chlorate in the match head is mixed with a number of other ingredients, including glue. On the striker, the other reactant, red phosphorus, is also mixed with a glue. When the match is rubbed on the striker, the glue coatings on the two reactants are broken and the reactants come into intimate contact and immediately react. The friction required to break the coatings is far less than that which can produce high temperature, even with a grit. In the case of the safety match, the mechanism of initiation is really that of hypergols, mixing two chemicals that instantly react with each other. The "strike anywhere" (SAW) match, on the other hand, operates by thermal ignition: friction raises the temperature of the fine grits that are included. Then these grits, as hotspots, cause the initiation of burning of the reactants. The major reactants of the tip of the SAW or wooden kitchen match are potassium chlorate and phosphorus sesquisulfide; the grit is powdered glass. These same materials are also used in string and tab pull friction ignitors used in some military and commercial systems as well as in the striker assemblies on *fusee* flares.

23.3 Stab Initiators

This type of initiator is among the most mechanically sensitive of all the non-electric types. A typical stab detonator is shown in Figure 23.3. Most stab ini-

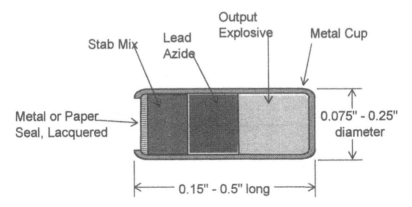

Figure 23.3. Stab detonator.

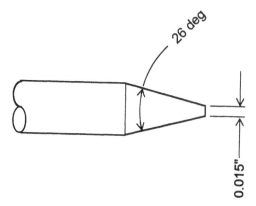

Figure 23.4. Typical stab initiator firing pin.

tiators are fairly similar in size and sensitivity; they use the same standard firing pin (with a few rare exceptions) shown in Figure 23.4.

During activation, the firing pin pierces the closure disc and penetrates the priming mix. This causes heating of the mix by compression of the mix in front of the pin, and friction on the mix by the conical sides of the pin. For the same primer mix at the same load density, the minimum firing energy increases approximately linearly with the thickness of the input closure disc. For the same closure disc thickness, the input firing energy increases with decreasing loading density of the primer mix (Table 23.1).

The firing energy for stab initiators is determined by a drop-weight test. A given weight is dropped from various heights onto a centered firing pin that then pierces the initiator. Many tests are conducted at various heights, and the data are reduced to form a chart of firing energy (height times weight) versus probability of function. Most stab initiators function at high reliability at input energy levels between 0.5 and 5 in. oz (0.0035 to 0.035 joules).

Table 23.1   Effect of Loading Pressure on Stab Initiator Sensitivity[a]

Loading Pressure (psi)	50% Function Probability Drop Test Height (in.)
15,000	1.3
25,000	0.91
40,000	0.77
60,000	0.68
80,000	0.57

[a] NOL primer mix in MK102 cups, 2-oz. ball drop weight. Ref. 10.

Stab initiators are used in military systems such as small mechanical fuses where very little mechanical energy is available because of weight limitations and the small dimensions of the springs. Stab detonators use the same ignition mix, or priming composition, as many percussion primers do. Some of these are listed Table 23.2.

The energy required to fire a given stab detonator is constant at higher firing-pin velocities and increases as firing-pin velocity decreases below a certain level. The reason for this behavior, which is common to all types of initiators, will be dealt with in a later chapter. Figure 23.5 shows this behavior characteristic for an M55, a common military stab detonator.

23.4 Percussion Initiators

Percussion primers are different from stab-type initiators in two ways; first, they are not punctured by the firing pin, and second, they are not made as integral detonators in themselves. The two major mechanical types of percussion primers are the rimfire, Figure 23.6, and the centerfire, Figure 23.7.

In the rimfire primer, an integral part of the small-calibre cartridge, the ignition mix, or primer composition is slurried into the cartridge base, crimped, then oven dried at relatively low temperatures (<110°F). The base of the cartridge serves as the striking surface, and the top of the crimp, supported by the weapon breech lip, serves as the anvil.

Table 23.2 Common Priming Compositions

Ingredients[a]	FA-956	FA-982	PA-100	PA-101	NOL-60	NOL-130	M31 Igniter Mix
Lead styphnate, basic	—	—	—	53	60	40	—
Lead styphnate, normal	37	36	—	—	—	—	—
Barium nitrate	32	22		22	25	20	—
Lead azide	—	—	5	—	—	20	—
Tetracene	4	12	—	5	5	5	—
Lead dioxide	—	9	—	—	—	—	—
Calcium silicide	—	—	—	—	—	—	—
Aluminum powder	7	—	—	10	—	—	—
Antimony sulfide	15	7	17	10	10	15	—
PETN	5	5	—	—	—	—	—
Zirconium	—	9	—	—	—	—	—
Potassium chlorate	—	—	53	—	—	—	55
Lead thiocyanate	—	—	25	—	—	—	45

Reference 10.

[a] FA, Frankford Arsenal; PA, Picatinny Arsenal; NOL, Naval Ordnance Laboratory.

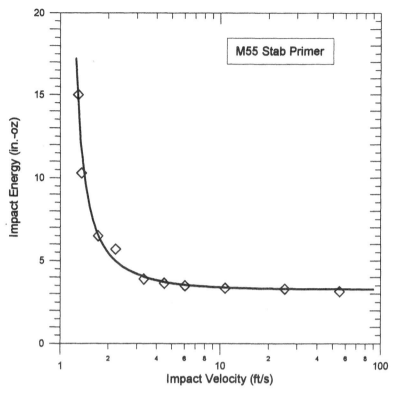

Figure 23.5. Energy-velocity relationship for firing M55 stab primer (data from Ref. 10).

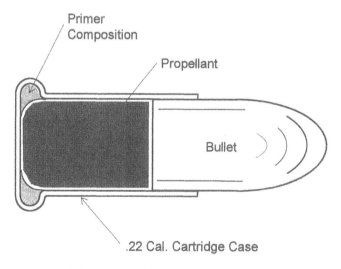

Figure 23.6. Typical rim-fired cartridge.

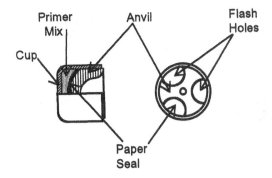

Figure 23.7. Center fire percussion primer.

The center-fire percussion primer is loaded with slurry mix, as is the rimfire, or dry pressed by some manufacturers. Certain center-fire percussion primers are sealed into another, inverted, cup during fabrication. These types are called shotgun or battery cup primers, and are shown in Figure 23.8.

Primer compositions fall into two main categories: corrosive mix types based on potassium chlorate/lead thiocyanate; and noncorrosive mix types based on lead styphnate/tetracene. Some of these mixes were listed in the previously in Table 23.2. High-temperature-resistant mixes (designated by a ''G'' following the mix name or number) are a varient of the corrosive mix types. These are based on potassium chlorate /antimony trisulfide/calcium silicide.

Firing pins for percussion primers have hemispherical tips with radii from 0.02 to 0.05 in., depending upon the particular primer with which they are to be

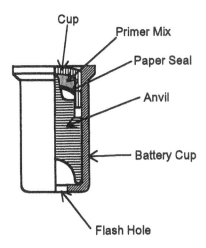

Figure 23.8. Battery cup percussion primer.

used. The pin does not penetrate the cup, and so the primer continues to pressure seal the cartridge after firing. This is important in their uses in small arms ammunition as well as in nonvented delay trains.

The firing energy requirements of percussion primers are somewhat higher than those for stab initiators. Firing energies range from 18 to 60 in. oz for the more common types. It is common for percussion primer manufacturers to quote the recommended, or all-fire, energy for a primer as the value of the mean plus five standard deviations, along with the drop-ball weight with which their acceptance tests were run. Like the stab initiators, the percussion primer exhibits an energy/firing-pin velocity curve in which the firing energy increases at lower velocities. This is shown for three primers in Figure 23.9.

When percussion primers are installed in the primer cavity of the next assembly, they are slightly compressed, or squashed down. This "recompression" is critical in their installation and is specified by each manufacturer. If not recompressed properly, the primers will not fire at the specified energies; they may also fail to seal the cartridge after firing.

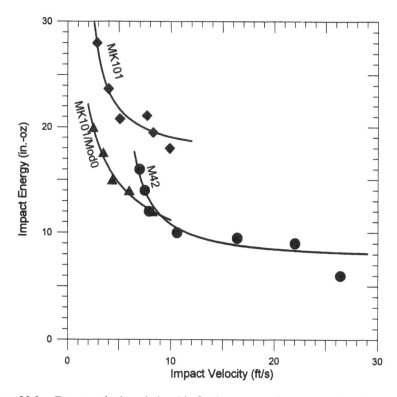

Figure 23.9. Energy-velocity relationship for three percussion primers (data from Refs. 10, 11, and 12).

The physical size of percussion primers is quite small. Size ranges and firing energy requirements, along with typical uses for a number of common percussion primers, are given in Table 23.3.

23.5 Energy-Power Relationship

Referring back to the Frank-Kimenetskii (*FK*) equation in the previous chapter, we had a relationship where the rate of energy produced by the decomposition reaction was equal to the rate at which the explosive material was heated plus the rate at which heat was lost to the surroundings. This same relationship also holds when the explosive material is heated from an external source, not by slow self-heating due to decomposition. The rate at which work is done, or energy is transferred, is called power:

$$P = dE/dt \tag{23.1}$$

where P is power, E is energy, and t is time.

Similar to *FK*, the power balance can be expressed as

$$(dE/dt) = P = \rho C(dT/dt) + \lambda T \tag{23.2}$$

where ρ is density, C is heat capacity, λ is heat transfer, and T is temperature.

If we recall the conditions that defined the critical temperature: the heat produced in the explosive due to the reaction at the rate corresponding to T_C, was transferred away at the exact rate so the temperature could not rise above T_C in an infinite time. If the external temperature was the least bit higher than T_C, then the reaction rate would increase, increasing the internal temperature to runaway reaction or explosion. Let us now consider a temperature of the explosive analogous to T_C but is, as we saw in the work of Afanas'ev and Bobolev, for the diameter of a hot spot, T_{ign}. If we could get any tiny spot in the explosive up to that temperature, then we would get the runaway reaction or explosion. In this case, if we integrate Eq. (23.2), we would get the energy required for ignition,

$$E = \rho C(T_{ign} - T_0) + \lambda \, Tt \tag{23.3}$$

where T_0 is the ambient temperature.

Now let us envision a power level so low that the steady-state heat transfer occurs at some temperature just below T_{ign}, and no matter how long we apply power, we transfer it away just fast enough that T cannot rise any further. In this case, we could not initiate the explosive. We will call the power at this condition P_0. Since we are at steady-state heat transfer, the temperature is constant and dT/dt in Eq. (23.2) must equal zero; therefore $P_0 = \lambda T$.

If the input power is any higher than this, then the explosive must eventually ignite. The higher the input power, the faster we will reach ignition.

Now let us consider the other end of this spectrum. Picture a condition at

Table 23.3   Common Centerfire Percussion Primers

Primer Name	Primer Mix	Dia. (in.)	Length h (in.)	Firing Energy (in. oz)	Drop Ball Wt. (oz)	Firing Pin Rad. (in.)	Typical Uses
108	282 C	0.1755	0.119	20	2	0.05	Pistol, revolver ammo
116	864 B	0.1755	0.119	36	2	0.05	Small rifle ammo
116-M	282 C	0.1755	0.119	36	2	0.05	.30 cal. Military Carbine ammo
116-D	257 W	0.1755	0.125	48	4	0.05	5.56 mm Military ammo
M42-G	530 G	0.1752	0.115	20	2	0.02	Fuse, delay powder trains
M42-C2	793	0.1752	0.115	26	2	0.02	"
M42-C1	PA101	0.1752	0.115	26	2	0.02	"
M29	FA70	0.2043	0.122	18	2	0.045	Ammunition fuses
M29-A1	257 W	0.2043	0.122	18	2	0.045	"
M35	706 A	0.2098	0.122	32	2	0.05	Mortar ignition cartridges
111	864 B	0.2110	0.119	28	2	0.05	Pistol, revolver ammo
111-M	295	0.2110	0.119	28	2	0.05	Military pistol ammo
120	864 B	0.2118	0.128	60	4	0.05	Rifle ammo
120-M	257 W	0.2118	0.128	60	4	0.05	Military rifle ammo
8-1/2	FA70	0.2118	0.129	60	4	0.05	Military ammo
3*	793	0.217	0.311	30	2	0.05	M1 firing device, demolition eq.
EX2926A*	793	0.217	0.311	30	2	0.05	Delay det/C12 riot hand grenade
M27*	257 W	0.217	0.311	24	2	0.05	Military signal flares, grenades
5*	548	0.227	0.335	26	2	0.05	Fuse powder train igniter
MLK-119*	FA70	0.227	0.335	36	2	0.05	"
M39A1C*	548	0.227	0.335	30	2	0.05	"
209*	955	0.2403	0.304	24	2	0.05	Comm. shot shells, gen. purpose
209-B*	772	0.2403	0.304	30	2	0.05	Military incendiary bombs
209*	981	0.2403	0.304	30	2	0.05	M2A2 ignition cartridge

Reference 13.

Notes: Firing energy is MEAN + 5 SIGMA
2. * designates battery cup type primer
3. Lead styphnate/tetracene, noncorrosive-type mixes:
  282C, 864B, 257W, 955, 981, 295, PA101, FA956, NOL60, NOL130.
4. Potassium chlorate/lead sulfocyanate type mixes:
  530G, 548, 760A, 772, 793, PA100, FA70, M31(ign. mix)
5. Mixes prefixed PA, from Picatinny Arsenal
              FA, from Frankford Arsenal
              NOL, from Naval Ordnance Laboratory
              M, from Rockford Arsenal
nonprefixed numbered mixes in table above from Olin Powder Co.

very high power. T_{ign} would be reached so quickly there is virtually no time available to transfer heat to the surroundings. In this case essentially all the input energy would go into heating the explosive to T_{ign}, at which point the runaway reaction starts; therefore, no additional external energy is required; so we will cut off the power. This condition represents the minimum energy required for ignition, E_0. Here, since time is so short, $t \to 0$, therefore, λTt [from Eq. (23.3)] $\to 0$, and $E_0 = \rho C(T_{ign} - T_0)$.

If we have a graph of input energy required for ignition versus input power, these two conditions would represent limits for both quantities, as shown in Figure 23.10. Regardless of type, all initiators behave in this manner. The required energy for ignition versus input power for all initiators is a hyperbolic relationship where E_0 and P_0 are the asymptotes. This is shown ideally in Figure 23.11.

A more realistic plot of energy versus power is shown in Figure 23.12, where the ignition characteristics are given in terms of mean firing energy and reliability.

Prudent engineering practice is to design firing systems to operate at power levels above those where the high-reliability curve flattens into or approaches a

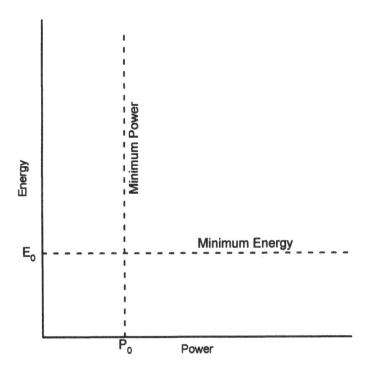

Figure 23.10. Energy versus power limits.

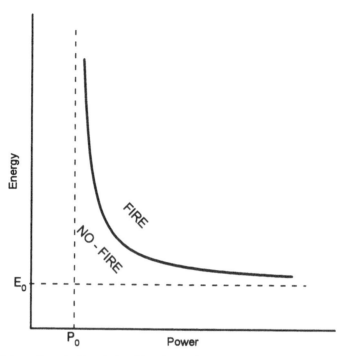

Figure 23.11. Hyperbolic relationship of P and E.

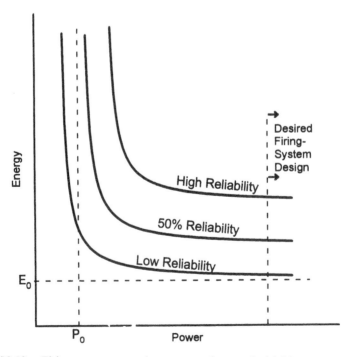

Figure 23.12. Firing energy versus input power for a typical initiator.

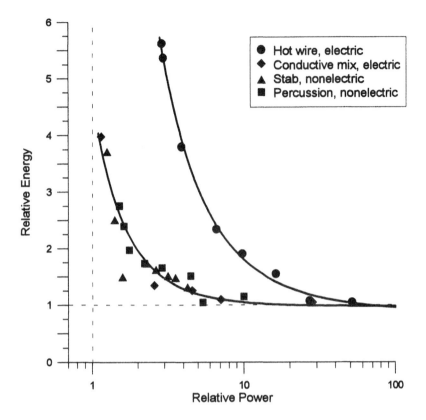

Figure 23.13. Reduced energy versus reduced power for different types of initiators (Refs. 10 and 12).

constant-energy characteristic. The safety considerations and procedures that must be established for a particular initiator are based on the minimum power levels for the low-reliability curve. This condition is often called the "no-fire" power.

It is interesting that some types of initiators are so similar in this hyperbolic relationship, that when plotted as relative (or reduced) energy and power (E/E_0 and P/P_0), they fall on the same line. This is shown in Figure 23.13. On this plot the firing pin velocity, not the firing pin power, has been plotted for stab and percussion primers. This is because the actual input power in these devices is proportional to the rate of crushing of the explosive, which in turn is proportional to the input firing-pin velocity. It is also interesting to speculate as to why hot-wire bridge initiators do not fall on this same curve. Is it possible that the volume of the bridgewire forces or determines an "artificial" hot spot size that is much larger than the "natural" or "free" hot spots that are formed in the other devices? In the next chapter, we will see that is the case.

24

Hot-Wire Initiators

The vast majority of electrical initiators are of the hot-wire type, in which a small wire is imbedded in, or in contact with, an ignition material, either a primary explosive or a pyrotechnic. Electrical current heats the wire, which in turn heats the ignition material to its ignition temperature and thus starts a burning reaction. This reaction is then propagated to the next element in the device, either another pyrotechnic (in the case of an ignitor) or an explosive (in the case of a detonator). The general arrangement is shown in Figure 24.1

All hot-wire initiators are similar in that they have a header, a bridgewire, and some kind of ignition charge. They differ considerably in size, shape, and construction details. Among the simplest are dipped electric matches.

24.1 Electric Matches

These initiators were first developed in the early 1900s in Germany. They are inexpensive because they are designed to be manufactured in simple, high-volume processes. The header of the dipped electric match (or fuse head) is made of cardboard, covered on both sides with metal foil. The bridge is soldered across the foils as shown in Figure 24.2.

The bridged fuseheads are dipped into an ignition-mix slurry or paste, then air dried, and dipped into an output-mix paste and dried again. Some matches are then coated by a third dip into nitrocellulose lacquer to give the bulb a

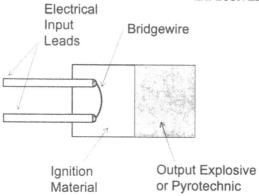

Figure 24.1.   Generic hot-wire initiator.

strengthening glaze and also to aid in moisture resistance. See Figure 24.3. The combs are then sawed apart, forming individual fuse heads, or matches; then lead wires are soldered on as electrical leads or leg wires. The finished match is shown in Figure 24.4.

Electric matches are used today in such diverse applications as thermal batteries, small rocket motors, and blasting cap ignition elements.

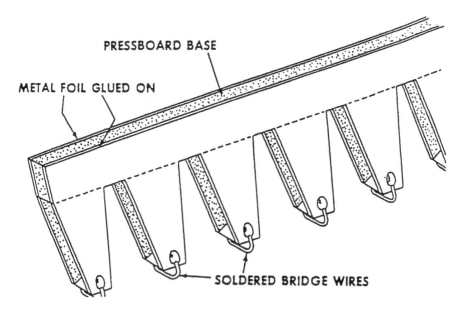

Figure 24.2.   Fuse head comb, bridged (w/permission from Ref. 14).

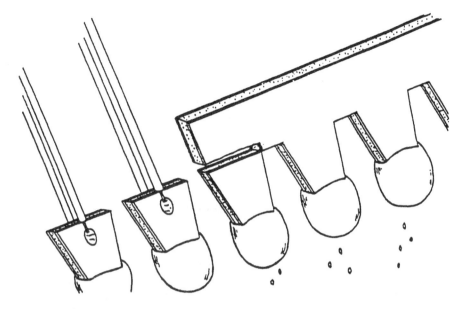

Figure 24.3. Loaded fuseheads (w/permission from Ref. 14).

24.2 Electric Blasting Caps

The most common, and most voluminously produced, hot-wire detonator is the electric blasting cap. The history of these devices seemed to start in England in 1745, when a certain Dr. Watson of the Royal Society of England demonstrated that black powder could be ignited by the spark discharge from a Leyden Jar.

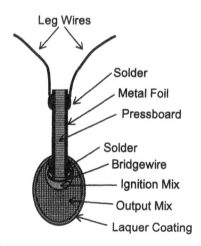

Figure 24.4. Finished fusehead.

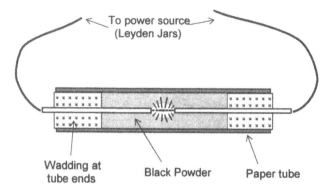

Figure 24.5. Benjamin Franklin black powder spark ignitor.

Five years later, in the British American Colony of Philadelphia, Benjamin Franklin (the printer with the funny eyeglasses and a penchant for flying kites in thunderstorms) made an electric black-powder ignitor based on this same principle, but prepackaged in a paper tube (Figure 24.5).

In 1822, Dr. Robert Hare developed the first hot-wire ignitor for black powder, shown in Figure 24.6. The ignition mix in this device is believed to be potassium chlorate/metallic arsenic/sulfur. The bridgewire was a single strand of fine wire separated out from a multistrand cable.

By 1867, Alfred Nobel was on the market with a nonelectric blasting cap filled with mercury fulminate, for use with his new dynamite a replacement for black powder. The fulminate-filled cap became a spark actuated or ''high-tension'' blasting cap under the development of American inventor, H. Julius Smith in 1868 (Figure 24.7).

A few years later, in 1875, both Smith and another inventor, Perry ''Pell''

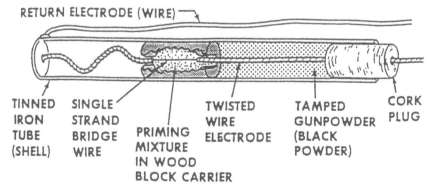

Figure 24.6. Dr. Robert Hare's hot-wire black powder ignitor (w/permission from Ref. 14).

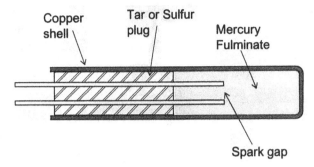

Figure 24.7. H. Julius Smith high-tension blasting cap.

Gardiner, independently introduced an almost identical hot-wire blasting cap (Figure 24.8). This blasting cap, which could operate at low voltages, obtainable from new-fangled storage batteries, and over long wire, opened up the era of big blasting.

The Smith-Gardiner blasting cap is essentially identical to electric blasting caps used today. Around the turn of the century, the mercury fulminate was "goosed-up" in output by the addition of 10 to 20% potassium chlorate. This was later (beginning around 1917) replaced by lead azide as the ignition or "priming" charge. About this same time, the Germans started using electric matches as the ignition element in blasting caps. This ignitor became fairly standard throughout Europe. In the United States, solid-pack-type ignition elements were used (until the 1980s). In these, the raised bridgewire was imbedded in the ignition mix pressing, just as in the early Smith-Gardiner cap. During World War II, pyrotechnic mixes became popular as the ignition mix, and the lead azide was loaded further down the cap as the "priming" or "boosting" charge. The last lead azide, on-the-wire caps disappeared from the American

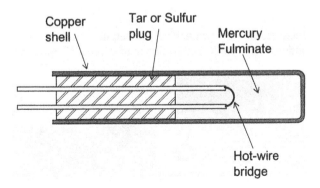

Figure 24.8. Smith-Gardiner hot-wire cap.

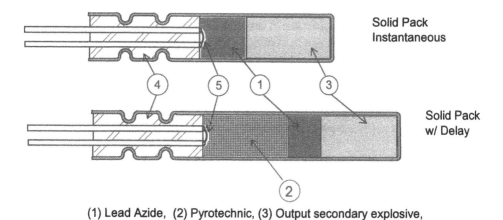

(1) Lead Azide, (2) Pyrotechnic, (3) Output secondary explosive,
(4) Insulating header plug, (5) Bridgewire

Figure 24.9. Older and current solid pack electric blasting caps.

market in the early 1950s. Figure 24.9 shows the early LA ignition and pyro-ignition solid-pack-type caps.

During the mid-1950s, Atlas Powder Co. was bought by ICI International and became at that time, by adopting their European standards, the only American manufacturer of the match initiated electric blasting caps. These caps (a delay-version is shown in Figure 24.10) are now pretty much the standard throughout the American as well as the world market.

24.3 Short Lead and Connectorized Initiators

Since the early part of World War I, a multitude of hot-wire ignitors and deto-nators have been developed for specialized uses by the military, aerospace, and nuclear weapons industries. In principle, these devices are like electric blasting

(1) Lead Azide, (2) Pyrotechnic, (3) Output
secondary explosive, (4) Insulating header plug, (5) Electric match

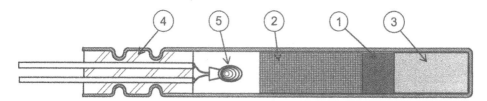

Figure 24.10. Match (or fusehead) initiated electric blasting cap.

caps; in appearance they are as different as horse chestnuts are from chestnut horses. What all these other initiators have in common is that they are all solid-pack types. As with blasting caps, a number of them have raised bridges. A typical example is the M36A1 detonator shown in Figure 24.11.

In a raised bridge device, the explosive or initiating mix completely surrounds the bridgewire. Most raised bridges are made of fine, round cross-section wires, but some are in the form of ribbons. The bridges in these devices are either soldered or welded to the conductor pins.

The other class of bridges is called flush bridges; they lie along, and are flush to, the surface of the header. Flush bridges are in the form of wires, ribbons, foils, and deposited films. They are connected to the flush pins by soldering and welding, and by the deposition process for deposited films. Figure 24.12 shows several typical arrangements of flush bridges.

Bridge materials are usually metals, either pure or alloyed. Common pure metals used for bridges include gold, platinum, tungsten, and chromium. Typical alloys include various nichrome types, platinum/iridium, gold/iridium, and gold/rhodium, and platinum/rhodium. Some initiators are fabricated as integrated connector devices; that is, instead of lead wires for the electrical inputs, the detonator or ignitor is built into an electric connector. An example of this is shown in Figure 24.13.

Integrated-type initiator's bodies are usually an alloy of steel. The header's insulating material can be anything from plastic to ceramic to glass.

24.4 Energy-Power Relationship

We saw earlier that all initiators behave in a hyperbolic manner when their required firing energy is plotted against the input firing power. In the case of

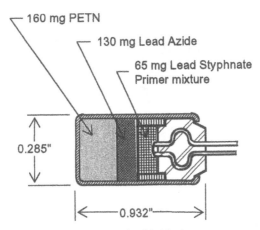

Figure 24.11. M36A1 detonator (note raised bridge).

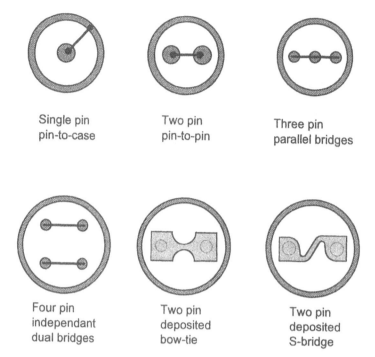

Figure 24.12.   Typical bridges and bridge configurations of flush bridge initiators.

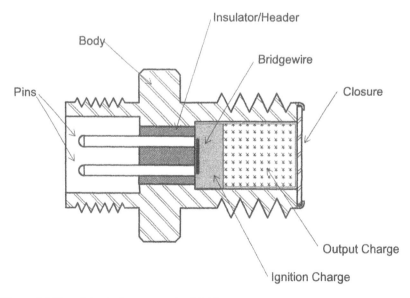

Figure 24.13.   A typical connectorized initiator.

hot-wire initiators, the input power is the product of input voltage times input current, $P = iV$, where i is the current, and V the voltage. Since voltage equals current times resistance (Ohm's Law), we can say input power equals i^2R, where R is the resistance. Then the power balance equation for hot-wire initiators becomes

$$\rho C_p \frac{dT}{dt} + \lambda T = i^2R \qquad (24.1)$$

For convenience, let us lump ρC_p together and in this case make it equal C, the total heat capacity; substituting this into Eq. (24.1) yields

$$C \frac{dT}{dt} + \lambda T = i^2R \qquad (24.2)$$

The term C now includes the mass of the bridgewire, and λ refers to the rate of heat transfer away from the bridgewire. This heat transfer in a raised bridge system is toward the pins or lands, axially, and to the ignition mix, radially. For a flush bridge, the heat transfer is from the bridge to the pins, or lands, toward the ignition material radially on one side, and to the header radially on the other side.

The integral of Eq. (24.2) is

$$C(T_{ign} - T_0) + \lambda Tt = i^2Rt \qquad (24.3)$$

and the firing energy is, of course,

$$E = i^2Rt \qquad (24.4)$$

At the minimum energy condition, high input power, where $t \to 0$, $\lambda Tt \to 0$, and

$$E = i^2Rt = C(T_{ign} - T_0) \qquad (24.5)$$

At high power, the minimum energy requirement is essentially the energy to heat the mass of the bridge to the ignition temperature.

If the bridge volume mass were changed, then one would expect the minimum firing energy to change proportionately. In Figure 24.14 we see that this is indeed the case. The data in Figure 24.14 are for lead azide buttered over a raised bridgewire. The units for bridge volume used in this figure are cylindrical mils. One cylindrical mil is the volume of a cylinder of 0.001-in. diameter by 0.001 in. long.

And, as before, at the heat transfer steady-state condition with minimum power, at constant bridge temperature the power equals the heat transfer,

$$P_0 = i^2R = \lambda T \qquad (24.6)$$

The relationships in Eqs. (24.5) and (24.6) are the limits, where the initiation behavior approaches the minimum power and minimum energy asymptotes. The

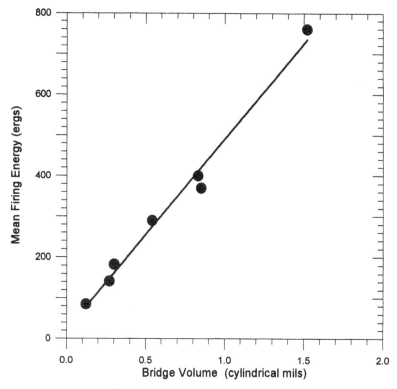

Figure 24.14. Mean minimum firing energy as a function of bridgewire volume (Ref. 10).

full behavior of the relationship of power to energy between the asymptotes is approximated by the Rosenthal model (Refs. 15, 16, and 17), which is expressed as

$$E = i^2 R_0 t \left(1 + \alpha K_2 + \frac{\alpha K_2}{K_1 t} (e^{-K_1 t} - 1) \right)$$

$$K_1 = \frac{kA - \alpha i^2 R_0}{m C_p}, \qquad K_2 = \frac{i^2 R_0}{kA - \alpha i^2 R_0}$$

$$t = \frac{1}{K_1} \ln\left(\frac{K_2}{K_2 - \Theta} \right), \qquad \Theta = K_2 (1 - e^{-K_1 t}) = T - T_0$$

where A is the bridgewire surface area; α, the bridgewire temperature coefficient of resistivity; m, the bridgewire mass; k, the effective explosive material thermal conductivity at the bridgewire interface; R_0, the bridgewire resistance; T_0, the bridgewire ambient temperature; T, the bridgewire temperature; Θ, the bridgewire temperature rise above ambient; i, the bridgewire current; t, the time; E, the energy; and C_p, the bridgewire specific heat. Using the physical character-

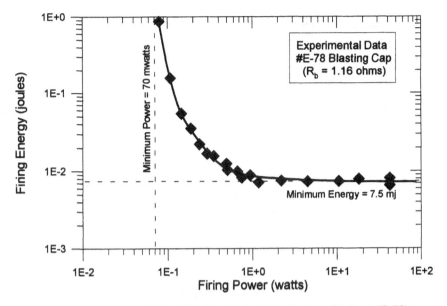

Figure 24.15. Energy-power relationship for a typical blasting cap (DuPont #E-78).

istics of the initiator header-bridge-explosive system, this model closely estimates the full hyperbolic relationship. Experimentally derived data of this relationship are shown in Figure 24.15 for a DuPont No. E78 blasting cap.

As stated earlier, it is desirable to design the firing system to fire the initiator in the high-power region, where the energy is at minimum. This gives us several advantages, as we will soon see.

24.5 Firing at Minimum Energy

At the minimum energy condition losses due to heat transfer are negligible, and differences that affect transfer, such as header surface irregularities, small density variation, etc., are also negligible. Since the required energy (at the minimum energy condition) is a constant, the equations that relate detonator firing to electric circuit parameters are tractable, engineering-wise. At low power levels, where firing energy is a system variable, the equations relating energy, heat transfer, firing time, and circuit parameters are nonlinear, and data for the various required constants are hard to come by.

24.5.1 Function Times

The total function time, t_f, of an electric initiator is defined as the time from the beginning of the input electrical signal until the "breakout" time at the business end of the device. This function time is made up of two parts, the time it takes to deliver the firing energy and start a burning reaction, t_e, and the transit time, t_t, which is the time it takes from start of burning until breakout at the end of the device.

$$t_f = t_e + t_t \tag{24.7}$$

The transit time is independent of the electrical firing characteristics, strictly a function of initiator design and construction; it depends upon the burning or detonation rates of the explosive materials as loaded, and on the distance, or length of the loaded explosive column.

The time it takes to deliver the firing energy, t_e, is a function of both initiator design and electrical circuit parameters. The initiator characteristics that affect this time are: the minimum firing energy, the bridge resistance, the particular ignition mix and its contact intimacy with the bridge, and the bridge material. The circuit parameters that affect the commit time are the source voltage and source impedance, pulse shape, circuit impedances, and, of course, circuit design and firing lead lengths.

For ease of explanation, let us confine our discussions here to constant-current or constant-voltage sources. A schematic of a simple firing circuit is shown in Figure 24.16.

The voltage source has some internal impedance that limits the current it can supply. For AC/DC power supplies, the current limitations are crucial, and over-loading, or attempting to draw current above the supply rating, can result in smoking it. In Figure 24.16, the term R_c is the general combined circuit resistance seen by the initiator. This includes the firing-line wire resistance.

Long lines add considerable resistance to the circuit. The resistance of the initiator leg wires, in the case of a blasting cap typically 25 to 30 feel long, is part of the circuit resistance. The term R_b is the bridgewire resistance of the initiator. So, considering all the above, let us correct the circuit in Figure 24.16 so that R_c includes all these elements. R_c is the source resistance, plus circuit resistance, plus firing lead resistance, plus legwire resistance.

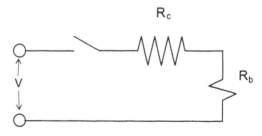

Figure 24.16.   A schematic of a simple firing circuit.

The current that will pass through the initiator is then

$$i = V/(R_c + R_b) \tag{24.8}$$

and the firing energy is

$$E = i^2 R_b t_e \tag{24.9}$$

Typical values of E and R_b for several types of hot-wire detonators are given in Table 24.1.

24.5.2 Series and Parallel Firing Considerations

If we fire more than one initiator at the same time from the same circuit, then it is imperative that we know how these initiators behave dynamically in order to assure that each does not affect the other adversely. Different types of initiators behave very differently, even at high power levels. Detonators that have lead azide, or other primary explosives, pressed directly onto the bridge, will detonate essentially instantly once the critical ignition temperature, T_{ign}, is reached. The shock developed by the detonation will destroy the bridge, electrically opening it. In detonators with a pyrotechnic over the bridge, detonation does not occur until the burning reaction reaches the primary explosive (or priming) charge. At that time, the shock travels back to the header through the residue of the pyrotechnic and may or may not destroy the bridge. This depends upon the particular design of the detonator. In nondetonating ignitors, the bridge is usually not dramatically disturbed by the burning pyrotechnic.

In those cases where the bridge survives intact after reaction has started, after it has reached T_{ign}, the current will continue to flow through the bridge and firing circuit. As the current continues to flow, the bridge continues to heat up until it reaches the steady-state temperature for that current and power level, or reaches the melting point of the bridge material. In the latter case, the bridge will open at that time.

Relative to the time that temperature T_{ign} is achieved (at t_e, the time it takes to deliver the firing energy), different detonators will electrically open at differ-

Table 24.1 Typical Values of E and R_b for the Major Classes of Hot-wire Initiators

Initiator Type	Bridge Material	Range of R_b (Ω)	Range of E (mJ)
Old military types	Carbon	1K–10K	0.01–0.1
"	Tungsten	2–10	0.1–1
Standard basting caps	Varies	1.5–5	5–10
"1-amp/1-watt"	Ni/Cr alloys	1	30–50

ent times depending upon their design and upon the current or power level. It is this characteristic, which we can call the bridge-opening time lag, that determines whether we can fire multiple detonators in series or in parallel.

24.5.2.1 Series Firing

No two detonators can be fabricated exactly alike. There is a tolerance on the minimum firing energy as well as on the bridge resistance. These small differences, when related to the time it takes to deliver the firing energy, become very significant, best shown by an example.

Suppose we have a detonator that has lead azide pressed directly on the bridge. The detonator has the following firing characteristics:

$E = 40$ mJ $\pm$ 10% (4 mJ)

$R_b = 1.0$ Ω $\pm$ 10% (0.1 Ω)

(These, by the way, are fairly tight specifications). If we chose two detonators from this lot representing the one with the highest E and lowest R_b, and the other with the lowest E and the highest R_b, we would have the fastest and slowest detonators in the lot.

We can fire them at a relatively high-power level, say 100 watts. Since $P = i^2R$, and R_b is nominally 1 ohm, this means the current is around 10 amps. The time it takes to deliver the firing energy for the first detonator is

$t_e = E/i^2R_b = (0.044$ J$)/(10^2$ A $\times$ 0.9 $\Omega) = 0.000489$ s (489 μs)

The other detonator, fired at 10 A also, is

$t_e = E/i^2R_b = (0.036$ J$)/(10^2$ A $\times$ 1.1 $\Omega) = 0.000327$ s (327 μs)

So we see that from the same lot, within relatively tight specifications and at relatively high power, we can get a difference in the time it takes to deliver the firing energy (for our example detonators) as high as 162 μs. For this type of detonator, lead azide on wire, it would cause a failure to fire of the slower detonator if they were fired in series. When this type of detonator fires, the shock from the detonation destroys the bridge within a few microseconds. By this example we see that we can identify one specific type of detonator that cannot be fired in series with another.

It is common practice in blasting operations to fire as many as 50 blasting caps in series. This is made possible because the blasting cap, which has *pyrotechnic* over the bridge, will not open electrically until the priming charge detonates. But there are limits here also. At high power levels where all the time jitter (difference in time from slowest to fastest detonator) is less than the minimum burning delay time (and this is subject to jitter also) then these 50 caps can be reliably initiated in series. However, at lower power levels, the firing times and time jitter get larger. When the jitter of the time it takes to deliver the firing energy exceeds the minimum burning delay time (delay time tolerance), then failure will occur. Typical tolerances for commercial blasting caps are

$$E: \pm 20\%$$

$$R_b: \pm 25\%$$

t_t (transit or delay time): $\pm 15\%$

Tolerances for high-quality hot-wire ordnance items for the military, aerospace, and nuclear weapons industries are better, but not that much better. Typically tolerances are

$E: \pm 10\%$

$R_b: \pm 10\%$, and pressed delay column rates:

$\pm 10\%$ (over temperature extremes)

Reliable series firings can be achieved if these considerations are taken into account. The advantages of series firings are that circuit current is lower than in parallel firing. This means that switches, voltage dividers, reference diodes, etc., can all be at lower current ratings, hence, smaller and probably less expensive. Also electrical cable and other conductors can be smaller, and usually hook-up is simpler and fewer connectors can be used.

24.5.2.2 Parallel Firing

All detonators can be fired reliably in parallel as long as sufficient firing current can be provided. This is not a trivial statement. When many initiators are to be fired in parallel, remember that the current provided must be at least the individual firing current times the number of devices.

Some reliability problems in parallel circuits stem not from bridgewire opening but from very low bridge resistance after firing. Certain types of ignitors with thermite-type ignition mixes will form highly conductive metallic burning residue. This residue can randomly short-circuit the bridge to resistances as low as several mΩ. This is more true of some mild mixes, such as Al/CuO, than of the more violent types, such as Al/KClO$_4$. When a bridge is short-circuited in this manner, it suddenly draws significantly higher current. If this occurred in the faster unit in a parallel hook-up, it would decrease the current available to the other bridges and possibly cause an ignition failure in the slower ones.

24.6 Safety Considerations in Design

By design, initiators are power and energy amplifiers. They take in a small power level or small amount of energy and produce large energy outputs at tremendously high power. Because they are designed to be sensitive to small inputs, they must be protected from inadvertent ones. Beside the obvious problems of protection from heat, flame, and impact, much attention has been given to protection from inadvertent or accidental electrical input. The two areas of concern with accidental electrical input are from coupling electrical current into the bridgewire and firing circuit, and from high potential electrostatic charges between the bridge circuit and the initiator case.

24.6.1 Protection from Coupled Current

Small, yet significant, electric current can be coupled into the bridge circuit in a number of ways. Assuming this current is not due to faulty circuit design, we can focus on some of the others. The most prevalent is from the behavior of detonator leads and circuit loops to act as antennas and pick up various signals just as a radio might. Microwave radiation from radar transmitting antennas is probably the worst case, especially aboard naval vessels, where ordnance is stored and at the ready in close proximity to large radar transmitting arrays. Portable radio transmitters are often inadvertently operated near initiators, and these may couple into the initiator leads also. In any case, because of the inability to control the radiation environment, it is necessary to make the initiators as immune as possible to the relatively low electrical currents. As you recall at low current, hence, low power input, the heat transfer from the bridge takes up a large portion of the input energy. The larger we can make λ, the heat transfer coefficient, the higher the minimum power level for firing will be. Increasing λ will not affect the minimum firing energy as long as we maintain the same bridge volume or mass. We accomplish this by giving the bridge the largest possible contact area with the header. Modern high "no-fire" initiators achieve this by utilizing deposited film bridges. In addition to increasing the heat-transfer area, attempts are made at increasing the thermal conductivity of the header material. The present specifications for minimum power or "no-fire" level acceptable to the military and aerospace and nuclear weapons industries are 0.01% fire probability at 95% confidence at 1 A and 1 W input current (this specification therefore implies that bridge resistance must be 1 Ω).

24.6.2 Protection from Electrostatic Charge

High electrostatic potential between the bridge circuit and the initiator case can cause an arc to jump from one to the other. This electrical arc can ignite the ignition material through which it is passing. Designs to prevent or minimize this phenomenon include loading the initiation material into an insulating cup inside the initiator. This forces an arc to have a longer path and therefore a higher breakdown potential. Also, ignition materials are developed that have higher tolerance to spark or arc ignition. On the outside of the initiator or built into the header, some designs incorporate a parallel breakdown path that could arc over at lower potential than any path through the ignition mix. Bleeder resistors, both external as well as built into the header, are incorporated to prevent slow static charge buildup.

There are no universal specifications for resistance to electrostatic hazards for hot-wire initiators. Each agency or service has somewhat different operating environments and hence different specifications. The more common specifications are those for protection from electrostatic discharge from personnel. This specification states that the initiator will not fire from the discharge of a 600 pF capacitor, charged to 20,000 V, through a 500 Ω resistor in series with the initiator such that the arc path is from the shorted bridge to the case. Other specifications are similar, only changing the values of the capacitor, resistor, and charge voltages.

24.7 Quality Control Testing

Aside from the expected physical inspections of piece parts and materials, powder pressings, etc., destructive tests are conducted on samples drawn from each manufacturing lot. These tests, usually firings done in Bruceton or Langlie fashion, test or establish data for firing energy mean and standard deviation. Systematic process changes or material variations usually can be spotted by changes in either or both of the mean or standard deviations. Because these destructive tests use up so many of the lot items, the average price of the lot yield is higher than it would be if so many units did not have to be sacrificed.

One way around this problem is to find a nondestructive test that would still give good indication of firing characteristics. This has been done, using the Rosenthal model (seen earlier) as the basis, with transient-pulse testing or electrothermal response. This procedure can be used on 100% of the units fabricated and find individual bad actors to be weeded out as well as to find systematic or lot to lot shifts at the same time. This procedure is based upon the same heat transfer and energy balance equation we saw earlier

$$C\frac{dT}{dt} + \lambda T = i^2 R$$

All metals have the characteristic that their electrical resistivity changes with temperature. For most materials the resistivity increases with temperature, and for most metals and alloys that are used as bridges in hot-wire initiators, this relationship is linear.

$$R = R_0(1 + \beta T) \tag{24.11}$$

where R is the resistance at temperature T; R_0, the resistance at reference temperature; β, the thermal resistivity coefficient; and T, the temperature above reference T_0.

We realize that as we heat the bridgewire, its resistance rises, so during the firing process R is changing and we can replace it in the heat-balance equation by Eq. (24.11) yielding

$$C\frac{dT}{dt} + \lambda T = i^2 R_0(1 + \beta T) \tag{24.12}$$

Suppose now that we test a hot-wire device at a very low current level, well below the minimum power condition. During this test we pulse the bridge with a square-current pulse at low amplitude. Using the magic of instrumentation, we measure the voltage through the bridge. Since the current is constant, the voltage will change proportionately with the resistance, which in turn changes linearly with temperature. Thus the voltage-versus-time data can be converted to temperature versus time (this assumes we know the value β). We are using the bridge itself as a thermometer.

Looking back quickly to Eq. (24.12), we see that knowing T versus t for a given i, will allow us to solve for the values of C and λ for the device being tested. These values are sensitive to changes in the initiator, such as

1. Initiator mix characteristics and density and contact with the bridge;
2. Welding or solder joints from bridge to lands;
3. Bridge parameters including dimensions;
4. Presence of foreign contaminants such as cleaning solvents; and
5. Corrosion or physical changes.

As long as this test pulse is low enough, no damage is done to the device under test, and we have a 100% screening process. For expensive ordnance items, the costs of this type of testing is more than offset by the increased lot yields.

The testing is done with digital processors that automatically sample and analyze the data. Complications, such as situations where the ignition mix is electrically conductive and forms a circuit path in parallel with the bridge, have been treated and analyzed as well for these (Ref. 18). Modern instrumentation techniques and equipment are now in use that give excellent experimental values of C and λ at previously untenable low current levels (Ref. 19). The data from such tests are applicable not only to lot screening for quality control, but are forming a new and powerful database for use in the design of next-generation hot-wire initiators.

25

Exploding Bridgewire Detonators

In mechanical details, the exploding bridgewire (EBW) detonator looks like any other simple electrically actuated or hot-wire initiator. The difference is that the explosive material loaded over the bridgewire cannot be ignited by the wire, even if the wire is heated to its melting temperature.

25.1 Construction of EBWs

A typical EBW detonator is shown in Figure 25.1. The bridgewires in EBWs are generally pure gold or gold alloys, and sometimes platinum or platinum alloys. These materials offer excellent corrosion resistance, but the main reason for their use is their relatively low electrical resistivity and their high density. The bridges are generally between 1 and 3 mils (0.001 to 0.003 in.) in diameter, with 1.5 mil being by far the most common. Bridgewire length varies from 10 to 100 mils, with 40 mils being the most prevalent.

Any secondary explosive can be used over the bridge, but in practice only PETN is used, usually pressed to 50% crystal density ($\sim$0.88 g/cm^3). In most cases, an output pellet of higher density is added over the PETN "initial pressing" to enhance the breakout pressure.

The EBW detonator functions differently from the hot-wire initiators we have seen, in which the ignition material is brought to a critical temperature and then starts to burn. Instead, the EBW produces a shock wave by means of actually exploding the bridgewire and, in turn, initiating detonation directly by the impulse of the shock.

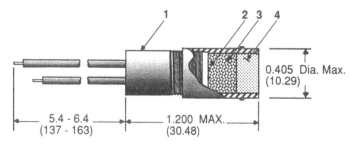

PARTS DESCRIPTION

1. MOLDED HEAD: Diallyl phthalate per MIL-M-14 type SDG.
2. BRIDGEWIRE: 99.9% gold, 6.5 ohms per foot, 0.0015 inches in diameter, 0.040 inches long.
3. INITIATING EXPLOSIVE: 251 mg of PETN.
4. HIGH DENSITY EXPLOSIVE: 375 mg of RDX with binder.

Figure 25.1. Typical EDB detonator (Reynolds Industries Systems Inc., RP-1).

As you will recall, there must be sufficiently high-energy fluence (or P^2t) and high enough pressure such that the explosive can be initiated in a reasonably short distance from a shock wave input. Electrical constraints limit the practical size (pressure, temperature, time) of the shock obtainable from the bridgewire, and these are such that we require an explosive with very low critical energy fluence, E_C (or P^2t_{crit}), and short run distance. PETN, at low density and small particle size, is the only explosive that will respond to these practical needs. However, superfine, high-surface-area RDX has recently been used successfully.

25.2 Explosion of the Bridgewire

For a bridgewire to be exploded with sufficient output to detonate the initial pressing, high input power and high current are required. Current of several hundred amps at very fast rise rate, in the neighborhood of 200 A/μs, are the nominal values required. The only practical source with low enough internal impedance to supply such needs is low-inductance, high-voltage capacitors. When a capacitor is discharged through a low-impedance load, the current-versus-time signal through the load looks like that shown in Figure 25.2.

When such a discharge flows into a small bridgewire, a series of events occurs culminating in the bridgewire explosion. First, the bridgewire heats up to near-melting temperature, during this time the resistance rises somewhat due to the thermal resistivity response. The bridgewire then melts, but this happens in such a short time that inertia prevents it from moving away from its position. The resistance increases further as the melted bridgewire continues to heat toward vaporization or boiling point. Upon vaporizing, the resistance rises rapidly, sufficiently so as to cause the current to suddenly dip at this point. So much current

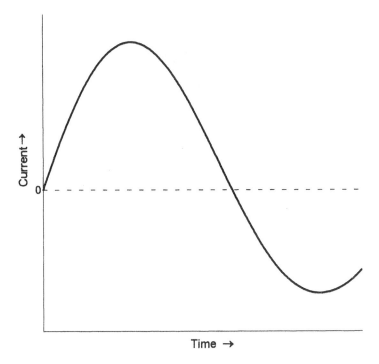

Figure 25.2.   Typical current trace through a low-impedance load.

is passing through this dense vapor, however, that it is further heated. Finally overcoming the inertia, the vaporized bridge material expands explosively. As an electrical arc is stabilized through the now expanding metal vapor, the resistance drops and the current increases, driving the shock of the exploded bridgewire even harder.

These last stages, melting to vaporization to shock expansion, take only a few tens of nanoseconds. Instead of the shape of the current trace we just saw, into a fixed load, the current trace up to and during the wire explosion appears ideally as shown in Figure 25.3. All the activity just described, melting through shock formation, occurs during that dip on the current trace. Figure 25.4 shows the resistance during this time. The voltage trace, of course, would look very much like the resistance trace, also peaking at the same time. This peak time of the voltage is called the bridgewire burst time, t_b. The current at this time is called the burst current, I_b.

25.2.1 Energy and Action

As you recall, in the case of a hot-wire initiator, at high power levels there was so little time for heat transfer that the heat transfer term, $\lambda T t$, could be neglected

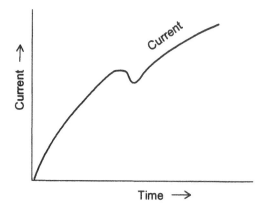

Figure 25.3.   Current trace of bridge explosion.

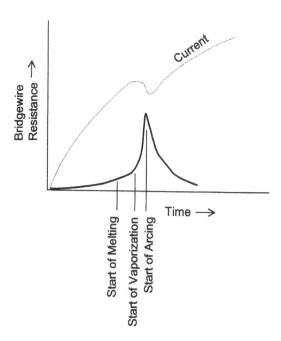

Figure 25.4.   Electrical resistance of exploding bridgewire.

in the energy balance equation. The energy required to fire that particular device became a constant, just sufficient to heat the bridge mass to some critical temperature. The same argument applies herein that, since the bridgewire explosion occurs so fast, there is no appreciable heat transfer; thus the energy required to bring the bridgewire, not to a critical temperature but to a critical explosion state, is also a constant. We see the same energy balance applied (less the λ term).

$$i^2R(E)\ dt = m\ dE \qquad (25.1)$$

where $R(E)$ is dynamic resistance, m is the mass of bridgewire, and E is energy.

The dynamic resistance is essentially the resistance change with temperature but includes also the phase changes from solid to liquid to vapor. We can express this resistance as dependent upon energy because, of course, the internal energy of the bridge is dependent upon temperature.

The resistance of a wire is equal to the resistivity of the metal of which it is made times the length of wire, divided by the cross-sectional area of the wire; so we can also say

$$R(E) = \gamma(E)L/A \qquad (25.2)$$

where $\gamma(E)$ is dynamic resistivity, L is the length of wire, and A is the cross-sectional area of wire.

By replacing $R(E)$ in the energy balance equation, and rearranging the terms, we get

$$i\ dt = \frac{mA\ dE}{L\gamma(E)} \qquad (25.3)$$

Further, the mass of the bridgewire is equal to its density times its volume. The bridgewire volume is the length times the cross-sectional area; so

$$m = \rho AL \qquad (25.4)$$

where ρ is the density of the bridgewire. Substituting this into the previous equation yields

$$i^2dt = \rho A^2\ \frac{dE}{\gamma(E)} \qquad (25.5)$$

The integral of Eq. (25.5), from the start of the current until bridge burst time, is

$$\int_0^{t_b} i^2\ dt = \rho A^2 \int_0^{E_b} \frac{dE}{\gamma(E)} \qquad (25.6)$$

where t_b is the bridgewire burst time, and E_b is the bridgewire burst energy.

The quantity expressed as the integral of energy deposited in the bridge divided by dynamic resistivity is essentially single valued (Ref. 20), and can be considered, within practical limits, to be a constant that depends only upon the bridgewire material. The density is also only a function of bridgewire material.

Since area is directly proportional to the square of diameter, we can lump these terms together and get

$$\int_0^{t_b} i^2 \, dt = K_b D^4 \tag{25.7}$$

where D is bridgewire diameter, and K_b is a constant we will call *burst action coefficient*, a property only of the bridgewire material.

The integral above in eq. (25.7) is called the *burst action* and is designated as G_b; it is the area under the curve of current squared versus time, from the start of the current pulse up to burst time. This is shown in Figure 25.5.

Again, the burst action, G_b, is a function only of the bridgewire material and the bridgewire diameter.

$$G_b = K_b D^4 \tag{25.8}$$

For gold, the most common bridgewire material, $K_b = 0.022$ A^2 s/mil^4. It was mentioned earlier that the most common bridge diameter is 1.5 mils; therefore, the burst action for all EBW detonators that use 1.5-mil gold bridgewires is

$$G_b = 0.022 \, (1.5)^4 = 0.11 \text{ A}^2 \text{ s}$$

The maximum current as well as the rise rate of current discharge with time from a given source, in this case a capacitor discharge fireset, is limited by the impedance of the source. Thus, for a given charge voltage on the capacitor, the maximum attainable current will be higher for a low-impedance source, and conversely, the current will be lower from a higher-impedance source. The impedance of the source capacitor is determined by its resistance and inductance. Figure 25.6 shows the current trace that could be obtained from three different firesets, all with the same capacitance charged to the same voltage, but with

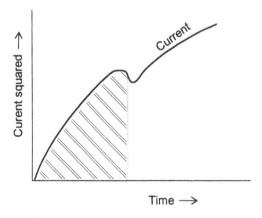

Figure 25.5. Burst action.

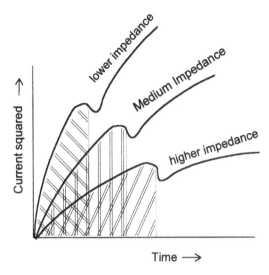

Figure 25.6. Burst point changes with changes in source impedance.

different internal impedances. We are discharging into an identical bridgewire on each trace; we get bridgewire burst at the same burst action, but at different burst times and at different burst currents.

The attainment of burst does not depend on the shape of the current trace, but occurs at the time and current where the area under the current squared versus time curve equals the burst action. Attainment of burst is necessary but not sufficient to form the shock that will detonate the initial pressing.

25.3 Detonation of Initial Pressing

As stated earlier, the critical energy fluence, E_C (or $P^2 t_{\mathrm{crit}}$), must be exceeded to detonate an explosive from a shock wave. The pressure of the shock wave as well as its duration from a bursting bridgewire are functions of the burst current (or of the peak burst power). There is a minimum burst current below which we cannot detonate the initial pressing. This minimum burst current is dependent not only on bridgewire parameters, but also upon the properties of the explosive used for the initial pressing. The critical properties of the explosive that affect minimum burst current are those that affect E_C, namely, the density, particle size, and specific surface area of the particles.

As with all initiators, we cannot state an exact firing energy, or in this case burst current, because of the statistical nature of these devices. We must describe the firing requirement in terms of mean and standard deviation. The mean burst current, for a given EBW detonator, is called the threshold burst current, and its symbol is i_{bth}.

Tucker (Ref. 20) has shown empirically that

$$i_{bth} = \frac{D}{L^{1/2}} (F1)(F2)(F3)$$

where

$$
\begin{aligned}
(F1) &= \left(850 + 35.5 \, \frac{(LDS_0^p \times 10^{-3} - 120)^2}{(LDS_0^p \times 10^{-3})^{3/2}} \right) \\
(F2) &= \left(\frac{1}{(1.88 - \rho)^3} \right) \\
(F3) &= \left(1 - \frac{2 \times 10^{-2} \, (T - 24)}{L^{1/2}} \right)
\end{aligned}
\qquad (25.9)
$$

S_0^p is the specific surface area (cm^2/g) of the PETN initial pressing, D is bridge-wire diameter (mils), L is bridgewire length (mils), ρ is PETN density (g/cm^3), and T is temperature (°C).

For the same PETN specific surface area, bridgewire, and assuming that the burst temperature is constant, we can find that the effect of density alone is

$$i_{bth} = K_2 \frac{1}{(1.88 - \rho)^3} \qquad (25.10)$$

where K_2 is a constant proportionality factor.

We mentioned earlier that a normal PETN density for the initial pressing is approximately 0.88 g/cm^3. If we increase this by 10%, we would find from Eq. (25.10) that the i_{bth} would be increased by 32%.

Since most EBW detonators utilize approximately the same PETN particle size and surface area, and are loaded at approximately the same initial pressing density, Eq. (25.9) can be approximated by

$$i_{ibth} = 850D/L^{1/2} \qquad (25.11)$$

Thus, for a 1.5-mil x 40-mil-long bridged detonator (standard PETN), $i_{bth} \approx 202$ A; for a 1.5 x 30-mil bridge, $i_{bth} \approx 233$ A.

Table 25.1 i_{bth} For Some Commercial EBW Detonators

Name of Detonator	Bridge D and L (mils)	i_{bth} (A)	Standard Deviation (A)	Calculated i_{bth} from Eq. (25.11) (A)
RP-1	$1\frac{1}{2} \times 40$	190	20	202
RP-2	$1\frac{1}{2} \times 30$	220	20	233
RP-80	$1\frac{1}{2} \times 40$	180	25	202
RP-87	$1\frac{1}{2} \times 20$	210	25	285

Data available for some commercial EBW detonators (taken from Ref. 21) are shown in Table 25.1.

The discrepancy of the calculated to measured i_{bth} in Table 25.1 is probably due to differences in initial pressing density, as well as possible differences in particle size and or specific surface area of the PETN.

We have seen that two criteria must be met in order to detonate the initial pressing: (1) we must have a bridge explosion that depends only upon bridge material and diameter, and is determined by the burst action; and (2) we must have burst current that is sufficiently high. The sufficiency of the burst current is determined by the bridge dimensions and the explosive properties.

25.4 Effects of Cables

Earlier, it was stated that the current supply was limited, at any given voltage, by the internal impedance of the fire set. Typical capacitor discharge fire sets have the following impedance characteristics:

capacitance: 1–10 microfarads (μF)
resistance: 50–150 milliohms (mΩ)
inductance: 50–150 nanohenries (nH)

Unless we have a system where the detonator is attached directly to the fire set, we must use a cable to transmit current to the detonator. Cables, of course, also have impedance. The impedance of cables is a function of both their construction and their length. Typical cables and their impedance characteristics are shown in Table 25.2. These are the most common cables used for EBW firing systems.

Comparing the values in Table 25.2 to those of typical fire sets, we see that for cables longer than a few feet, the cable impedance quickly becomes much larger than that of the fire set. The cable, therefore, becomes the limiting factor in both current level and current rate of rise.

Consider the current-time trace shown in Figure 25.7; this trace is the current measured at the detonator end of the cable. When the fireset capacitor is switched on, the current at that instant is limited to the charge voltage divided by the

Table 25.2 Impedances of Common EBW Detonator Cables

Cable Type	Resistance (mΩ/ft)	Capacitance (pF/ft)	Inductance (nH/ft)	Approximate Transmission Time (ns/ft)
C	10	50	66	1.5
L	50	30	90	1.5
RG213	3	30	80	1.5
No. 20[a]	20	20	200	1.7

[a] Solid copper, twin lead.

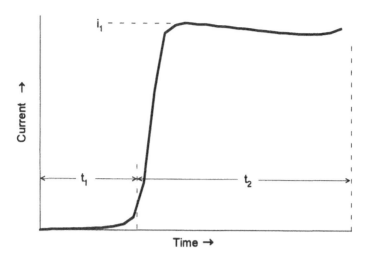

Figure 25.7.   Current trace on first round trip.

system impedance (mainly that of the cable). Since the time t_1 required for the pulse to travel down to the detonator end of the cable is set by the cable characteristics, the capacitor is initially "unaware" of what the load at the end of the cable may be. The initial current from the capacitor is set by the impedance of the cable, not the load. After the pulse reaches the end of the cable at time t_1, a reflection dependent on the load occurs that travels back up the cable to the

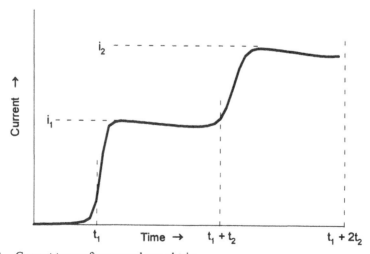

Figure 25.8.   Current trace after second round trip.

fireset. For an EBW load, this reflection arriving at the fireset causes an increase in the capacitor current that is equal to the remaining voltage on the capacitor divided by the system impedance plus initial current pulse. This current now travels down the cable for its round trip, and reaches the output at time $t_1 + t_2$, as shown in Figure 25.8.

This process continues, producing another "stem" each round trip of the current over the length of cable. After a while, the voltage on the capacitor drops enough to reach a peak current and let the signal "step" down. This is typical of all current traces we observe with EBW firing systems. The longer the cable, the longer the steps. The higher the voltage, the higher the current per step. We see also that the burst action is accumulated over longer time and lower current in long cable systems because longer cables produce longer steps. This effect is shown in Figure 25.9, where we see bridge burst on current traces for different cable length for the same fire set and capacitor-charge voltage.

From Table 25.2, we saw that different cable types have different impedance characteristics and different transmission rates. Therefore, for the same fireset at the same charge voltage (and the same detonator), we should see different step amplitudes and step widths (and hence different burst points) for the same

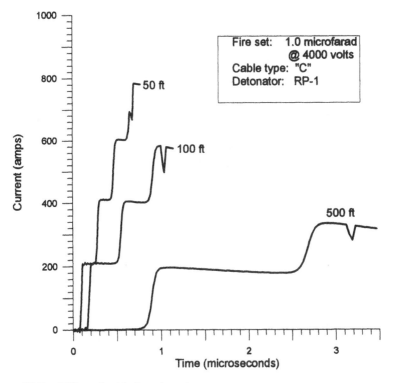

Figure 25.9.   Effect of cable length on burst.

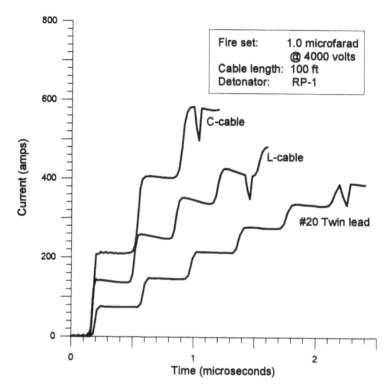

Figure 25.10.   Effect of cable type on burst point.

length of different cable types. This is shown in Figure 25.10 for three different cable types.

Now we have seen that the burst action is constant for a particular wire material and diameter; that the shape of the current-time trace is not a factor in achieving burst; that the current-time trace is determined by the fire set and cable parameters; that as long as we get burst, and as long as that burst current is sufficiently above threshold, we will detonate the initial pressing. To emphasize this, Figure 25.11 shows current traces and burst points that fired detonators from a wide variety of fireset/cable values.

25.5 Function Time

The function time, defined as the time from the start of discharge of the firing capacitor until break-out at the end of the detonator, is the sum of two values; these are the time until burst, t_b, and the transit time, the time from burst until breakout.

We saw in the earlier discussions that time to burst is a system variable; that

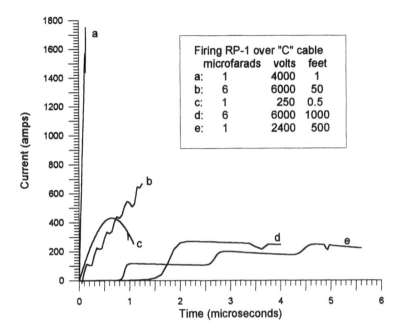

Figure 25.11. Potpourri of firing conditions.

is, it depends upon the firing system (fire set, charge voltage, cable type, cable length) and the detonator bridgewire (material, diameter).

The transit time is mainly a function of the particular detonator design, that is, the type, density, and length of the explosives loaded into the detonator. The transit time is equal to the length of each explosive element or pressing, divided by the detonation velocity of that element, plus the excess transit time due to the buildup of run distance to steady-state detonation. Recall that the run distance, and hence excess transit time, is a function of the initiating shock pressure. Also, the initiating shock pressure from an EBW is a function of the burst current. Therefore, the transit time of an EBW detonator is not independent of the system.

This relationship of excess transit time versus burst current is (Ref. 22) is seen in Figure 25.12. The excess transit time of a detonator is a function of burst current. The total function time is given by

$$t_f = t_b + \Sigma t_d + t_e(i_b) \tag{25.12}$$

where t_f is the total function time, t_b is burst time, Σt_d is the sum of length/detonation velocity for each explosive element, and $t_e(i_b)$ is excess transit time, a function of burst current.

Table 25.3 lists both i_{bth} and transit time (total of $\Sigma t_d + t_e(i_b)$, at $i_b = 400$–500 A) for several EBW detonators.

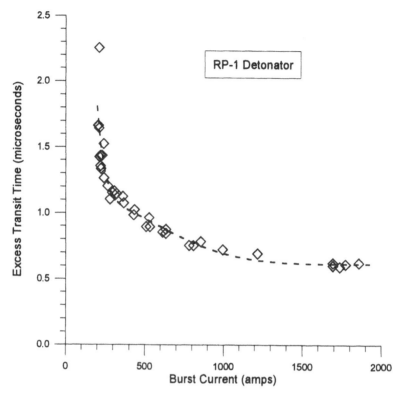

Figure 25.12.   Excess transit time of RP-1 detonator versus burst current.

25.6 Series and Parallel Firing Considerations

Series or parallel firing of several detonators at the end of a single trunk cable
is normally discouraged. Recommended practice is to use a separate cable from
the fire set to each individual detonator. The reason for this is that the trunk
cable, especially a long one, is the current limiter in an EBW firing system. If,

Table 25.3   Function Characteristics of Several Commercial EBWs

Detonator	i_{bth} (Amps)	t_t (μs)	Source
RP-1	190	2.95	RISI[a]
RP-2	220	1.90	RISI
RP-80	180	1.60	RISI
RP-83	180	5.15	RISI
RP-87	210	1.75	RISI

[a] Reynolds *Industries Systems, Inc.*

however, one must use a single trunk cable, then the following must be considered.

25.6.1 Series Firing

Since the current will be limited by the cable at the detonators' end of the cable, the normal firing current one would expect for a single detonator is the current that each in series will receive. However, the voltage drop across each detonator will be reduced to V_b/n (V_b, the burst voltage, n, the number of detonators). The peak burst power, $i_b V_b$, will be considerably lower because the voltage at burst is lower. Thus the number of detonators in series will be severely limited. It is best to test such an arrangement before fielding it. Testing should include measurements of bridge current, bridge voltage, and transit time to ensure that these are within the range one would expect for reliable function.

25.6.2 Parallel Firing

Again, the current, as with the series arrangement, will be limited by the trunk cable. In this case, however, the current to each detonator in parallel will be the normal current divided by the number of detonators. The voltage will be the same as for normal firing. Thus a cable type and length that would have provided, say, a 700-A burst current for a single detonator, will provide only a 350-A burst for each of two detonators in parallel. This system also should be tested for the same parameters mentioned earlier before being fielded.

25.7 Safety Considerations

The bridgewire in an EBW detonator will burn out at relatively low currents. For example, 1.4 A applied for less than 20 ms will burn out a $1\text{-}\frac{1}{2}$ mil gold bridgewire. The PETN initial pressing will not ignite at this condition. Therefore, stray currents are not a safety problem but may be a reliability problem. If one were to apply a 110 VAC normal household current directly across the bridgewire, the bridge would burst at a low current, significantly below i_{bth}. Under this condition, however, the arc caused inside the detonator would blow out the explosive powder and possibly ignite it (not detonate it!). If this detonator utilized a closed charge cup, the cup might burst and could produce minor fragment hazard. EBW detonators exposed to a sustained flame will burn out, not detonate. All in all, the EBW detonator is vastly safer to use than a hot-wire detonator. It is recommended, where costs permit (should that be a factor?), to use EBW detonators for all general explosives work.

References

1. Dobratz, B. M., *LLNL Handbook of Explosives*, UCRL-52997, Lawrence Livermore National Laboratory, March 1981 (updated Jan. 1985).

2. Bowden and Yoffe, *Initiation and Growth of Explosion in Liquids and Solids*, Cambridge Press, 1951.

3. Bowden and Yoffe, *Fast Reactions in Solids*, Academic Press, New York, 1958.

4. Kholevo, *Proceedings International Conference on Sensitivity and Hazards of Explosives*, pp. 5–26, London, 1963.

5. Andreev, (same as Ref. 4) pp. 47–151.

6. Afanas'ev and Bobolev, *Initiation of Solid Explosives by Impact*, Acad. Sc. USSR, 1968 (Translation 19   ).

7. Walker, *Derivation of the P^2t Criterion Based on Frey Shear Band Mechanism*, BRL Tech Report No. ARBRL-TR-02544, 1984.

8. Walker and Wasley, *Critical Energy for Shock Initiation of Heterogeneous Explosives*, Explosivstoffe 17, pp. 9ff, 1969.

9. Price, *Notes from Lectures on Detonation Physics*, Nav. Surf. W Cntr., White Oak, Maryland, NSWC-MP-81-399, Lecture No. 9.

10. *Engineering Design Handbook, Explosive Trains*, AMCP-706-179; Hq., U.S. Army Material Command, 1974.

11. E. R. Lake, *Percussion Primers, Design Requirements, Revision A*, McDonnel Aircraft Co. Report No. MDC A0514, June 30, 1970.

12. *M42G Primer Firing Curves*, Erwood Corp. (Chicago), for NBS Contract DAL-49-186-502-ORD(p)-199, May 1956.

13. *Percussion and Electric Primers*, 4th Edition, Winchester-Western Division, Olin Mathieson Chem. Corp., August 1967.

14. *Blasting Cap Recognition and Identification Manual*, Ernest W. Brucker, International Assoc. of Chiefs of Police, Inc., Research Division, 1973.

15. L. A. Rosenthal, *Electrothermal Equations for Electroexplosive Devices*, Naval Ordnance Report 6684, U.S. Naval Ordnance Laboratory, White Oak, MD, August 1959.

16. L. A. Rosenthal, ''Thermal Response of Bridgewires Used in Electroexplosive Devices'', *Review of Scientific Instruments* 32, 1033 (1962).

17. V. J. Menichelli and L. A. Rosenthal, *Electrothermal Characteristics of Electroexplosive Devices*, 7th Symposium on Explosives and Pyrotechnics, Franklin Research Institute, Philadelphia (1971).

18. O. L. Burchett, A. C. Strasburg, and A. C. Munger, *Electrothermal Response Testing of Electro-Explosive Devices with Electrically Conducting Explosives*, Proceedings of the 6th International Pyrotechnics Seminar, July 1980.

19. L. M. Moore, A. C. Strasburg, and J. J. Spates, *An Electrothermal Response Test System Using Sample-and-Hold Offset Balancing*, Sandia National Laboratory Report SAND85-2424, Albuquerque, NM, February 1986.

20. T. J. Tucker, *Explosive Initiators*, Behavior and Utilization of Explosives in Engineering Design, 12 Annual Symp., ASME/UNM Coll. Eng., Albuquerque, New Mexico, March 1972.

21. *Exploding Bridgewire Ordnance*, Technical Brochure of Reynolds Industries, Inc., 1982.

22. P. W. Cooper, R. N. Owenby, and J. H. Stofleth, *Excess Transit Time in EBW Detonators*, 14th Symposium on Explosives and Pyrotechnics, Burlingame, CA, 1990.

VI

ENGINEERING APPLICATIONS

Introduction

The title of this section may be misleading in that we will not go into specific design applications but rather look at methods that can be applied to design and design estimates. We will also look at some typical databases that will prove useful for many design problems. In the previous sections, we looked at how explosives behave internally. Now we will look at how explosives work on their surroundings. First we will discuss engineering units or dimensions, then look briefly at estimating methods. These will include empirical correlations, scaling by geometric similarities, and scaling by dimensional analysis. Then we will study several databases that will apply to the majority of engineering applications with which we normally deal. Among these are acceleration of metals (Gurney analysis), fragmentation of metal cylinders, flight or ballistics of fragments, air-blast waves, shocks in water, physiological responses to shock waves, and cratering. We will also look briefly at some speciality areas that involve the jetting behavior of metals such as shaped charges and explosive welding.

CHAPTER

26

Theories of Scaling

In this chapter, we will develop the theoretical background that is used in various scaling techniques. We will start with basic definitions of units and dimensions, and then apply these in two different techniques to scaling explosive phenomena. We shall also examine the source and form of the energy available from a detonating explosive.

26.1 Units and Dimensions

Unfortunately, we do not use a united or standard set of units in the field of explosives. All the following terms are in common use:

Length: mils, inches, feet, yards, meters
Mass: grams, grains, pounds, ounces, tons, Tonnes, slugs
Time: nano-, micro-, milliseconds, minutes, etc.
Temperature: °Celsius, °Fahrenheit, Kelvin, °Rankine.
Force: pounds, grams, poundals, dynes, newtons
Pressure: psi, atmospheres, bars, pascals, millimeters (and inches and feet) of mercury and water
Energy or work: calories, ergs, joules, BTU, foot pound, inch ounce, gram centimeters, newton meters
Power: horsepower, watts.

Most of the above can be resolved by use of the appropriate conversion tables. A common problem arises, however, in the confusion caused where we use the same word to denote force, mass, and weight. This problem is also due, in part, to the fact that we "mix and match" different sets of standard units. We must first remember that force is related to mass by Newton's equation $F = ma$, where F is force, m is mass, and a is the acceleration of that mass caused by that force. Weight is the force required to accelerate a unit of mass at the gravitational constant. So if we fix the definition of either force or mass, the other must agree in dimensions and units according to $F = ma$. Some of the extant systems are shown in Table 26.1.

In Table 26.1 we see that the dyne (CGS system) is the force required to accelerate 1 g of mass at 1 cm/s^2 and the erg is 1 dyne cm. The Newton is the force to accelerate a 1 kg mass at 1 m/s^2, and the joule is 1 newton meter (we see then that 1 joule = 10^7 ergs). In the MKS mass system we also get a definition of power, watts, where 1 watt is the rate of delivering 1 joule per second.

In the English language we get the problem that pound can be used for either force or mass (as we often blend or confuse the British and American systems). Where pound is force, mass is slugs, and 1 pound accelerates 1 slug at 1 ft/s^2. Where pound is mass, then force is poundals, and 1 poundal accelerates 1 pound mass at 1 ft/s^2.

When working a problem and checking the units, we find a discrepancy equal to some power of length per time squared; then we know that we have used force for mass or vice versa in our equations. This is easily remedied by multiplying or dividing by the appropriate power of g. We might remember that pressure is always defined as force per unit area and density as mass per unit volume.

26.2 Scaling by Geometric Similarity

Scaling by geometric similarity is a technique whereby the behavior of a given system can be determined by conducting experiments on a smaller model of that system. The smaller system is called the model, and its dimensions and properties are noted by the subscript "0." The larger system is called the prototype

Table 26.1 Systems of Defined Units

System	Force	Mass	Energy (or Work)	Power
CGS	dyne	gram	erg	
MKS (mass)	newton	kilogram	joule	watt
MKS (force)	kilogram			
British (mass)	poundal	pound		
U.S., Engineering	pound	slug		

and its dimensions and properties are noted by the subscript "1." The model and the prototype must be exactly similar in geometric design. Every dimension in the model is scaled down by the same factor, S. Thus, if we scaled a box with a scaling factor $S = 4$, then the prototype would be four times the height, four times the width, four times the depth, and four times the wall thickness of the model. If the model is to be tested to destruction, as when we are looking for the internal pressure that will burst open the seams, the nails or screws must be geometrically scaled also by a factor of four.

Geometric similarity also demands that material properties be exactly the same in both the model and the prototype, if those properties influence the behavior we are studying, for example, tensile strength or density.

Let us look at some examples and see how this works. First, consider a very simple system, a cube (Figure 26.1) The scaling factor is S. Therefore, $X_1 = SX_0$. How does the surface area of the prototype scale with that of the model?

Surface area of a cube $= 6X^2$

$$A_0 = 6X_0^2$$
$$A_1 = 6X_1^2$$
$$\frac{A_1}{A_0} = \frac{6X_1^2}{6X_0^2} = \frac{6S^2 X_0^2}{6X_0^2} = S^2$$

The areas in the prototype are scaled as the square of the scaling factor. How about volumes? Volume of a cube $= X^3$

$$V_0 = 6X_0^3$$
$$V_1 = 6X_1^3$$
$$\frac{V_1}{V_0} = \frac{6X_1^3}{6X_0^3} = \frac{6S^3 X_0^3}{6X_0^3} = S^3$$

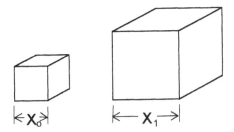

Figure 26.1. Model and prototype of a cube.

Volumes scale as the cube of the scaling factor. Simple. How about mass?

$$\text{mass} = \rho V$$

$$m_0 = \rho V_0$$

$$m_1 = \rho V_1$$

$$\frac{m_1}{m_0} = \frac{\rho V_1}{\rho V_0} = \frac{\rho S^3 V_0}{\rho V_0} = S^3$$

The mass scales as the cube of the scaling factor. Suppose we have as our system a spherical pressure vessel. We will burn a propellant inside the vessel and determine the final pressure. In both the model and the prototype (Figure 26.2) we use the same kind of propellant.

The propellant charge is also geometrically similarly scaled: $r_1 = Sr_0$, $r_{c1} = Sr_{c0}$, where r_c is the radius of the charge, and r is the radius of the vessel.

Propellant properties pertinent to the problem are density ρ_c, gas evolution V_g (cm^3/g), and isochoric flame temperature T_V. From the above, we know the volumes of the vessels V_0 and $V_1 = S^3 V_0$. We also know the masses of the propellants m_0 and $m_1 = S^3 m_0$. The volume of gas evolved from the propellant at standard temperature and pressure (STP) is

$$V_{STP} = V_g m \text{ (at STP)}$$

and the final temperature of the gas will be T_V. The initial temperature is T_{STP}, and the initial pressure is P_{STP}. Let us assume that the final pressure will not be very high; so we can invoke the ideal gas equation

$$PV = nRT$$

This allows us to find the final pressure in each vessel.

$$P_0 = \frac{V_g m_0 T_V P_{STP}}{V_0 T_{STP}}$$

$$P_1 = \frac{V_g m_1 T_V P_{STP}}{V_1 T_{STP}}$$

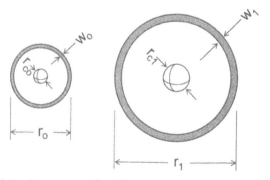

Figure 26.2. Model and prototype of a sphere.

then

$$P_0 = \frac{V_g m_1 T_V P_{\text{STP}}/V_1 T_{\text{STP}}}{V_0 m_0 T_V P_{\text{STP}}/V_0 T_{\text{STP}}} = \frac{m_1 V_0}{m_0 V_1}$$

We saw above: $m_1 = S^3 m_0$ and $V_1 = S^3 V_0$
  So

$$\frac{P_1}{P_0} = \frac{S^3 m_0 V_0}{m_0 S^3 V_0} = 1$$

We see that pressures are equal in both the model and prototype in geometric similarity modeling. Going a bit further, let us look at the stresses in the walls of these scaled vessels. The wall thickness was also scaled; so these are w_0 and $w_1 = S w_0$. The spherical hoop stress in a thin-walled ball is

$$\sigma = \frac{Pr}{2w}$$

where P is the internal pressure, r is the mean radius, and w the wall thickness.
  So

$$\sigma_0 = \frac{P_0 r_0}{2w_0} \quad \text{and} \quad \sigma_1 = \frac{P_1 r_1}{2w_1}$$

We saw above that $P_0 = P_1$ and were given $r_1 = S r_0$ and $w_1 = S w_0$,
so

$$\frac{\sigma_1}{\sigma_0} = \frac{P_1 r_1/2w_1}{P_0 r_0/2w_0} = \frac{P_0 S r_0/2 S w_0}{P_0 r_0/2w_0} = 1$$

So now we see that stresses are equal in both model and prototype.
  If both of these vessels were made of the same metal with the same heat treatment, they would both burst at the same pressure. Complex vessels that cannot be designed analytically can be modeled and tested in this manner very

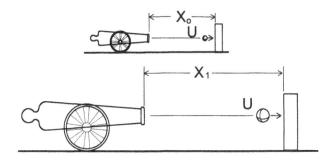

Figure 26.3. Model and prototype of a cannon.

economically. Let us look at another dimension, time. Suppose our system is now a gun firing over a short range into a target.

The velocity of the bullet is the same in both model and prototype. The time of travel from the muzzle to the target is $t = X/v$,

so

$$t_0 = X_0/v \quad \text{and} \quad t_1 = X_1/v$$

Then

$$\frac{t_1}{t_0} = \frac{X_1/v}{X_0/v} = \frac{SX_0/v}{X_0 v} = S$$

So time scales with the scaling factor. Without belaboring the point with more complicated equations, it can be shown that for scaled guns with scaled propellants as well as scaled bullets, the muzzle velocities would be equal. Geometrically similar scaled bullets with equal mass densities have the same drag coefficients. It will be shown later that the entire trajectory is scaled such that velocity loss is equal over scaled distances.

This technique has been successfully used to test many ordnance and explosive systems economically such as target damage in concrete walls from artillery projectiles. In that specific case, the aggregate in the concrete was also scaled so that the targets would be geometrically similar. The key word above is economics. Small-scale testing is less expensive than large-scale testing. Consider the cost of materials alone. A 5:1 scale model weighs 1/125 as much as its prototype, and materials are purchased by weight. A 10:1 scale model weighs 1/1000 the weight of a prototype.

Scaling by geometric similarity does have some limits, however; some processes do not scale. For instance, spalling of metal plates that were in contact with a detonating explosive is a process that is not conducive to scaling. The type of spall and its distance from the metal's first free surface are functions, among other things, of the slope of the Taylor wave behind the detonation front. This slope does scale somewhat for thin explosive charges (thin in the direction of detonation) but eventually becomes constant at a thickness greater that a few inches. So the spalling scales only to that thickness. The explosive in that case is not adhering to our stated rule of properties' similarity. Another case of this type is scaling a system containing a cylindrical explosive charge. Each explosive has some minimum diameter below which it does not propagate ideally. If the dimensions of the model are below this size and the prototype's above it, then the two systems are using effectively different detonation properties. This is also true of systems that use thin sheets of explosive.

As long as one is aware of such limitations, excellent results are obtained with the scaling technique. Another problem comes up, however, when one cannot keep all material properties constant. Perhaps a different metal must be used or a different explosive, or some other property changed. In that case, a different scaling technique must be used, and for that we turn to dimensional analysis and scaling by means of dimensionless groups.

26.3 Scaling by Dimensional Analysis

Very often in some design problem, we know which variables or system properties are important, but we do not know the precise analytical relationship between them. That is, we know the variables but cannot write an equation. We can determine the relationships experimentally by varying one parameter and measuring the effect on the dependent variable of interest. We will obtain a graph of this relationship when we plot the results. Let us say a particular problem has two variables, x and y. We run five experiments in which we vary x and measure the resultant change in y. Our data look like those in Figure 26.4.

We have developed a ''curve'' and we had to run five experiments to do this. Suppose our problem involved three variables: x, y, and z. In order to determine how they relate to each other, we would have to repeat our experiments of five shots for each of five values of z, thus conducting 25 experiments and developing a ''chart'' of data curves as shown in Figure 26.5.

If our problem involved 4 variables, then we would have to repeat our chart five times in order to relate x, y, and z to the fourth variable w, thus developing a

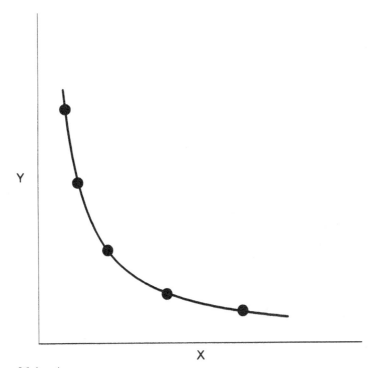

Figure 26.4.   A curve.

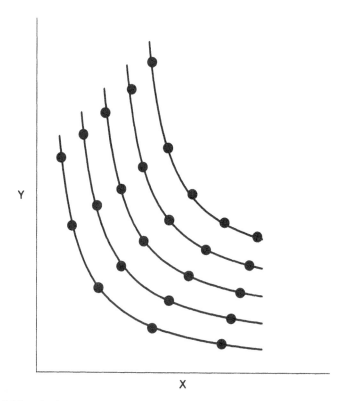

Figure 26.5.   A chart.

"set" of charts. We now have had to run 125 experiments. It is easy to see that this quickly gets excessive when several variables are involved.

In dimensional analysis we do experiments varying not one variable at a time, but a group of variables. Those groups are established so that they are dimensionless, and this allows us to reduce the number of experimental variables.

Before we get into these groups, let us first look at that term *dimensions*. To each property of matter that we can measure, we assign a dimension and a specific unit of measurement. Thus we have dimensions of length, mass, force, time, and temperature. Other entities or properties can be described by these above dimensions. For example, velocity is expressed as a length divided by time. Let us assign symbols for the basic dimensions.

Length, L;
Mass, M (for a mass-based system);
Time, T; and
Temperature, θ

In a system based on force, the dimension F is used and mass can be derived from Newton's law to have the dimensions FT^2/L or FT^2L^{-1}. Table 26.2 lists

Table 26.2 Dimensions of Entities

	Mass System	Force System
Length	[L]	[L]
Time	[T]	[T]
Temperature	[θ]	[θ]
Force	[MLT^{-2}]	[F]
Mass	[M]	[$FL^{-1}T^2$]
Specific weight	[$ML^{-2}T^{-2}$]	[FL^{-3}]
Mass density	[ML^{-3}]	[$FL^{-4}T^2$]
Angle	[1]	[1]
Pressure and stress	[$ML^{-1}T^{-2}$]	[FL^{-2}]
Velocity	[LT^{-1}]	[LT^{-1}]
Acceleration	[LT^{-2}]	[LF^{-2}]
Angular velocity	[T^{-1}]	[T^{-1}]
Angular acceleration	[T^{-2}]	[T^{-2}]
Energy, work	[ML^2T^{-2}]	[FL]
Momentum	[MLT^{-1}]	[FT]
Power	[ML^2T^{-3}]	[FLT^{-1}]
Moment of a force	[ML^2T^{-2}]	[FL]
Dynamic coefficient of viscosity	[$ML^{-1}T^{-1}$]	[$FL^{-2}T$]
Kinematic coefficient of viscosity	[L^2T^{-1}]	[L^2T^{-1}]
Moment of inertia of an area	[L^4]	[L^4]
Moment of inertia of a mass	[ML^2]	[FLT^2]
Surface tension	[MT^{-2}]	[FL^{-1}]
Modulus of elasticity	[$ML^{-1}T^{-2}$]	[FL^{-2}]
Strain	[1]	[1]
Poisson's ratio	[1]	[1]

derived dimensions for most of the more common engineering variables or entities (Ref. 1).

So we see that for a particular problem or system we can, because we are bright engineers, specify all the pertinent variables, and can determine the dimensions of those variables. We do not know the analytical relationship between these variables, but we do know that they are the pertinent ones.

As a rule of thumb, the number of discrete and independent dimensionless products that can be formed from a given set of variables is equal to the difference between the number of variables and the number of dimensions of those variables.

This can best be described by an example: Suppose that we are interested in the drag force on a sphere that is submerged in a moving stream of fluid. The velocity of the fluid stream some distance ahead of the sphere is V. The diameter of the sphere is D. The density of the fluid is ρ and the dynamic coefficient of viscosity is μ. We know that these are the only pertinent variables that will affect the drag force F. However, we do not know the relationship of these variables. We can write an "almost" equation.

$$F = f(V, D, \rho, \mu) \tag{26.1}$$

This equation means that F is a function of V, D, ρ and μ, but does not give any hint of that functionality.

In this problem, we have five variables: F, V, D, ρ, and μ. The dimensions of each variable are: $F(F)$, $V(L/T)$, $D(L)$, $\rho(FT^2/L^4)$, and $\mu(FT/L^2)$. We see that we have only three basic dimensions (I have chosen to use a force system here). Therefore, we should be able to form $(5 - 3 = 2)$ or two independent dimensionless groups. We find that these are

$$\frac{F}{\rho V^2 D^2} \quad \text{and} \quad \frac{VD\rho}{\mu}$$

Now, Eq. (26.1) can be rewritten as

$$\frac{F}{\rho V^2 D^2} = f\left(\frac{VD\rho}{\mu}\right) \tag{26.2}$$

If it took five experiments to develop a data relationship curve, in this problem if we varied one parameter at a time and have four independent variables, we would need a set of sets of charts representing 625 experiments. Instead, we run five experiments to develop the curve of the functionality relationship of the dimensionless groups in Eq. (26.2) (see Figure 26.6).

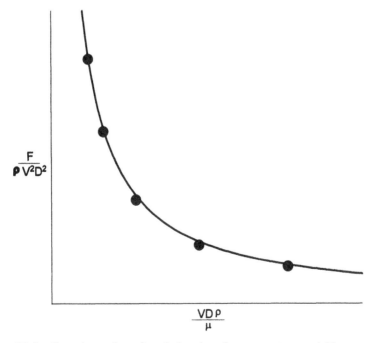

Figure 26.6. Experimental results relating drag force to system variables.

Regardless of how we vary the values of the individual original variables, for each value of $(VD\rho/\mu)$ there is only one unique value of $(F/\rho V^2 D^2)$. So our problem is greatly simplified. We have reduced a five-variable problem to a two-variable problem. We still do not know the analytical or physical relationship of all the variables, but we now know enough to be able to determine the drag force on a sphere for any given combination of the other four independent variables.

The most common variables in fluid mechanics problems are: force, length, velocity, mass density, dynamic coefficient of viscosity, acceleration of gravity, speed of sound, and surface tension. Along with their dimensions they are as follows:

F (F)   force

L (L)   length

V (LT^{-1}) velocity

ρ (FT^2L^{-4}) mass density

μ (FTL^{-2}) coefficient viscosity

g (LT^{-2}) acceleration of gravity

c (LT^{-1}) acoustic velocity

σ (FL^{-1}) surface tension

These eight variables have only three basic dimensions; thus we should be able to find five independent dimensionless groups for this general set of fluid mechanics variables. These groups are so common that there are names for each.

$(VL\rho/\mu)$ Reynolds number, N_{RE};

$(F/\rho V^2L^2)$ Pressure coefficient, P, because F/L^2 = pressure, this is

            also written as $(P/\rho V^2)$;

(V^2/Lg) Froude's number, F;

(V/c) Mach's number, M; and

$(\rho V^2L/\sigma)$ Weber's number, W.

Actually, an infinite number of dimensionless groups can be found for any set of variables; they will not all be independent, but they will all be some product of the basic set. The Buckingham Pi Theory deals with this in strict mathematical terms:

$$\pi_i = \pi_1^a \pi_2^b \pi_3^c \ldots \pi_n^x$$

where π is a dimensionless group. This says that any dimensionless group can be formed from the products of a set of dimensionless groups raised to any powers.

26.4 Work Functions or Available Energy

In the following chapters we will see how explosives, when detonated, work on their surroundings to perform some useful function. We will use databases that have been established by experimental data utilizing the techniques of geometric similitude scaling as well as dimensional analysis. Most of these databases will require that we know how much energy the explosive has available for use. Let us review how an explosive behaves.

26.4.1 The Hydrodynamic Work Function

You will recall from the previous sections on shock and detonation that upon detonation an explosive turns into a hot, highly compressed gas whose density in the detonation-wave front is greater than the original density of the unreacted explosive. This gas now is available to push on (or expand against) its surroundings. This pushing is the work it is doing or energy it is transferring. This work process is considered to be isentropic and the expansion is represented by the expansion isentrope on a P-v diagram of the explosive. In Figure 26.7, we see the process plotted for a typical explosive.

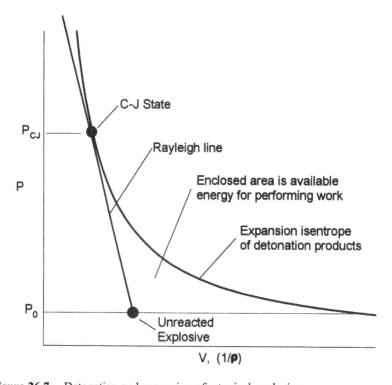

Figure 26.7.   Detonation and expansion of a typical explosive.

The area under this isentrope is

$$E = \int_{P_{CJ}}^{P_{amb}} P(V)_s\, dV$$

which is the energy available for work on the surroundings. This integral is called the hydrodynamic work function. We must know the equation of the isentrope to calculate the total energy or value of the work function. This work function derives from the equation of state of the detonation products. Here is the rub. The composition of the detonation products is in dynamic equilibrium; it is changing continuously from the CJ state down to some modest pressure and temperature where it finally becomes constant. The best of the extant computer codes can only roughly estimate the composition equilibria. Also, these equilibria have only been estimated for a few explosives. The equation of state of the gases, even knowing the composition, is a rough estimate at best. So we do not have readily available values for the hydrodynamic work function; those we have are only for a few of the most studied explosives. The hydrodynamic work function can be estimated by

$$E = P_{CJ}\, v_{CJ}/2 = 0.5 P_{CJ}/\rho_{CJ} \tag{26.3}$$

An alternate approach to the above is to use the thermodynamic work function (Refs. 2, 3).

26.4.2 The Thermodynamic Work Function

The thermodynamic work function is the Helmholtz free energy, ΔA, where

$$\Delta A = E = -\Delta H_R^0 + T\,\Delta S^0 \tag{26.4}$$

where ΔH_R^0 is the heat of detonation at a given standard state, T is the absolute temperature of that standard state, and ΔS^0 is the change in entropy for the reaction at the same standard state.

The thermodynamic work function considers only the starting and ending states of the detonation-expansion process. It has some problems also. First, when the detonation-product gases expand down to ambient conditions, they do not come to the standard-state temperature (usually specified as 298 K), but are at some higher but unknown temperature. Therefore, the T in Eq. (26.4) is in error. Nor do we really know the final expanded equilibrium composition; therefore, the estimates of the and ΔH_R^0 and ΔS^0 for the reaction are somewhat questionable. It should be noted that ΔS^0 also contains the terms for the entropy of mixing of each of the detonation products that are in the gas phase. The entropy of mixing for a given gaseous product is (from Ref. 2) $S_{mix} = R \ln y$, where R is the universal gas constant, and y is the mole fraction of this particular product in the gas phase.

The data for heats of formation and relative entropy for the above calculations are not always readily available, except for the more commonly studied explosives. These can be estimated using the method of group additivity that was shown in Section II. The work function for TNT, calculated from thermodynamics, as above, is 1160 cal/g (Ref. 3), calculated by Eq. (26.3) is 1080 cal/g, and determined experimentally (Ref. 3) is 1120 cal/g.

Therefore, we can only roughly estimate the energy available from a detonation. We will also see, as we examine each particular database, that each process (like expanding a cylinder, throwing a fragment, shocking air) has some limiting condition that prevents us from using all the energy available. For this reason, each of the databases uses a somewhat different way of handling the available energy in the form of empirical correlations that pertain to that particular system.

27

Acceleration, Formation, and Flight of Fragments

In this chapter, we will examine how fast explosives can throw pieces of metal, dependent upon the geometry of the charge and metal piece to be thrown; how a cylinder breaks up when it is explosively expanded; how methods are used for estimating the resulting sizes of the fragments from that breakup; and finally, how the velocity of a piece of metal will be affected by the air through which it is traveling after it is thrown.

27.1 Acceleration of the Gurney Model

In the early 1940s, R. W. Gurney developed a model that described the expansion of a metal cylinder driven by the detonation of an explosive filler charge (Ref. 4). The model closely predicts the initial velocity of the fragments produced by the breakup of the expanding cylinder. His model relied on a partition of an explosive's energy between the metal cylinder and the gases driving it. He assumed a linear velocity gradient in the expanding gases. The model yielded a deceptively simple relationship between the final metal velocity, the explosive energy, and the ratio of the mass of the driven metal to that of the loaded explosive charge.

For cylindrical charges, as shown in Figure 27.1, the Gurney equation is

$$\frac{V}{\sqrt{2E}} = \left(\frac{M}{C} + \frac{1}{2}\right)^{-1/2} \tag{27.1}$$

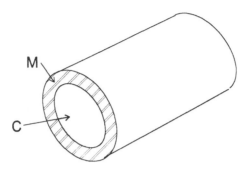

Figure 27.1. Cylindrical configuration.

where V is the metal velocity; M, the mass of metal; C, the mass of charge; and $\sqrt{2E}$, a constant, the Gurney velocity coefficient, which is specific to a particular explosive.

When the same analysis was applied to a spherical metal shell loaded with explosive (Figure 27.2), the relationship was only slightly changed. For center-initiated spherical charges,

$$\frac{V}{\sqrt{2E}} = \left(\frac{M}{C} + \frac{3}{5}\right)^{-1/2} \tag{27.2}$$

The third symmetrical geometry (Figure 27.3) to which the analysis was applied was a "symmetric sandwich," where

$$\frac{V}{\sqrt{2E}} = \left(2\frac{M}{C} + \frac{1}{3}\right)^{-1/2} \tag{27.3}$$

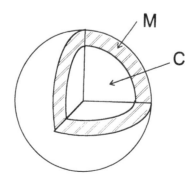

Figure 27.2. Spherical configuration.

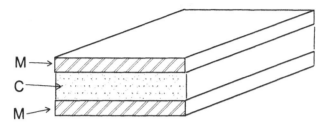

Figure 27.3. Symmetrical sandwich configuration.

The value of E or $\sqrt{2E}$ is determined experimentally for each explosive. It has also been correlated to several other explosive or detonation properties. Among the extant correlations,

$$\sqrt{2E} = 0.6 + 0.54(1.44\Phi\rho_0)^{1/2} \tag{27.4}$$

This is the Kennedy-Hardesty correlation (Ref. 5), where Φ is the Kamlet-Jacobs characteristic value (Ref. 6), and ρ is the density of the unreacted explosive.

Another correlation using these same variables is by Kamlet and Finger (Ref. 7).

$$\sqrt{2E} = 0.887\Phi^{1/2} + \rho_0^{0.4} \tag{27.5}$$

A simplier and also more accurate correlation is

$$\sqrt{2E} = D/2.97 \tag{27.6}$$

where D is the detonation velocity of the explosive. This correlation is derived from data shown in Table 27.1.

The values and correlations for $\sqrt{2E}$ in Table 27.1 are for the case where (for cylinders and spheres) the metals involved have very high ultimate strain. That is, the explosive gases can drive or work on the metal for a large portion of the gas expansion. This is not always the case, however. Metals that are brittle or have low ultimate strain will fracture at smaller expansion ratios. The detonation gases will then stream around the fragments or bypass them, and the acceleration process stops there. This is typical of warhead metals such as the various cast-iron alloys, where the final fragment velocities are lower than would be predicted by the Gurney model. Typically, the fragment velocities of exploding spheres and cylinders of brittle metals are 80% of the predicted value.

This correction does not apply when the metal does not fragment. These cases would be the various "sandwich" configurations. For the case of the unsymmetric sandwich, the Gurney model was modified (Refs. 8, 9) by inclusion of a momentum balance along with the energy balance and the assumption of an ideal EOS for the detonation gases.

Example 27.1   Let us suppose that we have two large sheets of $\frac{1}{4}$-in. thick steel (density is 7.87 g/cm³). Sandwiched between them is a layer of $\frac{1}{2}$-in. thick Datasheet™ explo-

Table 27.1 Correlation of $\sqrt{2E}$ to Detonation Velocity

Explosive	ρ_0 (g/cm³)	D (mm/ms)	$\sqrt{2E}$ (mm/ms)	$D/\sqrt{2E}$ (mm/ms)
Composition A-3	1.59	8.14	2.63	3.095
Composition B	1.71	7.89	2.70	2.92
Composition B	1.717	7.91	2.79	2.84
Composition B	1.717	7.91	2.71	2.92
Composition B	1.72	7.92	2.68	2.96
Composition B	1.72	7.92	2.70	2.93
Composition B	1.72	7.92	2.71	2.92
Composition B	1.72	7.92	2.77	2.86
Composition C-3	1.60	7.63	2.68	2.85
Cyclotol (75/25)	1.754	8.25	2.79	2.96
H-6	1.76	7.90	2.58	3.06
HMX	1.835	8.83	2.80	3.15
LX-14	1.89	9.11	2.97	3.07
Octol (75/25)	1.81	8.48	2.80	3.03
Octol	1.821	8.51	2.83	3.01
PBX 9404	1.84	8/80	2.90	3.03
PBX 9502	1.885	7.67	2.377	3.23
PETN	1.76	8.26	2.93	2.82
RDX	1.77	8.70	2.83	2.97
Tacot	1.61	6.53	2.12	3.08
Tetryl	1.62	7.57	2.50	3.03
TNT	1.63	6.86	2.37	2.89
TNT	1.63	6.86	2.44	2.81
TNT	1.63	6.86	2.46	2.79
Tritonal (80/20)	1.72	6.70	2.32	2.89
				mean = 2.97

sive (the detonation velocity of Detasheet™ is 7.0 km/s and its density is 1.54 g/cm³). Upon detonation, what velocity is imparted to the two sheets of steel?

Solution This is obviously a symmetrical sandwich configuration where

$$M/C = (0.25)(7.87)/(0.5)(1.54) = 2.555$$

The Gurney velocity constant,

$$\sqrt{2E} = (7.0)/(2.97) = 2.357 \text{ km/s}$$

Therefore, the velocity of each plate is

$$V = \sqrt{2E}\left(2\frac{M}{C} + \frac{1}{3}\right)^{-1/2}$$

$$= (2.357)[(2)(2.555) + (1/3)]^{-1/2}$$

$$= 1.01 \text{ km/s}$$

The general unsymmetrical sandwich model (Figure 27.4) is somewhat more complex than the previous ones, but is still quite tractable.

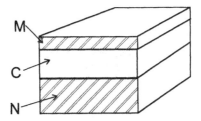

Figure 27.4. Unsymmetrical sandwich configuration.

Let

$$A = \frac{1 + 2\ M/C}{1 + 2\ N/C} \tag{27.7}$$

then

$$\frac{V}{\sqrt{2E}} = \left[\frac{1 + A^3}{3(1 + A)} + \frac{N}{C}A^2 + \frac{M}{C}\right]^{-1/2} \tag{27.8}$$

Example 27.2 Suppose we have the same situation as in the previous example except that one of the steel plates, which we will designate as the tamper, is 1 in. thick. What is the imparted velocity to the $\frac{1}{4}$-in.-thick flyer plate?

Solution This is an example of the unsymetrical sandwich configuration.

$M/C = (0.25)(7.87)/(0.5)(1.54) = 2.555$

$N/C = (1)(7.87)/(0.5)(1.54) = 10.22$, and therefore

$A = [(1 + 2(2.55)]/[1 + 2(10.22)] = 0.2845$, and

$V = (2.357)[(1 + 0.2845^3)/3(1 + 0.2845) + (10.22)(0.2845)^2 + 2.555]^{-1/2}$

$\quad = 1.234$ km/s

Note that increasing the tamping by a factor of four has increased the velocity by only 22%.

We can also derive two limiting cases for the unsymmetrical sandwich. First, where the tamper mass, N, approaches infinity (Figure 27.5), for $N \to \infty$, $A \to 0$, and

$$\frac{V}{\sqrt{2E}} = \left(\frac{M}{C} + \frac{1}{3}\right)^{-1/2} \tag{27.9}$$

Example 27.3 Following from the previous two examples, let us suppose that the tamper plate is now lying on top of another thick steel plate that is lying on a concrete pad, which is on the ground. We now have essentially infinite tamping. What will the initial flyer-plate velocity be now?

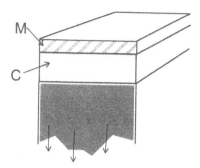

Figure 27.5. Unsymmetrical sandwich configuration where $N \to \infty$, also called *infinitely tamped configuration.*

Solution M/C is still 2.555, and the flyer plate velocity will be

$$V = (2.357)/(2.555 + 1/3)^{-1/2}$$

$$= 1.387 \text{ km/s}$$

This is the maximum velocity at which this charge could throw this plate.

And the very common case (Figure 27.6), where $N = 0$, which is also called the "open-face sandwich."

For $N = 0$, $A = 1 + 2(M/C)$

$$\frac{V}{\sqrt{2E}} = \left[\frac{1 + \left(1 + 2\dfrac{M}{C}\right)^3}{6\left(1 + \dfrac{M}{C}\right)} + \frac{M}{C} \right]^{-1/2} \tag{27.10}$$

Example 27.4 Following from the previous three examples, this time there will be no tamping at all behind the explosive. What will be the initial velocity?

Solution M/C is still 2.555, and in this open-face-sandwich configuration

$$V = (2.357)\{[1 + (1 + (2)(2.555)^3]/[(6)(1 + 2.555)] + 2.555\}^{-1/2}$$

$$= 0.65 \text{ km/s}$$

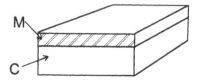

Figure 27.6. Unsymmetrical sandwich configuration where $N \to 0$, also called *open-face-sandwich configuration.*

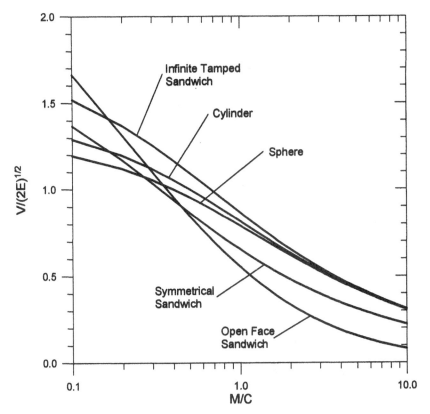

Figure 27.7. Comparison of various Gurney model configurations.

This is the minimum velocity with which this charge will throw the given flyer plate in a sandwich configuration. A comparison of these various configurations is shown in Figure 27.7.

The open-face-sandwich configuration will have considerable side losses when finite dimensions are imposed on its area and it no longer is an infinite plane. This is because there is no confinement to gas expansion in the direction parallel to the plate. Let us look at the example of a cylinder of explosive driving a plate off its end, Figure 27.8.

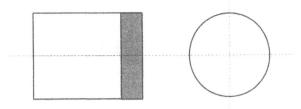

Figure 27.8. Cylinder charge driving plate.

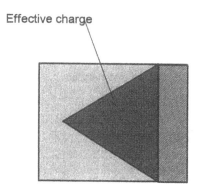

Figure 27.9. Effective charge volume.

The gases expanding to the sides will not exert any pressure on the plate; so their energy is lost. This can be thought of as effectively reducing the mass of the explosive charge. It has been found from numerous experimental observations that the effective charge weight, C_e, is that which would be contained within a cone with 60° base angle and base diameter equal to the charge diameter. This is shown in Figure 27.9.

All of the explosive outside of that cone is "wasted" as far as calculation of the model is concerned. In the model term M/C, the C should be C_e, the mass of charge *within the cone*. For charges shorter than the cone apex height, as shown in Figure 27.10, the effective weight is the mass within the truncated cone.

Example 27.5 Again, we have a flyer plate of steel $\frac{1}{4}$-in. thick, this time driven by a 1-in.-thick charge of Detasheet™. However, this time, the plate and charge have a finite diameter; they comprise a cylinder 1 in. in diameter. The charge is initiated at the end opposite the flyer. What will the flyer velocity be?

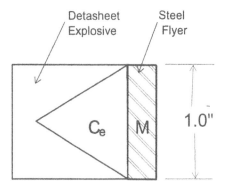

Solution This a case where we must take into account the side losses of the system.. Since this is an unconfined cylinder, all of the effect of the charge is lost outside of a 60°-base-angle cone in the explosive. The cone height is shorter than the charge length;

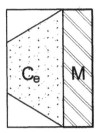

Figure 27.10. Effective charge mass for short cylindrical charges.

therefore, the entire cone is used. The volume of a cone is $(\pi/3)r^2h$, where r is the base radius and h is the perpendicular height from the base to the apex.

The height is $h = r \tan \theta$, where θ is the base angle of the cone. So the volume is found by

$$V = (\pi/3)r^2 \tan \theta,$$

$$= (\pi/3)(0.5 \text{ in.})^2 \tan(60°)$$

$$= 0.227 \text{ in.}^3 = 3.72 \text{ cm}^3$$

and we saw in Example 27.1 of this series that the density of Detasheet™ is 1.54 g/cm³ and so C_e the effective explosive weight is

$$W = (3.72)(1.54) = 5.73 \text{ g}$$

The weight of the flyer is $\rho_f \pi d^2 t/4 = 25.3$ g, and $M/C_e = (25.3)/(5.73) = 4.415$.

The flyer velocity can now be found from the open-face-sandwich configuration equation and

$$V = (2.357)\{[1 + (1 + (2)(4.415))^3]/[6(1 + 4.415)] + 4.415\}^{-1/2}$$

$$= 0.406 \text{ km/s}$$

If metal side tamping, or "barrel" tamping is provided, as shown in Figure 27.11, then the velocity of the gases to the sides is limited to that of the expansion rate of the metal-tamping cylinder.

Benham (Ref. 10) extended the Gurney model to include this factor. The 60°-cone-base angle is replaced in this model by the base-cone angle Θ. The mass of the tamping cylinder is Ψ.

$$\Theta = 90 - \frac{30}{(2\Psi/C + 1)^{1/2}} \tag{27.11}$$

Note that C in this expression is the full charge mass, uncorrected by the cone assumption.

Example 27.6 Let us now use the same flyer and charge as in the previous example, but we will asemble them into a $\frac{1}{4}$-in.-thick-walled steel cylinder. What will the flyer velocity be now?

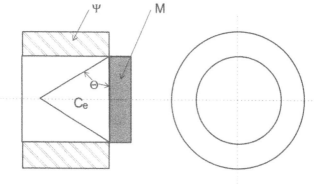

Figure 27.11. Barrel tamped charge.

Solution Before we can find M/C_e , we must first find Θ and then solve for C_e. The mass of the tamping cylinder is $\pi(r_0^2 - r_i^2)L\rho_\Psi = \Psi$, and the mass of the full charge of explosive is $\pi r_i^2 L\rho_{HE} = C$; so Ψ/C is $(7.87)(0.75^2 - 0.5^2)/(1.54)(0.5)^2 = 6.39$. Therefore, the base angle of the effective cone is [from Eq. (27.11)]

$$\Theta = 90 - \frac{30}{(2\Psi/C + 1)^{1/2}}$$

$$= 90 - 30/[(2)(6.39) + 1]^{1/2}$$

$$= 81.9$$

The height of the effective cone, $r \tan \theta$ is 3.51 in., and its volume is $(\pi/3)(0.5)^2(3.51)$ = 0.9189 cubic inches.

In order to determine the volume of the truncated cone, we can subtract from the full cone volume the portion of the cone outside of the charge. Its base radius is $(0.5)(3.51-1)/(3.51) = 0.35755$, and its volume is $(\pi/3)(0.35755)^2(3.51-1) = 0.336$ cubic inches. So the volume of the truncated cone, our effective charge volume, is $(0.9189 - 0.336) =$ 0.583 cubic inches and its mass is 14.7 g. So our M/C_e is $(25.3)/(14.7) = 1.72$, and the flyer plate velocity is now

$$V = (2.357)\{[1 + (2)(1.72)]^3/[6(1 + 1.72)] + 1.72\}^{-1/2}$$

$$= 0.886 \text{ km/s}$$

Here we see that barrel tamping has enabled us to more than double the flyer-plate velocity.

27.2 Fragmentation of Cylinders

The use of the Gurney model enabled us to predict the initial velocity of fragments that were produced by the explosion of a cased cylindrical charge. It is of great interest in relation to warhead design as well as to safety analyses, to be able to predict or estimate the number and size distribution of such fragments.

There are several theories of how fragments are formed and hence how large or small they would be; one of the earliest is from N. F. Mott (Refs. 11, 12). Mott explained the sizes of fragments as a function of the rate of cylinder expansion as compared to rate of a tensile relief wave around the cylinder's periphery. See Figure 27.12.

It is assumed that the cylinder is placed in greater and greater hoop stress (tensile) as it expands. A fracture eventually will occur at some point. The fracture presents a free surface, and a relief wave can now travel away from it.

Fracture can no longer occur in the relieved regions (shown in Figure 27.12 as shaded), but tensile stress and plastic flow are still growing in the unrelieved region where a new fracture is free to form. The size of the fragments then are determined by the balance between the rate of increasing strain and the rate of the relief wave.

Other theories give fragment size as a function of a hypothesized critical expansion velocity (Rinehard and Pearson, Ref. 13), or based upon a critical strain rate, a mechanism related to radial, not tangential, stress gradient across the cylinder wall (Garg and Siekmann, Ref. 14, and Taylor, Ref. 15). Mott, however, gives a more tractable mathematical treatment suited to first-order engineering, as compared to the others. Mott's equations are

$$N(m) = \frac{M_0}{2M_K^2} e^{-\left(\frac{m^{1/2}}{M_K}\right)} \tag{27.12}$$

where $N(m)$ is the number of fragments that are larger than mass m; m, the mass of a fragment (lb); M_0, the mass of the metal cylinder (lb); and M_K, a distribution factor (lb$^{1/2}$).

$$M_K = Bt^{5/16} d^{1/3}\left(1 + \frac{t}{d}\right) \tag{27.13}$$

where B is a constant that is specific for a given explosive-metal pair; t, the wall thickness (in.); and d, the inside diameter of a cylinder (in.).

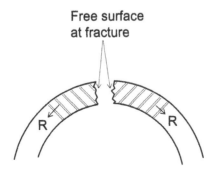

Figure 27.12. Stress-relief waves leaving a fracture.

An example of the size distribution predicted by the Mott formula along with experimental data are shown in Figure 27.13. These data and the calculational results are from Stromsoe and Ingebrigtsen (Ref. 16). As can be seen from this figure, the Mott formula tends to overestimate the larger fragment sizes. This is typical, but usually acceptable for first-order engineering purposes. The biggest problem with the Mott work is finding values of B for a variety of metals and explosives. Table 27.2 gives some values for mild steel and a number of explosives (Ref. 17). Of course a test can be run on a small model of a system of interest to find B for that system.

The values of the CJ pressure for each explosive are included because that was the only explosive parameter found that came close to correlating with B for this rather limited group of explosives. That correlation is shown in Figure 27.14. Often it is of interest to find or predict the largest fragment weight that

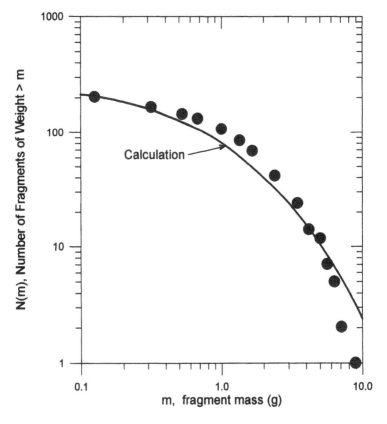

Figure 27.13. Mott calculation along with experimental data for a cast iron aerial bomb (data and calculation from Ref. 16).

Table 27.2 $\quad$ Values of Mott coefficient, B, for Mild Steel Cylinders

Explosive	B ($\mathrm{lb}^{1/2}$ in.$^{-7/16}$)	P_{CJ} (kbar)
Baratol	0.128	137
Composition B	0.0554	265
Cyclotol (75/25)	0.0493	316
H-6	0.0690	227
HBX-1	0.0639	227
HBX-3	0.0808	219
Pentolite (50/50)	0.0620	255
PTX-1	0.0554	249
PTX-2	0.0568	284
TNT	0.0779	190
Composition A-3	0.0549	305
RDX/Wax (95/5)	0.0531	279
Tetryl	0.0681	226

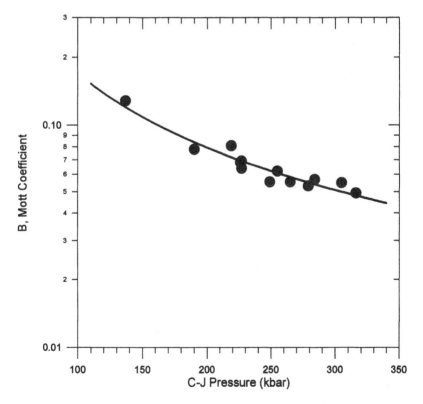

Figure 27.14. $\quad$ Correlation of Mott coefficient, B (for mild steel cylinders) with CJ pressure.

will be produced. To do this, we merely set $N(m) = 1$ and solve Eq. (27.12) for m.

$$m_1 = \left[M_K \ln\left(\frac{M_0}{2M_K} \right) \right]^2 \tag{27.14}$$

Example 27.7 Let us look at the steel cylinder we used for barrel tamping in the previous problem. We recall that it was one inch long, one inch inside diameter, and had a wall thickness of a quarter inch. It was filled with Detasheet™ explosive, which had $\rho_0 = 1.54$ g/cm^3, $D = 7.0$ km/s, and $P_{CJ} = 20.0$ GPa. We can assume that when the charge detonates the casing will fragment. How many fragments should we expect whose weights are are greater than 10, 7.5, 5, 1, and 0.1 g? Also, what will the weight of the largest fragment be? What would that fragment likely look like?

Solution First we will have to find values for B and M_K. From Figure 27.14 we can find B for this explosive since we know its P_{CJ}. We find $B = 8 \times 10^{-2}$. Using this in Eq. (27.13) we find M_K.

$$M_K = (8 \times 10^{-2})(0.25)^{5/16}(1)^{1/3}(1 + 0.25/1)$$

$$= 0.0315$$

Now we find M_0, which was the weight of the cylinder before we blew it up.

$$\text{Vol} = (\pi/4)(OD^2 - ID^2)L$$

$$= (\pi/4)(1.5^2 - 1^2)(1)$$

$$= 0.982 \text{ in.}^3$$

$$M_0 = \rho V = (0.982 \text{ in.}^3)(0.284 \text{ lb/in.}^3)$$

$$= 0.279 \text{ lb}$$

The number of fragments whose weight is greater than 10 g (10 g $= 0.022$ lb) is, from Eq. (27.12),

$$N(10) = \frac{(0.279)}{(2)(0.0315)^2} e^{-\left(\frac{0.022^{1/2}}{0.0315} \right)}$$

$$= 1.27 \text{ or } 1$$

There is one fragment whose weight is greater than 10 g. For the number of fragments whose weight is greater than 7.5 g,

$$N(7.5) = \frac{(0.279)}{(2)(0.0315)^2} e^{-\left(\frac{0.0165^{1/2}}{0.0315} \right)}$$

$$= 2.38 \text{ or } 2$$

Similarly, for the other size fragments, we find

Weight (g)	Number
10	1
7.5	2
5	5
2.5	13
1	31
0.1	87

The weight of the largest fragment (we now know that it is larger than 10 g) is found from Eq. (27.14),

$$m = \left[(0.315)\ln\left(\frac{0.279}{(2)(0.0315)^2} \right) \right]^2$$

$$= 0.0243 \text{ lb } (11 \text{ g})$$

The volume of this fragment is

$$\text{Vol} = (0.0243 \text{ lb})/(0.284 \text{ lb/in.}^3)$$

$$= 0.0856 \text{ in.}^3$$

The wall thickness of the original cylinder was 0.25 in.; if we assume that the wall was not significantly compressed, that would lead to a square fragment 0.585 in. on a side by 0.25 in. thick or a circular fragment 0.66 in. diameter by 0.25 in. thick. The former is more likely.

27.3 Flight of Fragments

Now we know the size of fragments and their initial velocity. As they travel through the air they are slowed down or decelerated. Two major forces act to slow down the fragments, the "face drag" due to dynamic pressure, and the "base drag" due to turbulence behind the fragment. The magnitude of these forces is a function of the projected face area of the fragment, the density of the air through which it is moving, the velocity of the fragment, and a shape factor expressed as the drag coefficient

$$F = C_D A_f \rho_a V_f^2 / 2 \tag{27.15}$$

where F is the force; C_D, the total drag coefficient; A_f, the face area of fragment; ρ_a, the local air density; and V_f, the fragment velocity.

The face drag coefficient, which is larger than the base drag coefficient, is generally relatively constant at low velocities, increases as velocity approaches

Mach 1, and then tends toward a higher constant value with further increasing velocity. The magnitude of the base drag coefficient is just the opposite, decreasing with increased velocity. The sum of these two, called the *total drag coefficient*, C_D, is used when calculating trajectories.

Figures 27.15 through 27.18 show Hoerner's data (Ref. 18) for five regular geometries, a rod (side-on), a sphere, a cube (face-on and corner-on), a cylinder (face-on), and a disc (face-on). These geometries represent the range of idealized fragment shapes most commonly found.

From classical physics, we know that force is equal to mass times acceleration:

$$F = ma \tag{27.16}$$

The drag forces on a fragment slow it down or impart a negative acceleration; therefore, Eq. (27.16) for this case can be written as follows:

$$a = -F/m \tag{27.17}$$

Replacing F, the force, in Eq. (27.17) with the drag force as expressed earlier in Eq. (27.15), yields

$$a = -\frac{C_D A_f \rho_a V_f^2}{2m} \tag{27.18}$$

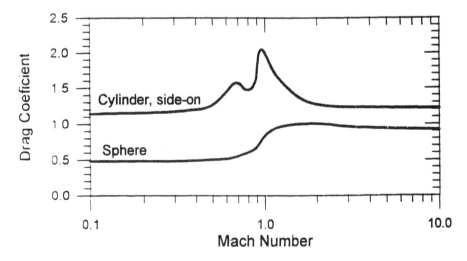

Figure 27.15. C_D versus Mach number, side-on cylinder and sphere.

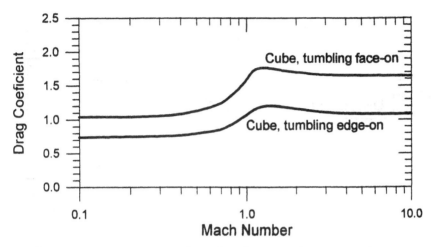

Figure 27.16. C_D versus Mach number, face-on, and corner-on cube.

From the basic definitions of acceleration and velocity, $a = dV/dt$ and $V = dx/dt$, where t is time and x is distance, we have

$$V \, dV = a \, dx \tag{27.19}$$

Combining Eq. (27.18) with Eq. (27.19), we obtain:

$$\frac{1}{V} \, dV = -\frac{C_D A_f \rho_a}{2m} \, dx \tag{27.20}$$

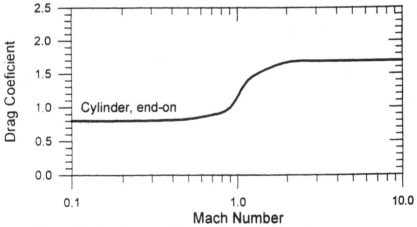

Figure 27.17. C_D versus Mach number, cylinder, end-on.

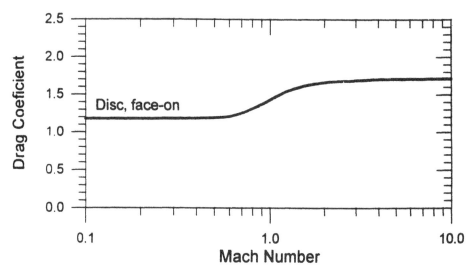

Figure 27.18. C_D versus Mach number, disc, face-on.

and integrating this expression between the limits V_0, x_0 and V, x, where $x_0 = 0$, yields

$$\ln\left(\frac{V}{V_0}\right) = -\frac{C_D A_f \rho_a x}{2m} \tag{27.21}$$

This is the velocity loss due to drag as a function of the distance traveled by a fragment along the path of its trajectory. Of course, this does not take into account the angle of launch and resultant slant range. Putting those factors in along with drop due to the acceleration of gravity complicates the mathematics by generating a set of nonlinear equations. The above Eq. (27.21) can be used for short ranges. The full trajectory calculation will involve either graphical or finite-difference calculations.

Example 27.8 In the last example we found that the $\frac{1}{4}$-in. wall steel cylinder had generated a fragment $\frac{1}{4}$-in. thick with a face area of 0.342 in.^{2}. The cylinder was 1-in. inside diameter and filled with Detasheet™. What was the initial velocity of this fragment? How far would it travel by the time it had lost half its velocity to drag forces?

Solution The initial velocity is found from the Gurney equation for a cylinder. The M/C is $\rho_{STL}(OD^2 - ID^2)/\rho_{HE}ID^2 = 6.39$, and the velocity is

$$V = \sqrt{2E\left(\frac{M}{C} + \frac{1}{2}\right)^{-1/2}}$$

$$= (2.357)(6.39 + 0.5)^{-1/2}$$

$$= 0.898 \text{ km/s}$$

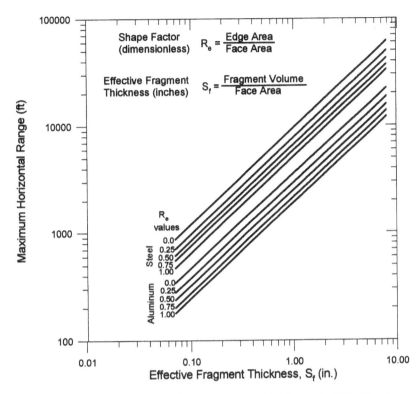

Figure 27.19.   Maximum ranges of fragments with initial speed of 10,000 ft/s.

The velocity of sound in air is around 0.335 km/s; therefore, the Mach number of the fragment is initially $(0.898)/(0.335) = 2.7$. At the point where it has lost half its velocity its Mach number is 1.35. From Figures 27.16 and 27.18, we see that the drag coefficient in this velocity range is around 1.6. From Eq. (27.21) we can now find the distance at which the initial velocity would be halved.

$$x = 2m \ln (V/V_0)/(C_D A_f \rho_a)$$

where $m = 0.0243$ lb (11 g); $V/V_0 = 0.5$; $C_D = 1.6$; $A_f = 0.342$ in.^{2}; $\rho_a = 4.645 \times 10^{-5}$ lb/in.^{3} (1.287×10^{-3} g/cm^3), and $x = 1325$ in. (110 ft).

An interesting analysis was made by Bishop (Ref. 19) for safety evaluations. His analysis took the full slant range and trajectory into account, and was made to estimate the furthest path of the worst fragment from an explosion. His results are shown plotted on Figure 27.19. They are recommended as a conservative safety guide for explosive test ranges.

28

Blast Effects in Air, Water, and on the Human Body

In this chapter we will look at shock or blast waves in air and water and learn how to predict their behavior as a function of both the magnitude of the explosion (weight of explosive) and the distance from the explosion. We shall also look at the effects that blast has upon the human body in respect to damage to hearing, lungs, and to life itself.

28.1 Scaling Air Shock

Although air shock is occasionally scaled against the linear dimensions of the explosive charge (Ref. 20), it is more often scaled to an equivalent weight of TNT. The TNT equivalency is based on energy of explosion obtained in various ways. The preferred method being calculation of either the hydrodynamic or the thermodynamic work function. (Recall section 26.4.) TNT weight equivalence is

$$\text{wt(TNT equiv)} = \text{wt(HE)} \times E_{\text{EXP}}(\text{HE})/E_{\text{EXP}}(\text{TNT}) \tag{28.1}$$

Other methods for estimating TNT equivalence are based either on correlation or empirical tests. One of the oldest of the correlation techniques goes back to the end of the nineteenth century and was developed by M. Berthelot, one of the earliest pioneers of detonation theory. Berthelot did not actually work with TNT, but developed his method to correlate gas-phase detonations with that of nitroglycerine. The conversion of his work to TNT was done later by other workers in the early part of this century (Ref. 21).

$$\%(\text{TNT equiv}) = 840 \times \Delta n(-\Delta H_R^0)/(\text{FM})^2 \qquad (28.2)$$

where Δn is the number of moles of gases produced per mole of HE (water included as a gas); ΔH_R^0, the molar heat of detonation (Kj/mole); and FM, the formula or molecular weight of the given explosive. The Berthelot calculation can be done at either the detonation condition or the fireball condition, whichever describes the specific application.

For comparison purposes, the total energy available from the detonation of TNT calculated from the thermodynamic work function is 1160 cal/g. Estimation of this energy from the hydrodynamic work function using $(\frac{1}{2}) P_{CJ}/\rho_{CJ}$ yields a value for TNT of 1080 cal/g. Calculation of the energy of explosion based upon a large number of air blast experiments using TNT give the value of 1120 cal/g (Ref. 22).

Using the hydrodynamic estimate of one-half of Pv at the CJ state and also assuming that $P_{CJ} = \rho_0 D^2/4$, we can derive a simple estimate of TNT equivalence.

Since $v = 1/\rho$, $\frac{1}{2}(Pv)_{CJ} = P_{CJ}/2\rho_{CJ}$
and at the CJ state: $P_{CJ} = \rho_0 D^2/4$ and then $\rho_{CJ} = \frac{4}{3}\rho_0$
combining these yields; $\frac{1}{2}(Pv)_{CJ} = 3D^2/32$
TNT equivalent:$= \frac{1}{2}(Pv)_{CJ}(\text{HE})/\frac{1}{2}(Pv)_{CJ}(\text{TNT})$
$= D^2 (\text{HE})/D^2 (\text{TNT})$

If we use TNT at 1.64 g/cm^3 as our standard of comparison, then $D(\text{TNT}) =$ 6.95 km/s, and TNT equivalent $= D^2(\text{HE})/48.3$.

Table 28.1 gives TNT equivalents for several explosives and for several meth-

Table 28.1 Comparison of TNT Equivalent Estimates with Test Data

	Ammonium Picrate	HBX-3	Military Dynamite (MVD)	Pentolite (50/50)	Torpex	Tritonal
Density (g/cm³)	1.55	1.81–1.84	1.1–1.3	1.66	1.81	1.72
Detonation velocity (km/s)	6.85	6.92–7.01	6.6–7.2	7.465	7.6	6.475–6.70
TNT equivalent	0.97	0.99–1.02	0.9–1.07	1.15	1.20	0.87–0.93
by $D^2/48.3$						
by Berthelot method	0.87	1.10	—	1.56	1.18	0.89
Air-blast tests	0.85	1.16	1.05	1.16 (1)	1.23	1.07
Ballistic mortar tests	0.99	1.11	1.22	1.26	1.38	1.24
Plate dent tests	0.91	—	—	1.21	1.20	0.93
Sand crush tests	0.82	0.94	—	1.16	1.24	—
Trauzl tests	—	—	1.10	1.22	1.64	1.25

Sources: References 22 and 23.

ods of obtaining the equivalence, where sufficient data could be found for comparisons.

All of the above gives us the energy that is potentially available from the explosive. We do not always recover all of this energy, however. The amount of energy available from the explosive may be partitioned between the air shock and other work that the explosive is doing at the same time. An example of this is a cased charge. Let us say that we have a cylinder of TNT encased in steel. The $M/C = 1$ (this is typical of fragmentation bombs). When the explosive detonates, it expands and fragments the steel casing. Energy is transferred to the steel in three modes: strain and fracture, shock heating, and kinetic energy of the fragments. The interaction of the detonation with the steel will produce a shock of about 32 GPa. This will heat the steel to about 300°C. This takes around 35 calories for each gram of steel. The strain energy (depending on the particular alloy) can be anywhere from 25 to 150 cal/g of steel. The fragments (from Gurney calculations) will be launched at around 1.87 mm/ms, and their kinetic energy would then be around 415 cal/g. Since the $M/C = 1$, for each gram of steel, there was 1 g of TNT. The energy available in the TNT was 1159 cal/g; we gave up a total of ~500 cal/g to the steel. This leaves only ~660 cal/g to work on the air in the form of an air-blast wave. In this case almost half of the energy is lost prior to the air shock. Other means of energy loss would be in crater formation for detonations on, in, or close to the ground.

While we do lose some energy to crater formation and ejecta kinetic energy in a ground blast, the effective yield of a ground blast is increased because of reflection of the shock from the ground surface. If no crater were formed and the ground was a perfect reflector, then the effective air blast yield of a ground-level explosion would be equivalent to twice that of the same charge fired in free air. Generally, the effective yield for ground-level detonation on undisturbed dirt is around 160 to 180%.

From all of the above, we see that at best we can only roughly estimate the energy available to form air-blast. Fortunately, as we will soon see, air blast parameters scale with the cube root of energy available. So large errors in energy estimates are cut down considerably when we predict air-blast parameters.

Let us look now at scaling air-blast pressure using dimensional analysis; see Figure 28.1. Let us picture an explosive charge at mass W. We will scale against mass of charge since we will later define this mass in terms of TNT weight equivalents. We will measure the peak overpressure, P^0, at a distance R from the center of the charge. The other parameters that will pertain to or affect this pressure are the ambient air pressure, P_a, and ambient air density, ρ_a.

We have five variables in this problem. They are W, R, ρ_a, P_a, and P^0. The dimensions are, respectively, FT^2/L, L, FT^2/L^4, F/L^2, and F/L^2. There are only three dimensions, therefore, we should be able to form two dimensionless products. These are

$$\left(\frac{P^0}{P_a}\right) \quad \text{and} \quad \left(\frac{R^3\,\rho_a}{W}\right)$$

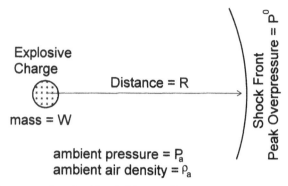

Figure 28.1. Parameters involved in air-blast scaling.

Now, rather than calculate the value of ρ_a for each condition we test, let us replace ρ_a with (P_a/T_a) since we know that from the ideal gas law $\rho_a = KP_a/T_a$. T_a, by the way, is the absolute ambient air temperature. So our scaling parameters are now

$$\left(\frac{P^0}{P_a}\right) \quad \text{and} \quad \left(\frac{R^3 \, P_a}{WT_a}\right)$$

It is customary to use the cube root of the second term (with permission from Dr. Buckingham) so we will have a scaling curve in the form:

$$\left(\frac{P^0}{P_a}\right) = f\left[R\Big/\left(\frac{WT_a}{P_a}\right)^{1/3}\right]$$

This is shown as Figure 28.2.

Example 28.1 Let us suppose that we are at an altitude of 5300 ft and the temperature is 75°F. We will detonate a charge of 20 kg of Cyclotol 77/23 (at $\rho_0 = 1.755$ g/cm³). What will the peak shock overpressure be at 10 m from the center of the charge?

Solution The ambient pressure at 5300 ft altitude is about 0.834 bar, and the Kelvin equivalent of 75°F is 297 K. If we are to scale according to Figure 28.2, then we need the scaled distance Z as

$$Z = R\Big/\left(\frac{WT_a}{P_a}\right)^{1/3}$$

and we need W as TNT equivalent weight. From Section IV, Table 20.1, we find D for Cyclotol 77/23 at $\rho_0 = 1.755$ is 8.29 km/s, and we will estimate the TNT equivalence as

$$\text{TNT equivalent} = D^2/48.3 = (8.29)^2/48.3$$

$$= 1.42$$

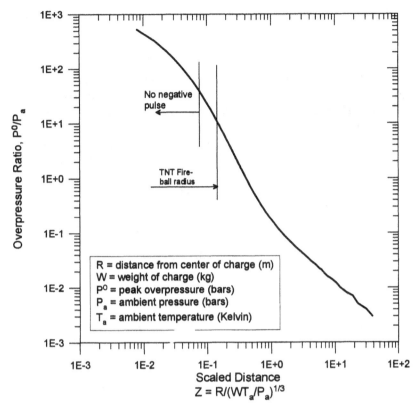

Figure 28.2.   Peak overpressure ratio versus scaled distance (based on data in Ref. 22).

Now we can find Z,

$$Z = 10 \left/ \left(\frac{(20)(1.42)(297)}{(0.834)} \right)^{1/3} \right.$$

$$= 0.46$$

From Figure 28.2, we see that for this value of Z, the value of $P^0/P_a = 0.8$, and

$$P^0 = (0.8)(P_a) = (0.8)(0.834) = 0.67 \text{ bar (9.7 psi)}$$

The procedure used by Kinney and Graham (Ref. 22) to find time, both shock time of arrival as well as shock positive pulse duration, is to first determine the scaled times from the reference explosion shown in Figures 28.3 and 28.5 using Z, the scaled distance as was done with pressure, above. The scaled time thus obtained is then corrected for explosive weight, ambient temperature, and ambient pressure

$$\text{time} = (\text{scaled time}) \times W^{1/3}/[(P_a/P_0)^{1/3}(T_a/T_0)^{1/6}]$$

where P_a and T_a are the actual ambient conditions and P_0 and T_0 are the reference conditions (1.01325 bar and 288 K). The time of arrival is in milliseconds, and P_a and T_a are in bars and Kelvin, respectively. The air-blast wave has a pressure-time profile similar to that shown in Figure 28.4.

The profile seen in Figure 28.4 is typical for waves some distance from the charge. The negative portion of the wave is caused by the inertia of the air that had been accelerated in the direction of the blast. When the positive phase has passed, this air mass is "pulled back" by the still air behind it. There is no negative phase very close to the charge. The scaled distance at which a negative phase can begin is shown on the various scaling curves. The curves also indicate the scaled radius of a TNT fireball.

The positive pulse duration is scaled in the same manner as the time of arrival,

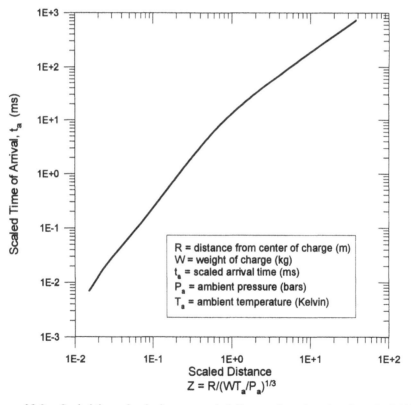

Figure 28.3.   Scaled time of arrival versus scaled distance (based on data from Ref. 22).

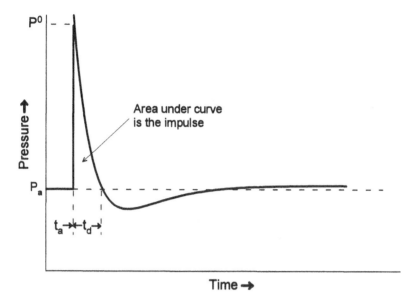

Figure 28.4. Typical profile of an air-blast wave.

and is shown in Figure 28.5. The dip in pulse duration is due to the phenomenon mentioned earlier about where the negative pulse can form. Notice that the minimum positive pulse duration occurs at that point.

Example 28.2 For the same charge and conditions as in the previous example, how long would it take for the shock to arrive at 10 m from the time of detonation?

Solution We can find the time of arrival using Figure 28.3. In Example 28.1 we found $Z = 0.46$, and for this value of Z we find from Figure 28.3 that scaled time $= 4.0$ ms. The actual time of arrival is then

$$\text{time} = (\text{scaled time}) \times W^{1/3}/[(P_a/P_0)^{1/3}(T_a/T_0)^{1/6}]$$

$$= (4.0)(20 \times 1.42)^{1/3}(0.834/1.01325)^{1/3}(297/288)^{1/6}$$

$$= 11.5 \text{ ms}$$

28.2 Scaling Shocks in Water

Since water is relatively incompressible, we do not get concerned with an "ambient pressure" or density term in the scaling. Also, because the water decelerates fragments so close to their origin, much of the kinetic energy as well as the heat imparted to casings is recovered. Also recovered is much of the thermal energy

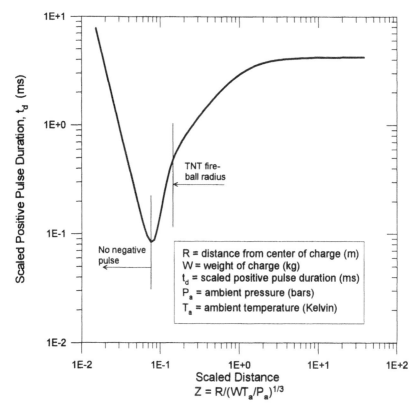

Figure 28.5. Scaled positive pulse duration versus scaled distance (based on data from Ref. 22).

of the gases that help form steam in the pressure driving bubble. A fairly simple empirical scaling law is used (Ref. 24) for scaling pressures.

$$P^0 = K\left(\frac{W^{1/3}}{R - R_0}\right)^a \tag{28.3}$$

where P^0 is the peak overpressure (psi), K is a constant that depends upon the particular explosive, W is the charge weight (pounds), R is the distance from the center of the charge (in.), R_0 is the radius of the charge (in.), and α is a constant that depends upon the particular explosive. Values for K and α are given in Table 28.2 for several explosives (Ref. 25).

The values for heat of detonation were included in this table to show that there seems to be decent correlation (among this very limited group) between the constant K and ΔH^0_{exp}, seen plotted in Figure 28.6. Reference 26 gives the constants for Pentolite as $K = 4.78 \times 10^5$ and $\alpha = 1.194$.

Table 28.2 K and α Values for Several Explosives

Explosive	K	α	ΔH^0_{exp} (kcal/g)
TNT	3.6×10^5	1.13	1.41
Composition A-3	3.9×10^5	1.13	1.58
Tetryl	3.75×10^5	1.15	1.51
PETN	3.85×10^5	1.13	1.65
Pentolite (50/50)	3.74×10^5	1.13	1.53
HBX-1	4.34×10^5	1.15	1.84
HBX-3	4.55×10^5	1.14	2.11

Example 28.5 Let us assume that we have an 8-pound sphere of cast TNT under water. When detonated, what peak pressure will be produced 12 ft away from the source?

Solution Since the TNT has been cast, we can assume that its density is 1.64 g/cm³. We can find the radius of the charge from the weight and density (which gives us the volume) and

$$\text{Volume} = \tfrac{4}{3} \pi R_0^3$$

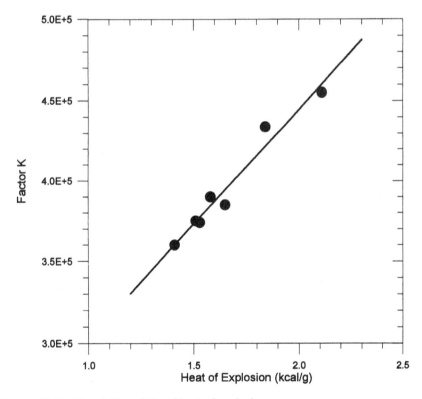

Figure 28.6. Correlation of K and heat of explosion.

We find that $R_0 = 3.18$ in. From Table 28.2 we find, for TNT, that $K = 3.6 \times 10^5$ and $\alpha = 1.13$. We can now find the pressure

$$P^0 = K\left(\frac{W^{1/3}}{R - R_0}\right)^a$$

$$= 2940 \text{ psi}$$

We can change Eq. (28.3) to a scaling form by replacing the weight term W with its equivalent density and volume. Assuming the charge is spherical

$$W = \tfrac{4}{3} \pi R_0^3 \rho_0$$

Replacing W in Eq. (28.3) with this yields

$$P^0 = K\left(\frac{(4\pi\rho_0/3)^{1/3} R_0}{R - R_0}\right)^a$$

which in turn can be reduced to

$$P^0 = K\left(\frac{(4\pi\rho_0/3)^{1/3}}{(R/R_0) - 1}\right)^a$$

The only serious problem with using Eq. (28.3) or (28.4) is for estimating pressure very close to a charge. As can be seen, the difference $(R-R_0)$ approaches zero and pressure approaches infinity as R approaches R_0. This artifact becomes noticeable when R/R_0 is less than ~ 1.6. For the region between the surface of the charge and $R/R_0 \approx 1.6$, another estimate can be made.

The pressure at the interface of the charge and the water can be found by solving the shock wave interaction between the explosive detonation products P-u isentrope and the P-u Hugoniot for water (for TNT this is around 120 kbar). This point (at $R/R_0 = 1$) can then be extrapolated back to the Eq. (28.3) curve of $P = f(R/R_0)$ using a power fit (a straight line on a log-log plot). This line should come tangent to the P versus R/R_0 curve. This is shown in Figure 28.7 for P vs R/R_0 for TNT under water.

Scaling of impulse underwater can be done by another simple correlation (Ref. 27).

$$I = BW^{1/3}\left(\frac{12W^{1/3}}{R - R_0}\right)^F \tag{28.4}$$

where I is impulse (lb s/in.^{2}), W is the explosive weight (lb), R, R_0 are in inches, and B, F are constants characteristic of each explosive. The problem with this relationship is the lack of data for a reasonable number of different explosives. Table 28.3 gives values of B and F for three explosives.

Example 28.6 For the same charge and location as in the previous problem, what impulse is delivered at 12 ft?

Solution The impulse scaling is given in Eq. (28.4), and the values of B and F are found from Table 28.3. For TNT, $B = 1.46$ and $F = 0.89$; therefore,

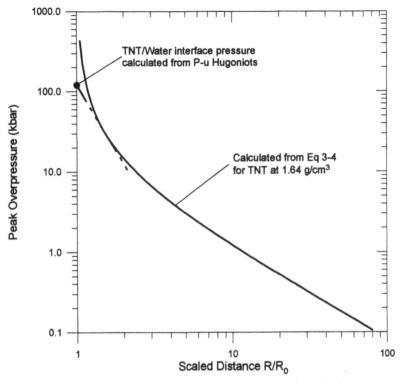

Figure 28.7. Peak pressure versus scaled distance for shock from TNT underwater.

$$I = (1.46)(8)^{1/3}\left(\frac{12(8)^{1/3}}{144 - 3.18}\right)^{0.89}$$

$$= 0.6 \text{ lb s/in.}^2$$

Time of arrival is easily calculated because below $\sim$2-3 kb, the wave velocity closely approaches the standard acoustic velocity ($\sim$1500 m/s). For close-in timing, velocities are found from the P-u and U-u Hugoniots for water or a convenient equation of state such as that from Rice-Walsh (Ref. 28)

$$U_s = C_0 + 10.99 \ln(1 + u_p/51.9) \tag{28.5}$$

Table 28.3   Values of Factors B and F for scaling impulse from underwater explosions

Explosive	B	F
TNT	1.46	0.89
Tetryl	1.73	0.98
Pentolite	2.18	1.05

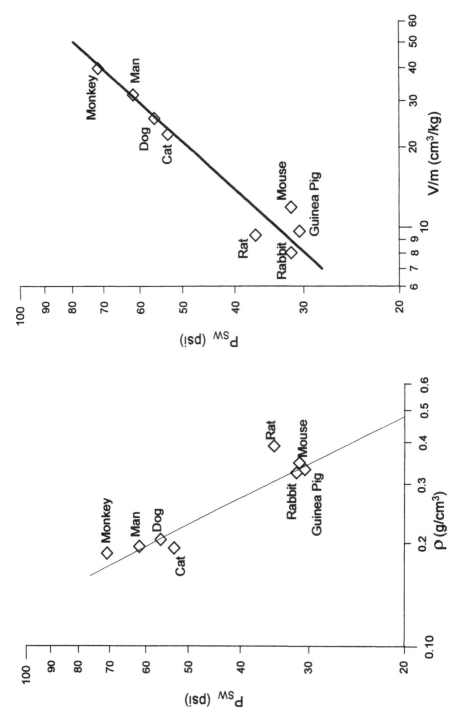

Figure 28.8.   Square-wave overpressure resulting in 50% survival (P_{SW}) as functions of average lung density an average lung volume per unit body mass for eight mammalian species.

where U_s is the shock velocity (km/s); C_0, the bulk sound speed (1.483 km/s); and u_p, the particle velocity (km/s). For lower pressures (< 100 kbar) the linear approximation $U_s = 2 + 1.4\, u_p$ is sufficient.

28.3 Physiological Response to Air Blast

The most vulnerable organs in our bodies in response to shock are the ears and lungs because these organs contain air or other gases. The damage is done at the gas-tissue interface, where spalling and tearing can occur. Testing of the response to shock would be rather difficult if we had to use people, especially if we wanted a good statistical sample. Scaling of biological responses can be done here. For lethality of shock trauma, it has been found that the lungs can be scaled from one species to another. Testing is done such that a statistical sampling can be made to find the 50% lethality and sigma as a function of incident peak pressure and pulse duration. The denser the lungs, the more vulnerable a particular mammalian species will be to shock. Also, the lower the specific lung volume (lung volume/body mass), the more vulnerable the species is to shock. These relationships, 50% lethality pressure level versus lung density and volume, are linear on log-log plots, as shown in Figure 28.8 (Ref. 29). Using this type of scaling data, experiments have been run on small species and extrapolated to determine response of humans to shock.

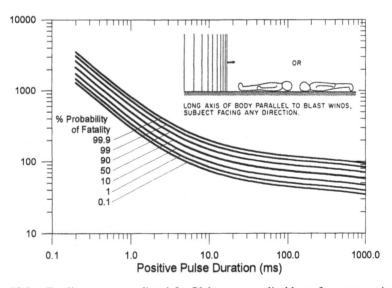

Figure 28.9. Fatality curves predicted for 70-kg man applicable to free-stream situations where the long axis of the body is parallel to the direction of propagation of the shocked blast wave.

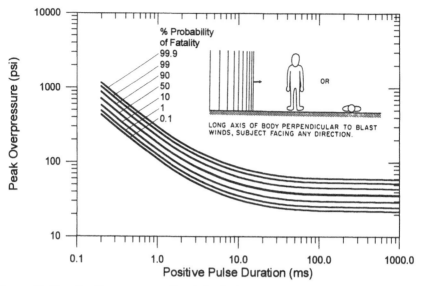

Figure 28.10. Fatality curves predicted for 70-kg man applicable to free-stream situations where the long axis of the body is perpendicular to the direction of propagation of the shocked blast wave.

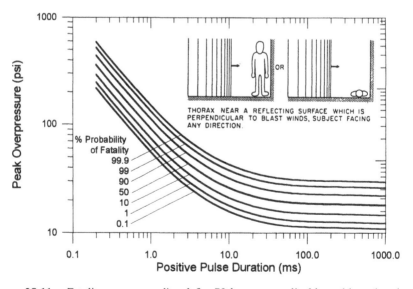

Figure 28.11. Fatality curves predicted for 70-kg man applicable to blast situations where the thorax is near a surface against which a shocked blast wave reflects at normal incidence.

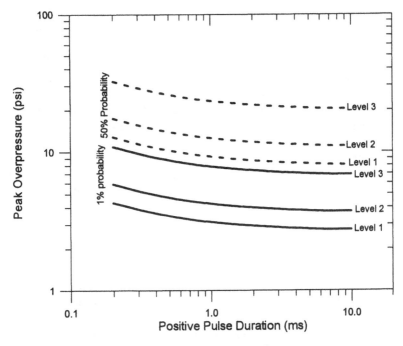

Figure 28.12.   Response of ears to a single pressure pulse.

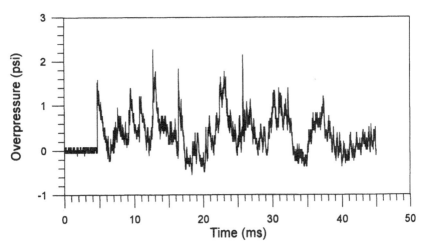

Figure 28.13.   Complex pressure signal generated by firing a single explosive charge in a closed room.

It has been found that both ear and lung responses are dependent not only on pressure but also upon impulse and body orientation. The shorter the pulse width, the higher the pressure the body can tolerate.

In Figure 28.8 the term P_{SW} is the square-wave overpressure resulting in 50% survival, ρ is the average density of lungs, and V/m is the average gaseous volume of lungs per unit body mass.

Figures 28.9, 28.10, and 28.11 show this relationship for humans. The figures differ in the body orientation relative to the incoming shock. Figure 28.12 shows the 1% and 50% probabilities of adverse ear responses in relation to pressure and pulse duration for three different levels of ear injury. Level 1 consists of minor damage and/or healable small tears or rupture of the tympanic membrane (ear drum); level 2 for tearing of the membrane that will result in permanent hearing loss; and level 3, which includes severe rupture of the membrane along with inner ear damage.

All of the preceding are for single-pulse response. This is typical of outdoor blast waves. When an explosion occurs indoors, the wave is reflected off walls, ceiling, and floor, forming a series of pressure pulses. The amplitude of the reflected pulses can often exceed that of a free-air explosion at the same distance. An example of this is shown in Figure 28.13. Ear and lung responses to repetitive pulses behave as if the peak pressure were delivered in a very long pulse. This effect must be accounted for when analyzing safety indoors with small explosive devices.

When dealing with ear damage potential, pressures are often expressed as ''db noise level.'' This is a log transform of pressure in normal units. The transform (relative to 2×10^{-4} μbar) is

$$db = 180 + 20\log_{10}\left(\frac{P}{2.900755}\right)$$

where P is peak overpressure (psi).

29

Scaling Craters

When an explosive is set off on, in, or near the ground, it makes a hole. We call such an explosively formed hole a crater. To describe the mechanisms that take place during crater formation, we first must divide the precrater conditions into two major categories: above-ground explosions and below-ground explosions. We make this distinction because the mechanisms are different in these two categories.

In this chapter, we will examine craters created from ground-level bursts, above-surface bursts, and buried bursts. We will find methods to scale from existing databases that will allow us to correct for the type of explosive as well as the type of ground medium.

29.1 Crater Formation Mechanisms

Above-ground explosions can be at zero "height of burst" (HOB), where the center of the charge is at ground level or higher. The height is often expressed in charge radii, where we are assuming a spherical or spherical-equivalent charge. When such a charge is fired, a shock wave is coupled into the ground. This shock shatters the ground medium, scouring it, and infusing detonation gases into it. The expansion of these gases ejects much of this material, called "ejecta," into the air. The ground, compacted and in plastic flow, has a rarefaction wave following into the compressed region, reversing the particle velocity direction and spalling back more ejecta. Some of the ejecta falls back into the hole, and this leaves us with a crater. The fall-back fill also contains subsidence material that sloughs off of the upper walls and lip. A typical crater is shown in Figure 29.1 (Ref. 3).

Below-ground or buried explosions also couple shock into the ground causing compaction, plastic flow, and shattering. The detonation products form a high-pressure cavity while this is occurring. When the shock reaches the surface, a relief wave is formed that travels back down into the ground. Shattered earth is spalled off into the air at the free surface and behind the relief wave. When the relief wave reaches the pressurized cavity, the gases can now heave the loose dirt (working against gravity) up into the air behind the spalled dirt. Eventually the fall back settles and again forms a crater. This process is shown in Figure 29.2 (Ref. 30). If the explosive is buried too deep, it does not have enough energy to lift the overburden, and no crater is formed. As burial depth is increased, the crater dimensions increase until a maximum is reached.

Burial deeper than this results in smaller and smaller craters until the point is reached where no crater forms. Figure 29.3 (Ref. 30) shows craters typical for various heights or depths of burial (DOB).

29.2 Surface Bursts

Because we cannot categorize and control all of the various states and kinds of dirt (earth, gravel, etc.) we cannot make precise analytical models of the cratering process, and so we resort to dimensional analysis.

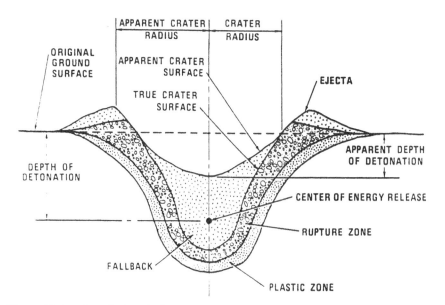

Figure 29.1.   A typical explosion crater.

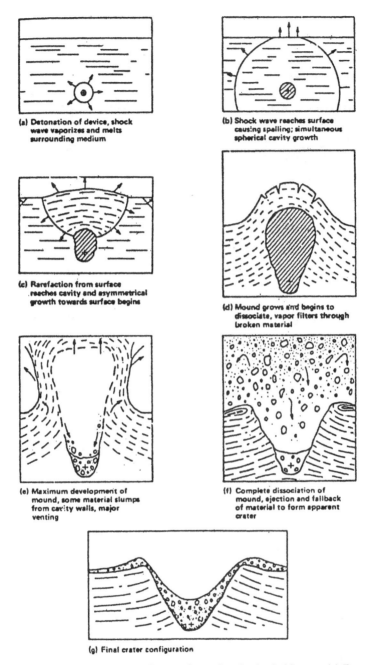

Figure 29.2. Sequential stages of crater formation for buried bursts. (a) Detonation of device, shock wave vaporizes and melts surrounding medium. (b) Shock wave reaches surface, causing spalling; simultaneous spherical cavity growth. (c) Rarefaction from surface reaches cavity and asymmetrical growth towards surface begins. (d) Mound grows and begins to dissociate; vapor filters through broken material. (e) Maximum development of mound; some material slumps from cavity walls; major venting. (f) Complete dissociation of mound; ejection and fallback of material to form apparent crater. (g) Final crater configuration.

423

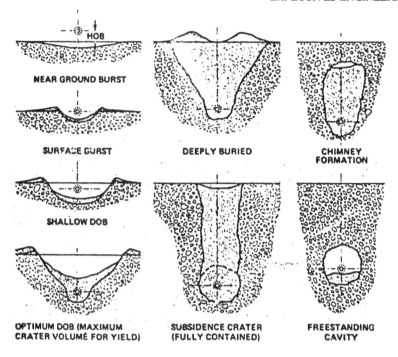

Figure 29.3.   Primary types of craters resulting from a variety of burst positions.

We do know that the variables or parameters that seem to be important in relation to surface blasts are as follows:

E is the energy in explosive (FL);

P_{CJ} is the detonation pressure of explosive (F/L^2);

Y is the strength of ground medium (F/L^2); and

R_a is the apparent crater radius (L) and the dependent variable, or crater parameter that we will attempt to predict.

That is four variables containing two basic dimensions. We can form two dimensionless products:

$$\left(\frac{R_a^3 Y}{E}\right) \text{ or } R_a\left(\frac{Y}{E}\right)^{1/3} \quad \text{and} \quad \left(\frac{P_{CJ}}{Y}\right) \tag{29.1}$$

The second product is only a function of the ratio of ground-to-explosive properties. If we held both the ground and explosive properties constant, then we would have only one dimensionless product that would always be equal to some constant value

$$R_a\left(\frac{Y}{E}\right)^{1/3} = \text{constant, or}$$

$$R_a = KE^{1/3}$$

(29.2)

where K contained the constant ground property, Y, as well as the explosive property P_{CJ}. In other words, for a fixed ground type and the same explosive, crater radius is proportional to the cube root of the energy of explosion. To check this out, Figure 29.4 shows data for apparent crater radius versus charge energy for charges varying from less than 1 g to over 10 megatons. Indeed R_a does vary proportionately with $E^{1/3}$. When we did our dimensional analysis, we neglected the viscosity of the dirt as well as density and acoustic or shock velocities. These variables would have added more dimensionless products. If these variables were important in the mechanism of surface-burst craters, then R_a versus $E^{1/3}$ would not be proportional over this huge range because we could not hold all of these other products of variables constant. Because R_a versus $E^{1/3}$ is proportional, however, we know that we were justified in neglecting these other variables.

You will notice that although the trend in Figure 29.4 is cube-root scaling, there is quite a bit of significant scatter. This is because these data are for various explosives as well as for various types of dirt. (Remember those other two dimensionless groups!)

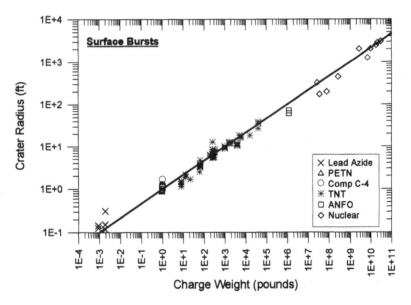

Figure 29.4. Crater radius versus charge weight for a broad range of explosives types and weights.

Let us examine the effect of dirt type first. A large number of explosive tests as well as a series of very near surface nuclear shots have been conducted on very different types of geology. From these experiments, statistical values for "cratering efficiency," E_{CR}, as a function of geological type were determined. The efficiency, E_{CR}, is related to crater *volume*. This same number can be used in relation to radius as well as depth in the form $E_{CR}^{1/3}$, since the volume of a crater is roughly proportional to the cube of its linear dimensions. Table 29.1 gives E_{CR} values for various geological materials.

The way we can apply this to crater dimension prediction is to use the ratio of the E_{CR} of the material in question to E_{CR} of a material for which you have known scaling data. An example follows:

We know that K [Eq. (29.1) for TNT shots in NTS alluvium is equal to 0.97 ft/lb$^{1/3}$. We want to know what R_a versus E is for TNT in Montana clay soil,

$$R_a = 0.97 \ E^{1/3} \left(\frac{E_{CR1}}{E_{CR2}} \right)^{1/3} \tag{29.3}$$

where E_{CR1} is the volumetric cratering efficiency in the Montana clay soil, and E_{CR2} is that for NTS alluvium. Simple. Figure 29.5 shows these same efficiency data plotted on a relative scale (Ref. 30).

If we use NTS alluvium as the reference or standard ground medium, then a general scaling relationship can be written as

$$R_a = K_{NTS} \left(\frac{E_{CR}E}{0.5} \right)^{1/3}$$
$$= K_{NTS}(2E_{CR}E)^{1/3} \tag{29.4}$$

where E now becomes TNT equivalent energy. Since explosive energy is proportional to explosive weight, we can express E as weight as long as the proportionality is included in K.

Now let us look at the effect of explosive type. Using data from Refs. 31 and 32, the soil type was adjusted and compensated for as just shown. These data were for four explosives: TNT, Composition C-4, PETN (in prima cord) and ANFO. Having corrected all of the data to equivalent craters in NTS alluvium, the ratio K was then calculated for each point [Eq. (29.2)]. The mean value for K for each type of explosive was then plotted against that explosive's CJ pressure. The results are shown in Figure 29.6. This resulted in the linear relationship:

$$K = 0.46 + 0.027 \ P_{CJ} \tag{29.5}$$

where K is in ft/lb$^{1/3}$ and P_{CJ} in GPa. This can then be substituted into Eq. (29.4).

$$R_a = (0.46 + 0.027 \ P_{CJ})(2E_{CR}W)^{1/3} \tag{29.6}$$

Table 29.1 HE Cratering Efficiencies for Various Earth Materials at a Zero Height of Burst

Material	Test Area (Project)	Cratering Efficiency E_{CR} (ft³/lb.)
Coral sand (saturated)	Eniwetok Atoll (Pace)	2.00
Clay soil/shale (saturated)	SE Colorado (Middle Gust)	2.00
Clay soil/shale claystone	Montana (Diamond Ore)	0.95
Glacial soil	Alberta, Canada (DRES)	0.75
Clay siltstone	Louisiana (Essex)	0.60
Clay soil/shale	SE Colorado (Middle Gust)	0.55
Alluvial soil	Nevada (NTS)	0.50
Sandy clay soil	New Mexico (CERF)	0.475
Playa	Nevada, New Mexico (NTS, PLEX)	0.45
Soil/sandstone	W. Colorado (Mixed Company)	0.25
Basalt-granite	Washington, Utah (Mice, Mine Shaft)	0.20

Reference 30.

ᵃ These numbers are not all of equal confidence (Ref. 30).

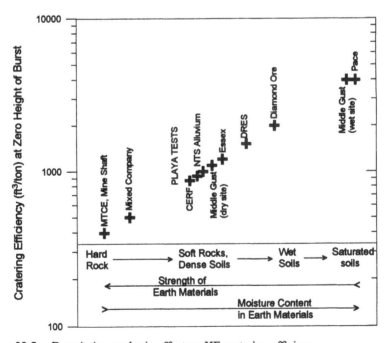

Figure 29.5. Descriptive geologic effect on HE cratering efficiency.

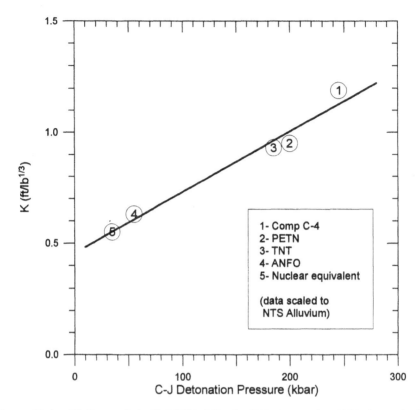

Figure 29.6. (1) Comp. C-4; (2) PETN (PC); (3) TNT (cast); (4) ANFO (bagged) (*) nuclear equivalent (data scaled to NTS alluvium).

where W is in TNT equivalents, pounds. This can now be used as the NTS alluvium scaling for radius versus energy$^{1/3}$.

Example 29.1 A thirty gallon shipping drum containing 215 pounds of picric acid is found half buried in glacial soil. Rather than move this material, it is decided to detonate it in place. What size crater will be formed?

Solution We must get the properties of the explosive first. In order to find the detonation velocity and the P_{CJ}, we have to first estimate the density of the explosive. It is normal to have about 10% spare volume when packing a drum with powder, so let us assume the volume of the explosive is about 27 gallons or about 98 liters. Then the density of the picric acid would be approximately 1 g/cm^3. The detonation velocity at this density can be found from $D = j + k\rho$ and we can find the values of j and k in Table 21.7 in Section IV.

$$D = 2.21 + (3.045)(1) = 5.255$$

The CJ pressure can now be estimated from $P_{CJ} = \rho_0 D^2(1 - 0.7125\, \rho_0^{0.04}) = 7.94$ GPa. The TNT equivalent weight can be estimated by $W_{TNT} = W_{HE}\, (D^2/48.3) = 123$ pounds.

The cratering efficiency found from Table 29.1 is 0.75, and now the crater radius can be estimated as

$$R = (0.46 + 0.027 \ P_{CJ})(2 \ E_{CR} \ W_{TNT})^{1/3}$$
$$= 3.8 \ \text{ft}$$

29.3 Above-Surface Bursts

When the explosive charge is raised above the ground surface, the air between the two acts as a cushion, reducing the pressure that reaches the ground surface. Rather small heights have very significant effects. The mechanism of crater formation is the same as that for surface bursts except that the energy coupled to the ground is lower.

At surface, HOB = 0, approximately 33% of the explosive energy is coupled to the ground to form the crater. At a HOB = 2.5 radii (of charge) only approximately 1% of the explosive energy is coupled to the ground. Figure 29.7 shows this coupling efficiency, percent E in ground versus HOB.

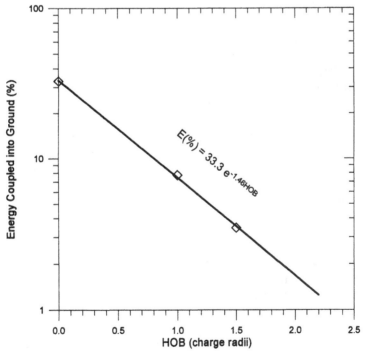

Figure 29.7. Percent energy coupled to ground as a function of height of burst (Ref. 30).

Use the scaling data for surface bursts to scale crater dimensions for above-ground bursts. Correct these data for ground medium and explosive type, but also correct the explosive equivalent weight by the ratio of percent energy coupled at height of burst to 33% (%E coupled at HOB = 0). Combined with the equation shown in Figure 29.6, this becomes

$$R_0 = (0.46 + 0.027\ P_{CJ})(2\ E_{CR}\ We^{-1.457\ \text{HOB}})^{1/3} \qquad (29.7)$$

Example 29.2 If the drum of picric acid in Example 29.1 had been sitting on a table (30 in. tall) when it was detonated, how big would the crater be in this case?

Solution Let us assume that the drum is a sphere and its equivalent radius is then

$$R_{\text{charge}} = (3V/4\pi)^{1/3}$$

We said the volume of explosive was about 98 liters or 5960 in.[3], so the charge radius is 11.25 in. The center of the charge is then this 11.25 + 30 in. and the HOB in charge radii is 3.67.

Solving Eq. (29.5) for crater radius with this HOB correction yields a radius of only 0.65 ft. Following a rough rule of thumb that crater depth is about one-half the radius, this crater would be only about 4 in. deep.

29.4 Buried Bursts

When we attempt to correlate or scale crater data for buried bursts using the cube root of weight or energy as a scaling parameter, we run into a problem because the data do not correlate well. Figure 29.8 shows crater radius divided by depth of burial versus the cube root of TNT equivalent weight divided by depth of burial. Notice that these data do not correlate well with these scaling parameters. Empirical scaling parameters have been developed that tighten up the data considerably (Ref. 33); primary among these is scaling weight of TNT to the 1/3.4 power. The same data from Figure 29.8 are shown replotted in this form, $R/_{\text{DOB}}$ versus $W^{1/3.4}/\text{DOB}$, in Figure 29.9. Why does the (1/3.4) power scaling correlate the data? Remember when we described the mechanisms of crater formation that buried bursts also had to lift the overburden. This difference in mechanism is quite profound. There are several theories on why the 1/3.4 scaling works. One, by Vesic (Ref. 35), postulates that due to the compressibility of the earth medium, the strength increases as depth increases, and therefore, the pressure developed in the cavity increases with depth. Vesic attributes the cratering scaling difference to the cavity pressure. His scaling factor varies with actual depth. Another theory, by Chabai (Ref. 36), shows that dimensional analysis predicts that buried crater dimensions should scale to the $\frac{1}{4}$ power to the $\frac{1}{3}$ power, depending upon depth.

Baker et al. (Ref. 37) also find the scaling from dimensional analysis to contain both the $\frac{1}{3}$ and $\frac{1}{4}$ power terms, but show that the product of the dimensionless

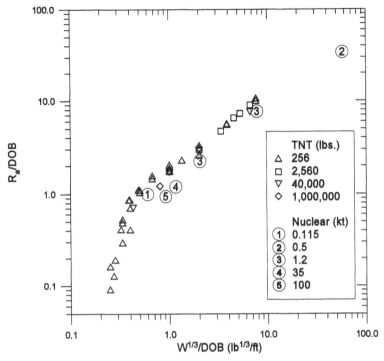

Figure 29.8. Experimental data from buried charges plotted as cube-root scaling.

groups containing these terms is what leads to the 1/3.4 power correlation. Let us examine that.

In addition to the parameters we found applicable to surface bursts, we must now add those that affect the lifting and containment effects of the overburden. These parameters include depth of burial and pressure gradient of the soil with depth (which is assumed linear), giving us

E, energy in the explosive (FL);
DOB, depth of burial (L);
Y, strength of soil (F/L^2);
K, pressure gradient in soil (F/L^3); and
R_a, apparent crater radius (L).

Notice that neither Chabai nor Baker includes the explosive CJ pressure among the important variables. That is reasonable since the work here is in lifting, *PV* work. Shattering and shock coupling with resulting surface spalling, which played an important role in the formation of surface blast cratering, are not major effects here.

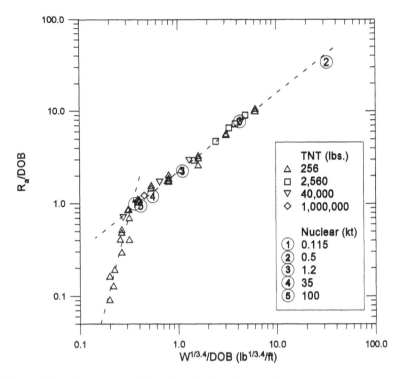

Figure 29.9. Experimental data from buried charges plotted at 1/3.4 scaling.

We have five variables and two dimensions; so we should need three dimensionless groups that are

$$\left(\frac{R_a}{DOB}\right), \left(\frac{W}{Y \times DOB^3}\right), \text{ and } \left(\frac{W}{K \times DOB^4}\right)$$

$$\text{or } \left(\frac{R_a}{DOB}\right), \left(\frac{W^{1/3}}{Y^{1/3}\, DOB}\right), \text{ and } \left(\frac{W^{1/4}}{K^{1/4}\, DOB}\right)$$

Therefore, our scaling relationship must take the form of an equation such as

$$\left(\frac{R_a}{DOB}\right) = f\left(\frac{W^{1/3}}{Y^{1/3}\, DOB}, \frac{W^{1/4}}{K^{1/4}\, DOB}\right)$$

Baker et al. showed that Y and K, although variables, were fairly constant over the real range of interest, that is, soils in which we bury explosives. They also showed that the particular relationship of these two independent variables, when plotted three dimensionally with experimental data, formed a rectangular hyperbola and thus yielded

$$\left(\frac{R_a}{DOB}\right) = f\left(\frac{W^{1/3}}{Y^{1/3}\, DOB} \times \frac{W^{1/4}}{K^{1/4}\, DOB}\right)$$

The terms could now be multiplied yielding

$$\left(\frac{R_a}{DOB}\right) = f\left(\frac{W^{7/24}}{Y^{1/6}K^{1/8}\ DOB}\right)$$

and $7/24 \approx 1/3.4$. So, what appeared to be a purely empirical correlating factor has been shown to be the product of rational dimensional analysis.

There is some additional interesting interpretation of Figure 29.8. The point $W^{7/24}/DOB \approx 0.3$ represents the depth for a given weight of explosive that will produce the maximum crater radius for that explosive weight. Further, that radius will be approximately equal to 0.8 times the depth of burial. Another interesting observation is that at that point the slope of the scaling relationship suddenly changes, telling us that the mechanism is changing. As the burial gets deeper for a given weight, the correlation should approach $R_a/DOB = 0$ as $W^{7/24}/DOB$ approaches a value of about 0.16 to 0.18, which represents the depth for a particular weight of explosives that will provide complete containment (the blast will no longer breach the surface).

Example 29.3   If we wanted to detonate the barrel of picric acid from the last example such that it would be completely contained under the soil, how deep should it be buried?

Solution   We have just seen that crater dimensions approach zero at a burial where $W^{7/24}/DOB$ approaches 0.16 to 0.18. Therefore, for full containment,

$$DOB = W^{7/24}/0.16$$
$$= 25\ \text{ft.}$$

CHAPTER

30

Jetting, Shaped Charges, and Explosive Welding

A jet may be formed when two surfaces collide at high relative velocity and at certain angles. The jet is made of material that has been stressed into plastic flow caused by high pressure at the collision interface. Jets, thus formed, are the critical element in two special areas of explosives applications: cutting with shaped charges and explosive welding.

30.1 Shaped Charges

30.1.1 Configurations of Shaped Charges

The shaped charge is generally a conical shape (Figure 30.1): The liner material is usually copper, aluminum, or mild steel, although glass is also sometimes used. The explosive is usually pressed or cast. The charge works by explosively collapsing the liner, which forms a high-velocity jet of liner material. The formation of the jet is rather complex to model mathematically; so let us look at the phenomenon qualitatively.

30.1.2 Qualitative Aspects of Conical Shaped Charges

30.1.2.1 Jet Formation

As the explosive detonation wave passes over the liner, the liner is accelerated at some small angle to the explosive liner interface (Figure 30.2).

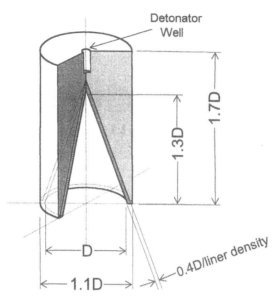

Figure 30.1. Conical shaped charge.

Nearer the apex of the cone, the M/C ratio is lower and the liner velocity is higher. When the liner material converges at the center line (or axis) of the charge, the surface material is squeezed out at high velocity (Figure 30.3). This "squeezed-out" material forms the jet. Since the material closest to the apex was at higher velocity, the portion of the jet that comes from that area is also highest in velocity; therefore, we have a jet with a velocity gradient; the leading tip is moving faster than the rear. This gradient is assumed to be linear. The remaining material, which is the bulk of the liner, forms a heavy "slug" that follows the jet at much lower velocity (Figure 30.4).

Since there is a velocity gradient along the jet, the further it travels the longer it gets. Because of minor inhomogeneities in the charge and liner, many of the

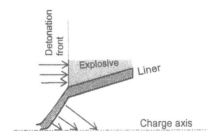

Figure 30.2. Acceleration of liner during passage of explosive detonation wave.

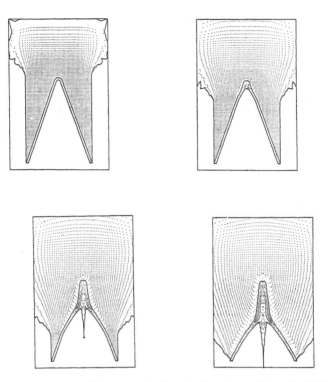

Figure 30.3. Converging of liner at axis of explosive charge (from Ref. 38).

particles that form the jet have slightly different directions of flight. The jet begins to break up after travel of several diameters (of the original charge).

30.1.2.2 Effect on Target

Erosion of a target by a penetrating jet is very similar to what occurs when a stream of water from a hose is squirted into a bank of dirt. Material dislodged at the deepest part of the hole turns to mud and flows back along the walls of the hole (Figure 30.5).

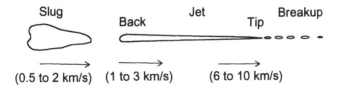

Figure 30.4. Jet configuration.

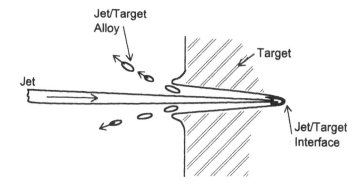

Figure 30.5. Erosion of jet through target.

Metal targets under shaped-charge jet attack behave like fluids because, at the impact velocities of the jet, both jet and target at the interface are at several megabars pressure, well into the plastic region for almost all materials. This erosion process continues until the entire jet has been used up or until the target has been perforated.

30.1.2.3 Effect of Standoff

"Standoff" refers to the distance of the base of the charge from the target. This is usually expressed as charge diameters (Figure 30.6).

At very short standoffs, the jet is still very short; it has not had time to form or "stretch"; therefore, the penetration into the target is less than optimal. At very long standoffs, the jet breaks up, and each particle hits further and further off center and is not contributing to the penetration at the center of the target (Figure 30.7). We can see that there is an optimum standoff (Figure 30.8).

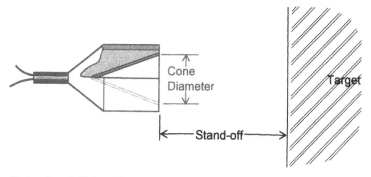

Figure 30.6. Standoff from the target.

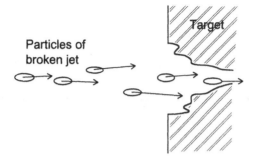

Figure 30.7.   Example of longer than optimal standoff.

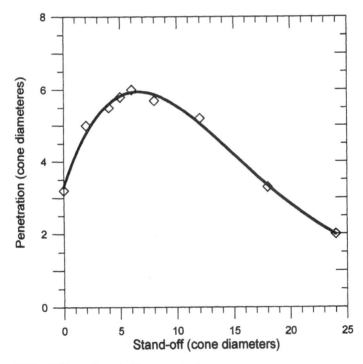

Figure 30.8.   Effect of standoff on penetration.

30.1.2.4 General Observations

1. Greatest penetration is obtained at cone angles of around 42°.
2. Optimum standoff is between 2 and 6 charge diameters.
3. Penetration is normally around 4 to 6 diameters and could go as high as 11 or 12.
4. Optimum liner thickness is about 3% of the cone base diameter for soft copper. This can be scaled for changes in density (change of material) by keeping weight constant. That is, lower-density liners should be thicker.

30.1.3 Penetration Model

We looked briefly and qualitatively at what shaped charges are and how they work. Now let us consider a simple model that will help quantify some of these observations. The model assumes that both the jet and the target are ideal liquids (that is, they do not exhibit any viscosity). This is not a bad assumption because at the impact pressure at the interface of jet and target (several hundred kilobars), most metals are far into the plastic region and do indeed behave like liquids.

The next assumption is that the jet is traveling at a constant, uniform velocity. We know this is not true, but it is surprising how well the model holds even with this oversimplifying assumption.

The last assumption is that the jet, a liquid, is in the form of a rod. Figure 30.9 illustrates a jet penetrating a target. It does not show the jet or target material flowing backward and out of the hole, just the progress of the jet and the hole.

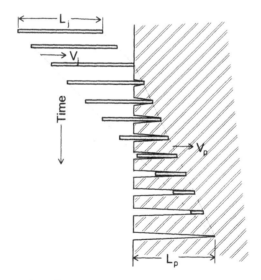

Figure 30.9.   Jet penetration of a target.

From the Bernoulli theory, the pressure of the jet at the jet-target interface is

$$P = \frac{1}{2} \rho_j V_R^2 \tag{30.1}$$

where V_R is the relative velocity of the jet and the end of hole (the end of the hole is receding from the jet at the penetration velocity, V_p); that is,

$$V_R = V_j - V_P \tag{30.2}$$

Equation (30.1) becomes

$$P = \frac{1}{2} \rho_j (V_j - V_P)^2 \tag{30.3}$$

The pressure in the target at this same interface is, of course, the same, and is

$$P = \frac{1}{2} \rho_T V_P^2 \tag{30.4}$$

The time it takes to complete the penetration is found from the penetration velocity and the depth of penetration.

$$t = \frac{L_p}{V_P} \tag{30.5}$$

This time is the same as that required to "use up" the jet, or the time it takes for the complete travel of the back of the jet (see Figure 30.9) past the front surface of the plate.

$$t = \frac{L_j + L_p}{V_j} \tag{30.6}$$

Let us combine Eqs. (30.3) and (30.4):

$$\rho_j (V_j - V_P)^2 = \rho_T V_P^2$$

and then

$$\left(\frac{\rho_j}{\rho_T}\right)^{1/2} = \frac{V_P}{V_j - V_P} \tag{30.7}$$

Now combine Eqs. (30.5) and (30.6):

$$\frac{L_p}{V_P} = \frac{L_j + L_p}{V_j}$$

so

$$L_P = L_j \left(\frac{V_P}{V_j - V_P}\right) \tag{30.8}$$

Now, combining Eqs. (30.7) and (30.8), we get

$$L_P = L_j \left(\frac{\rho_j}{\rho_T} \right)^{1/2} \tag{30.9}$$

This is the idealized penetration equation; it allows us to predict penetration performance in various targets and to help analyze shaped charge designs.

Notice that the equation implies that the depth of penetration, L_p, is independent of the jet velocity and depends only on the length of the jet and the relative density of jet to target. Experiments show that this is true and holds fairly well for jet velocities above 3 km/s.

30.1.4 Shelf Hardware

Many shaped charges, both commercial and military, are available "off the shelf." Some are far more efficient than others in relation to depth of penetration as a function of size or explosive weight. This spread of penetration efficiency is due not to poor design, but that each charge was designed for a particular application, and not all were optimized with penetration alone in mind. Figure 30.10 presents the performance of a compendium of over a hundred different

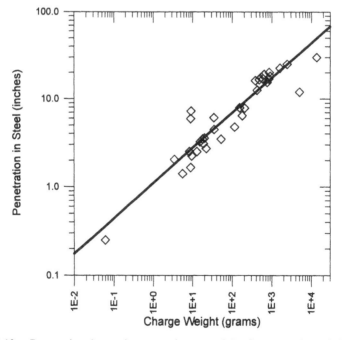

Figure 30.10. Penetration in steel versus charge weight for a number of different shaped charges.

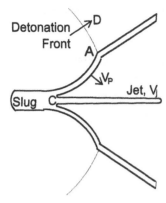

Figure 30.11.   Shaped charge cone during collapse.

shaped charges, showing the ranges one should expect of penetration versus weight of charge based upon what is available off the shelf.

30.1.5 Jet Formation Parameters and Mechanism

Let us consider now how the jet is formed. To do this, picture the cone at the time the detonation wave has progressed about halfway down the length of the cone axis (Figure 30.11). As seen in Figure 30.11, the detonation has progressed

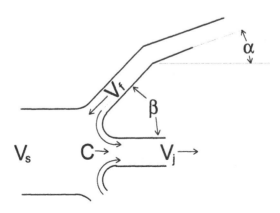

Figure 30.12.   Jet formation near collision point.

to point A. It has accelerated the cone inward at velocity V_P. The velocity, $\vec{V}_p$, is a function of cone density, cone thickness, explosive parameters (detonation velocity, or Gurney energy, as well as the local explosive weight), and the direction of grazing or cone angle. The collision point, C, is moving to the right at velocity V_C. In Figure 30.12 we see more detail in the collision region. We see that a particle in the liner traveling at $\vec{V}_p$ is also, in effect, traveling toward the collision point at $\vec{V}_f$. The angle β, or "collapse angle," is a function of the liner mass and explosive mass and detonation properties, as well as a function of the detonation direction relative to the original cone angle, α. At the collision point, the pressures are sufficiently high to cause the metal to be in the plastic state. Part of the liner is squeezed out forward as a jet whose velocity is $\vec{V}_C + \vec{V}_f = \vec{V}_j$ and part of the liner squeeze together to form a slug whose velocity is $\vec{V}_C + \vec{V}_f = \vec{V}_s$.

The angle β and flow velocity V_f are critical design parameters. V_f must be subsonic relative to the plastic material into which it is flowing. The angle β must be greater than some minimum angle that is a function of the liner materials, usually $\beta_{min} \approx 5°$ to $10°$. At these low βs, even though material can jet, the jet is not coherent, and only a spray of particles is formed. At higher values of β ($25°$–$50°$), the result is a coherent rod-shaped jet. This is the desirable jet for most purposes. At higher angles, sprays are again obtained, and finally above some value β_{max}, no jet is formed at all. At very high β, the broad cone can invert and form an aerodynamic slug, but that is another story. The jet has a velocity gradient from tip to back because V_P varies as the local M/C varies. Usually M/C increases in the direction of the mouth or base of the cone. This also causes a variation in β.

That the jet breakup (or particulation) is due to stretching of the jet is only partially true. Jet breakup is also due, in part, to high oscillatory stresses that are built into the jet at the collision point. The jet material, plastic at the collision point, rapidly returns to its elastic state as it leaves the high-pressure collision-point environment. Flow oscillations are then "frozen" in place in the form of local stresses.

30.2 Explosive Welding

If one were to take two pieces of metal, meticulously clean their surfaces, and press them together under a vacuum environment, the two pieces would weld to each other. This would occur without benefit of high temperature. The critical factor would be the cleanliness of the mating surfaces.

30.2.1 Structure of the Interface

This, in essence, is what is done in explosive welding. The technique looks very simple, but in mechanism is quite complex. Figure 30.13 shows the two most common welding setup configurations, for welding or "cladding" flat plates.

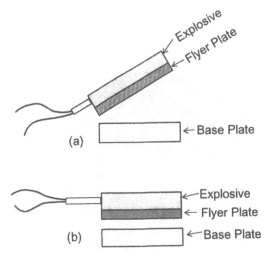

Figure 30.13.   Plate welding setups

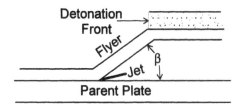

Figure 30.14.   Flyer plate collision at β forms jet.

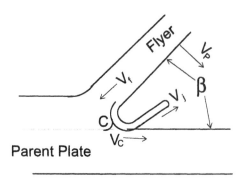

Figure 30.15.   Collision and flow.

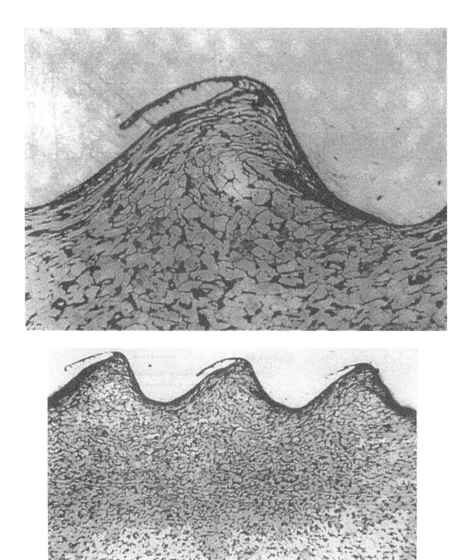

Figure 30.16. Copper welded to mild steel (Ref. 39).

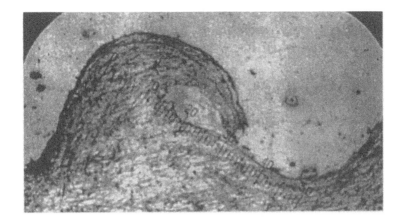

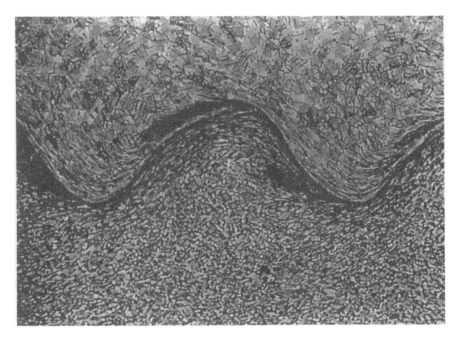

Figure 30.17. (a) Photomicrograph (100X) of stainless steel (Type 304L) explosion welded to carbon steel (SA-516-70). (b) Photomicrographs of the interface of explosively welded metals. Mild steel welded to mild steel (Ref. 39).

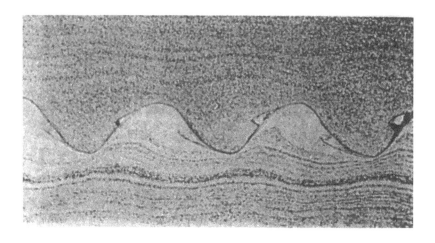

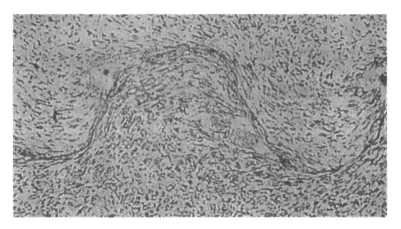

Figure 30.18. (a) 4130 Steel explosion welded at 4130 steel. (b) Explosion weld of 1018 steel to 1018 steel. (c) Explosion weld interface of sample made of alternate layers of Cu and Ni electroplate. Each square 0.0005 in. × 0.0005 in. (d) Explosion weld of Ni to Co (Ref. 40).

Configuration A is called ''angled standoff'' and B is called ''constant stand-off.'' In both cases, the flyer plate is accelerated toward the parent plate at an angle by the explosive charge. An angle of attack or collision is determined by the flyer standoff angle (if any) and the explosive/flyer plate properties. The collision angle is analogous to the collision angle for conical shaped charges, and forms a jet at the proper angles (Figure 30.14). The details at the collision point are shown in Figure 30.15.

The same behavior (and criteria) are observed as with a shaped charge. A jet is formed above some critical β. The flyer is essentially ''peeled'' of its face

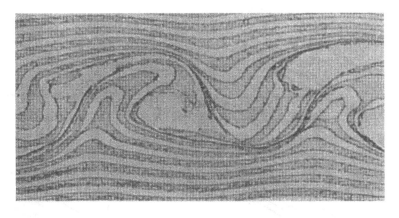

Figure 30.18. *continued*

layer of material, thus cleaning it for the weld interface. The parent plate is "scraped" of its surface, thus cleaning it. When β is just right, flow oscillation is set up at the collision point, and a "wavy" weld interface is created. This wavy nature, because it adds additional contact surface, supposedly adds to the weld strength. This is seen in Figures 30.16 through 30.18.The mechanism that forms the waves is shown in the sequence of sketches in Figure 30.19. At (a) the impact or collision is shown stagnated at point s, and the jet going to the right is dragging up parent-plate material with it. At (b) the parent-plate material dragged up by the jet has formed a dam that diverts the jet up into the flyer. At (c) the flyer flow toward the collision point drags the jet backwards, flowing over the parent dam. At (d) the hump formed by the dam now diverts flyer material downward in the form of a new jet that stagnates in front of the damn in sketch (e), and the process now repeats itself. This same behavior is typical

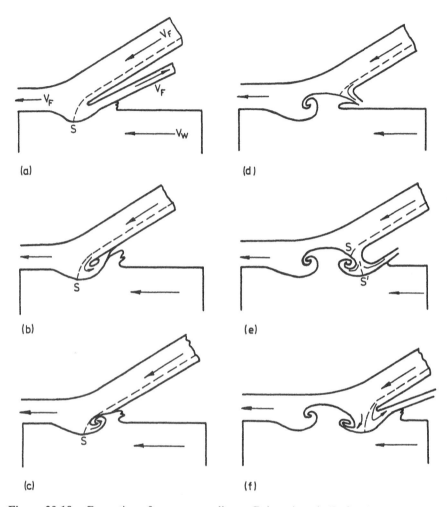

Figure 30.19. Formation of waves according to Bahrami et al. (Ref. 41).

at the shear interface of any two fluid streams moving and impacting at different velocity vectors.

30.2.2 Critical Welding Parameters

As with shaped charges, optimal jetting occurs at a critical range of angle β, and flow velocities V_f and V_C must be below sonic velocities at the collision point pressures. For design purposes this is taken as normal longitudinal sound velocities for each of the metals. Figure 30.20 shows curves that relate critical β and collision-point velocity, V_C. Note that this oscillatory behavior is probably an

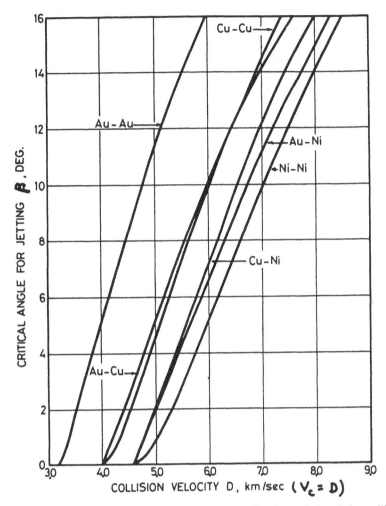

Figure 30.20. The relationship between the calculated values of β and the collision point velocity for initially parallel plate (Ref. 42).

exaggerated case of the stress oscillation that are set up in conical shaped-charge jets.

References

1. Langhaar, L. H., *Dimensional Analysis and Theory of Models*, John Wiley & Sons, New York, New York, 1951.

2. Kiefer, P. J., et al., *Principles of Engineering Thermodynamics*, John Wiley & Sons, New York, New York, 1954.

3. Kinney, G. F., and K. J. Graham, *Explosive Shocks in Air (Second Edition)*, Springer-Verlag, Berlin, 1985.

4. Gurney, R. W., *The Initial Velocities of Fragments from Bombs, Shells, and Grenades*, Ballistic Research Laboratory, Aberdeen, Maryland, BRL-405, 1943.

5. Kennedy, J. E., and D. E. Hardesty, *Thermochemical Estimation of Explosive Energy Output*, Compustion and Flame, Vol. 28, 45–59, 1977.

6. Kamlet, M. J., and S. J. Jacobs, *J. Chem. Physics* **48**, 23–35, 1968.

7. Kamlet, M. J., and M. Finger, *Combustion and Flame* **34**, 213–14, 1979.

8. Jones, G. E., Kennedy, J. E., and L. D. Bertholf, *Am. J. Physics* **48** 264–69, 1980.

9. Kennedy, J. E., *Explosive Output for Driving Metal*, Behavior and Utilization of Explosives Symposium (12th), ASME/UNM, March 1979.

10. Benham, R. A., *Analysis of the Motion of a Barrel-Tamped Explosive Propelled Plate*, SAND78-1127, Sandia National Laboratories, September 1978.

11. Mott, N. F., *Fragmentation of High Explosive Shells, A Theoretical Formula for the Distribution of Weights of Fragments*, Army Operationsl Group Research Memo No. 24, March 1943, NOS-AC-3642.R 87/NFM, United Kingdom.

12. Mott, N. F., *Fragmentation of Shell Cases*, Proc. Royal Soc. **A189** 300–8, 1947.

13. Rinehart, J.S., and J. Pearson, *Metals under Impulsive Loading*, Cleveland (ASM), 1954.

14. Garg, S. K., and J. Siekmann, *Experimental Mechanics*, Vol. 39, January 1966.

15. Taylor, G. I., *Scientific Papers of G. I. Taylor*, Vol. III, No. 4, Cambridge Univ. Press, Cambridge, MA, 1963.

16. Stromsoe, E., and K. O. Ingebrigtsen, *Modification of the Mott Formula for Prediction of Fragment Size Distribution, P. E. P. Journal* **12**, 175–78, October 1987.

17. *A Manual for the Prediction of Blast and Fragment Loadings on Structures*, SWI for U.S. Dept. of Energy/Albuquerque Offices, DE82-000536, DOE/TIC-11268, Amarillo, TX, November 1980.

18. Hoerner, S. F., *Fluid Dynamic Drag*, 1965.

19. Bishop, R. H., *Maximum Missile Ranges from Cased Explosive Charges*, SC-4205(TR), Sandia National Laboratories, Albuquerque, NM, July 1958.

20. AMCP 7-6-180, *Engineering Design Handbook, Principles of Explosive Behavior*, HQ. U.S. Army Materiel Command, Washington, D.C., 1972.

21. Berthelot, M., *Explosives and Their Power* (translated from French to English by Hake and Macnab), John Murray Publ., London, 1892.

22. Kinney, G. F., and K. J. Graham, *Explosive Shocks in Air*, Springer-Verlag, Berlin, 1985.

23. *Engineering Design Handbook*, AMCP-706-177, U.S. Army Material Command, January 1971.

24. Cole, R. H., *Underwater Explosions*, Princeton Univ. Press, Princeton, 1948.

25. Ezra, A. A., *Principles and Practice of Explosive Metalworking*, Garden City Press Ltd., Hertfordshire, 1973.

26. Liddiard, T. P., and J. W. Forbes, *Shock Waves in Fresh Water Generated by the Detonation of Pentolite Spheres*, NSWC TR 82-488, Dahlgren, Virginia, 1983.

27. Rinehart, J. S., and J. Pearson, *Explosive Working of Metals*, MacMillan Co., New York, 1963.

28. Rice, M. H., and J. M. Walsh, *Equation of State of Water to 250 kbar*, *J. Chem. Physics* **26** (4), 824–30, April 1957.

29. Bowen, I. G., Fletcher, E. R., and D. R. Richmond, *Estimate of Man's Tolerance to the Direct Effects of Air Blast*, HQ DASA, DA-49-146-XZ-372, Washington, D.C., October 1968.

30. *Nuclear Geoplosics Sourcebook*, Vol. IV, Part II, DNA 6501H-4-2, Defense Nuclear Agency, Washington, D.C., March 1979.

31. Vortman, L. J., *Ten Years of High Explosive Cratering Research at Sandia Laboratories*, Nuclear Applications & Technology, Vol. 7, September 1969.

32. Stephenson, D., *Small Scale Cratering Studies in Sand/Clay Soil*, E. H. Wang C.E.R.F./UNM for Defense Nuclear Agency and Air Force Weapons Laboratory, AFWL TR-73-99, August 1979.

33. Nordyke, M. D., *An Analysis of Cratering Data from Desert Alluvium*, *J. Geophysical Research* **67** (5), May 1962.

34. Nordyke, M. D., *Nuclear Cratering Experiments: U.S. and U.S.S.R.*, Impact and Explosion Cratering, pp. 103–24, Pergamon Press, New York, 1977.

35. Vesic, A. S., *Cratering by Explosives as an Earth Pressure Problem*, 6th Intl. Conf. on Soil Mechanics and Foundation Engineering, Montreal, Canada, 1965.

36. Chabai, A. J., *Influence of Gravitational Fields and Atmospheric Pressures on Scaling of Explosion Craters*, Impact and Explosion Cratering, (pp. 1191–214), Pergamon Press, New York, 1977.

37. Baker, W. E., Westline, P. S., and F. T. Dodge, Similarity Methods in Engineering Dynamics, Theory and Practice of Scale Modeling, Southwest Research Institute, San Antonio TX, 1973.

38. Sedgwick, R. T., Gittings, M. L., and J. M. Walsh, *Numerical Techniques for Shaped Charge Design*, Behavior and Utilization of Explosives in Engineering Design, 12th Annual Symposium, ASME and UNM, Albuquerque, NM, March 1972.

39. Schwartz, M. M. (Ed.), *Source Book on Innovative Welding Processes*, Amer. Soc. for Metals, Metals Park, Ohio, 1981.

40. Ezra, A. A., *Principles and Practice of Explosive Metal Working*, Vol. I, Garden City Press, Hertfordshire, United Kingdom, 1973.

41. Crossland, B., *Explosive Welding of Metals and Its Application*, Clarendon Press, Oxford, 1982.

42. Blazynski, T. Z., *Explosive Welding, Form, and Compaction*, Applied Science Publ., London, 1983.

Index